DATE DUE

AC 5 '00			
AP 24 03			
NV 17 05			

DEMCO 38-296

NOVELTIES IN THE HEAVENS

NOVELTIES IN THE HEAVENS

Rhetoric and Science in the Copernican Controversy

Jean Dietz Moss

THE UNIVERSITY OF CHICAGO PRESS
CHICAGO AND LONDON

...rofessor and director of The Rhetoric
...Catholic University.

The University of Chicago Press, Chicago 60637
The University of Chicago Press, Ltd., London
© 1993 by The University of Chicago
All rights reserved. Published 1993
Printed in the United States of America

02 01 00 99 98 97 96 95 94 93   1 2 3 4 5

ISBN (cloth): 0–226–54234–3
ISBN (paper): 0–226–54235–1

Library of Congress Cataloging-in-Publication Data

Moss, Jean Dietz.
  Novelties in the heavens : rhetoric and science in the Copernican
controversy / Jean Dietz Moss.
     p.   cm.
  Includes bibliographical references and index.
  ISBN 0-226-54234-3. — ISBN 0-226-54235-1 (pbk).
  1. Astronomy, Renaissance.   2. Copernicus, Nicolaus, 1473–1543.
3. Galilei, Galileo, 1564–1642.   I. Title
QB29.M67   1993
520′.9′03—dc20
                                                            92-21608
                                                                CIP

⊚ The paper used in this publication meets the minimum requirements of
the American National Standard for Information Sciences—Permanence
of Paper for Printed Library Materials, ANSI Z39.48–1984.

CONTENTS

CONTENTS

PREFACE

This study attempts to document the use of rhetoric as a means of furthering scientific claims where previously that art had been employed in the domain of science merely to present, clarify, or amplify a topic. The writings examined herein testify to an important change, a change that might be said to constitute a second revolution in science, following closely on the first begun by Copernicus. They reveal the expansion of rhetoric into astronomy, one of the disciplines of natural philosophy, which traditionally recognized only dialectical argumentation and demonstrative reasoning as appropriate methodology for reaching conclusions. Dialectic could yield probable truths and demonstration certainty. But, in the case of the Copernican thesis, when certainty was unattainable and probability the next best recourse, the likely story declaimed by rhetoric found an opening. The supporters and opponents of Copernicus invoked "persuasible" proof to tip the balance created by dialectical arguments that seemed equally probable.

The reasons for this revolution in argumentative practice become clearer when we consider the historical context in which the issue was debated. The time, the place, and the public affected by the debate on the Copernican hypothesis are all significant factors. The attitudes of those who read Copernicus—Galileo, Kepler, Scheiner, and Grassi—were greatly determined by their religious beliefs, for the Scriptures seemed to underwrite the older opinion, the stable earth of Ptolemy, which was the center of gravity and of celestial orbits. The fact that Galileo found corroborating evidence for the Copernican universe, and that he made these discoveries and announced them in Italy where the campaigns of the Counter-Reformation were mounted, helps to explain why the issue became a matter of importance not only for astronomers but for a larger public as well. The interest of that public in the question was the major reason for the introduction of rhetoric into the debate. Rhetoric exists because an author wants to persuade a public of something. Unlike dialectic, where the aim is to find what seems to be true, the object of rhetoric is to induce assent by whatever means the orator thinks appropriate.

The treatises, letters, disputations, and dialogues that furnish the raw materials for this study were composed at a time when knowledge of rhetoric was second only to knowledge of grammar, and awareness of the rhetorical dimensions of writing and speaking was pervasive. Rhetorical principles of organization and presentation of discourse, argumentative strategies, figures of speech, the use of commonplaces were all second nature to the authors involved in the debate over the Copernican thesis. Rhetoric, however, had traditionally been associated with politics and literature rather than with natural science, so that its increasing use in the realm of astronomy justifies both an examination of that usage and speculation about the cause of that change. Such a focus has the advantage also of enlarging our understanding of the audiences' response. But the extensive and conscious use of rhetoric in the Renaissance, which was to assume unusual importance in this debate, was an ascendancy that proved to be short-lived. Ironically, one of the reasons for the decline of the academic discipline appears to be the simultaneous rise of interest in experimental science and the desire of scientists to prevent the incursion of rhetoric into the "objective" communication of its findings. Rhetoric, however, was to prove difficult to eradicate. Whenever scientists thought it necessary to convince a public of the "truth" of an investigation, or of the significance of a scientific discovery, they turned to rhetoric, even though they were not always aware of it nor knew enough to enjoy its artful employ.

After an introductory chapter describing the practice and province of dialectic, demonstration, and rhetoric, the focus of the discussion turns to the writings relevant to the Copernican thesis. Part One, The Celestial Revolution, contains a rhetorical analysis of Nicolaus Copernicus's overview of a sun-centered cosmology in the first book of *De revolutionibus*. This analysis is preceded by a brief look at the text it proposed to replace, Ptolemy's *Almagest*. The chapters that follow treat selected responses to Copernicus from the turn of the sixteenth century through the middle of the seventeenth, when scholars began to argue their views with more and more vigor. These passages were chosen for analysis because of their relation to two of the most prominent protagonists in the controversy: Kepler and Galileo. Kepler's early mathematical arguments for the Copernican thesis provide a significant contrast to the startling evidence Galileo found to support the Copernican worldview when he first turned his telescope on the heavens. The discoveries announced by Galileo in *Sidereus nuncius* Kepler then ecstatically endorsed in his review of that work, but he included some mild correctives also. The publicity given to the Copernican question through these writings escalated tensions over the issue. The seemingly tangential problem of the nature of the sunspots, discussed in the fourth chapter, finds the parties in this dispute, Galileo and the Jesuit Christopher Scheiner, increasingly adamant and intent upon maintaining opposing positions.

All of the authors whose writings are discussed here worked within the frame of Aristotelian logic and dialectic, a point supported by the testimony of the arguments they advanced and also by their own statements regarding their methodology. The rhetorical techniques are also classical in their foundation and are consciously employed or delimited. Each applies rhetoric to scientific questions, but in very different ways. While Copernicus prefaces his treatise with rhetorical appeals, classical allusions, and some commonplaces, the arguments he uses to advance his hypotheses are in the main dialectical and demonstrative, not rhetorical. Kepler offers convincing computations and observations in support of some of the critical points of the system advanced by Copernicus, but he recognizes the need for physical evidence to prove the thesis. The rhetoric of his presentation is cosmetic and the partial proofs he offers in support of heliocentrism are in accord with the methodology of the day. Full proof that would satisfy the scientific canons of Aristotle was not available until the nineteenth century when Bessel was able to offer stellar parallax as proof that the earth moves around the sun and Foucault could show the earth's rotational movement by his experiments with the pendulum.[1]

When Galileo began to take up the Copernican cause after his discoveries with the telescope, he saw that he could provide only partial physical evidence for the system, and in the *Sidereus nuncius* he was content to keep his discussion within the bounds of accepted scientific discourse. In his debate with Scheiner in the letters on sunspots, both disputants employed dialectical arguments to develop the most probable explanation for the phenomena. When their arguments seemed not quite convincing, they turned to rhetoric in the form of enthymemic arguments that included ethical and pathetic appeals.

The movement of the weight of proof from dialectic to rhetoric is most apparent in the writings analyzed in Part Two, The Hermeneutical Crisis. Here the problem of the proper reading of Scripture in relation to the Copernican hypothesis is addressed by theologians, prominent Dominicans, and prelates of the Roman Catholic Church, and also by two astronomers: Kepler and Galileo. Before an official position on *De revolutionibus* was declared, a number of theological tracts seeking to harmonize scriptural exegesis with the work appeared; those of Diego de Zuñiga and Paolo Antonio Foscarini are significant among these. Kepler, a Protestant trained in theology at the University of Tübingen, also called for a new interpretation of Scripture in one of his works. Early in the controversy, the voices of two Dominicans, Tommaso Campanella and Giordano Bruno, were raised in support of the Copernican side. Campanella attempted to show that the Copernican system did not merit the condemnation its author feared the work would receive, and the Dominican, like Foscarini, found Galileo's discoveries important evidence for that position. Campanella's piece is essentially a scholastic argument, a disputation treating arguments for and

1. See the discussions of this point by Alexandre Koyré, *The Astronomical Revolution*, trans. R. E. W. Maddison (Ithaca: Cornell University Press, 1973), p. 108, n. 29.

against Galileo, which remains within the bounds of dialectic. Bruno's praise of Copernicus's treatise, on the other hand, made far more use of rhetoric. It was actually the first public espousal of the system, but the friar's arguments were more an embarrassment to the cause than an aid, since he used heliocentrism as the physical basis for the infinite, animistic cosmos he believed should embrace all religions.

Of major importance to our concern with the relation of rhetoric to science and theology is Cardinal Bellarmine's outline of the requirements for a new exegesis. He articulates these in response to the Carmelite friar Paolo Foscarini, who had sent him a copy of an exegetical piece he had written requesting the cardinal's advice. Writing in the period before the Church's position on Copernicus had been framed, Bellarmine stated that, unless a necessary demonstration of the thesis was forthcoming, the traditional reading of Scripture must stand. When they were communicated to him, Galileo grappled with the restrictions Bellarmine outlined and tried to show that these were abrogated by his new discoveries. In his *Letter to Madame Christina of Lorraine, Grand Duchess of Tuscany* Galileo appealed to the Church not to suppress *De revolutionibus,* but to consider instead another interpretation of the difficult passages in Scripture. In this piece the famous Tuscan carries rhetoric beyond the limits placed on it in the scholastic system, demanding recognition of the heliocentric system as an acknowledged truth.

Part Three, The Triumph of Rhetoric, traces the burgeoning of rhetoric in the debate through the period of its most impassioned use. Chapter eight treats a topic related to Copernicanism, the question of the nature and position of the comets, which engaged Galileo in an encounter with the Jesuit astronomer, Orazio Grassi. The rhetorical perspective of this chapter and the next gives a somewhat different view of the significance of the arguments advanced, and it also enables us to see the deepening emotional character of the dispute, which effectively prevented a meeting of minds. By this point the Copernican thesis had been condemned by the Church, and whatever bore upon that thesis took on greater importance.

The next to the last work discussed in this study is the most varied in its rhetorical character, Galileo's celebrated *Dialogue Concerning the Two Chief World Systems*. In his earlier disquisitions on scientific subjects Galileo employed a wide range of genres: treatise, disputation, letter, and dialogue. But the last is the form that allowed him the widest range of persuasive techniques. Galileo transforms the Ciceronian form of the dialogue, a favorite of Renaissance authors, into a scientific variety uniquely his. He enlivens what might have been a dreary rehearsal of mathematical-physical arguments by infusing his interlocutors with personality, and engaging these characters in a series of witty exchanges that take place over several days in the setting of a

Renaissance palace in Florence. Through three different personas he unfolds dialectical arguments, scientific experiments, and a variety of academic views. The choice of the dialogue form enables Galileo to imply that he is simply airing opinions without taking sides. He disclaims in the preface any attempt to convince his readers of the truth of the Copernican system but intends only a "pure mathematical caprice."

The last chapter considers the extension of the arguments for Copernicus by an English supporter of Galileo, John Wilkins, who was to become Bishop of Chester and a member of the Royal Society. His writings on the Copernican System provide a suitable ending to this study, for they testify to a paradoxical effect of rhetoric in the cause of science. Wilkens enthusiastically accepts Galileo's new philosophy, yet the Englishman acknowledges that the arguments for the superiority of the Copernican over the Ptolemaic system are only dialectical. Although these induce belief, they remain probable proofs, insufficient for attaining certainty. He recognizes the rhetorical arguments offered on either side, but declines to employ them himself in like manner. He must be seen as a conservative in the rhetorical revolution, but numbered among the vanguard in the scientific revolution.

The era of rationalism and empiricism then emerging made rhetorical arguments suspect. Yet, undoubtedly, the rhetorical skill of Galileo in the *Dialogue* and the *Letter to Christina* was of great importance in disposing Wilkins's mind and the minds of readers throughout the Western world to accept the thesis advanced by Copernicus, overcoming whatever religious or commonsense misgivings they may have had. The *ethos* of the Tuscan scientist and the *pathos* of his situation argued more strongly than the scientific proofs could. While the effects of his rhetoric on the attitudes of his public were extensive and enduring, within the scientific community itself rhetoric remained an unacceptable means of gaining assent to a scientific hypothesis. Thus, the rhetorical revolution begun so dramatically by Galileo rumbled beneath the surface for years, not quite able to overthrow the Aristotelian canons of scientific proof.

The central figure in my study is quite obviously Galileo. His involvement in the Copernican question has dictated the scope of the study and its focus. As the father of the scientific revolution, he is also the father of the revolution in the use of rhetoric I have attempted to describe. He was the most imaginative in his use of it and in his reliance on it when proof by the canons of the time eluded him. It was also ultimately to be one of the most important factors in his problems with the Church. Had he been less vigorous in wielding the weapons of rhetoric, he might have saved himself from the wrath of Pope Urban VIII. But then he would not have been the man he was, and we would not today find him so compelling a figure or his writings so persuasive.

The stance adopted by the Roman Catholic Church on the Copernican hypothesis was greatly influenced by its need to gain back some of the ground it had lost to the Reformers. Conservatism, a desire to conserve the faith, was the prevailing attitude of ecclesiastical leaders when faced with the alternative of a new interpretation of Scripture. In the forefront of the Counter-Reformation's campaign to reclaim the unfaithful, the Jesuits were persistent in their defense of the Pope's position. When they addressed issues such as the sunspots and the comets, therefore, the Jesuits cloaked their identity so as not to engage the Order in what might eventually become a problematic position. The rhetorical content of their discourse at times appears to have been prompted by the religious stance expected of them rather than by the scientific import of the question.

All of these circumstances tend to make a rhetorical analysis of the writings of the participants in the debate over the Copernican question a complex task. Since the term rhetoric is variously understood, an introductory clarification seems desirable here, though a fuller discussion follows in the first chapter. By rhetoric most people today mean the unfair manipulation of an audience, whether through honeyed words, emotional appeals, empty bombast, or even through a nonverbal posture. But this conception is more properly applied to the misuse of rhetoric. Rhetoricians themselves lack agreement on the nature of their discipline. Some consider it to be the art of persuasion applicable to any issue, especially to those of a political, judicial, or moral nature. Others think of it as communication in general, not limited to persuasion or to any arena, which is operative whenever verbal or written discourse affects an audience. Both of these senses have their counterparts in classical teachings, which were conveyed in the Renaissance through the original texts, commentaries, and contemporary derivatives.

For Aristotle the function of rhetoric was "to find in each case the means of persuasion."[2] These means included rational and emotional appeals undergirded with the speaker's ethos. While Cicero and Quintilian also accorded attention to the discovery of persuasive arguments, they dwelt more heavily on that other traditional concern of rhetoric: apt and eloquent expression. As Quintilian expressed it, rhetoric was "the art of speaking well."[3] "Well" for him had a moral connotation also, one that invoked the

2. Aristotle *Rhetoric* 1.1.14.1355b; cf. 1.2.1–3.1355b. Trans. John Henry Freese, Loeb Classical Library (Cambridge, Mass.: Harvard University Press, 1975).
3. Quintilian *Institutio Oratoria* 2.14.8–10. Trans. H. E. Butler, Loeb Classical Library (Cambridge, Mass.: Harvard University Press, 1980). The emphasis of Isocrates on the citizen-orator's responsibility to the state was also to have its effect in the Renaissance as his

responsibility of the citizen-orator to argue for the people's welfare. The two emphases are mutually reinforcing, but it was the Romans' interest in masterful expression and statesmanship that appealed to the Humanists. Their concomitant interest in persuasive argumentation was guided more by traditions of dialectic than by rhetorical doctrine, a point explained at length in the first chapter. The classical conception of rhetoric will dominate our discussion of the Copernican debate, since it is the one that was familiar to our authors.

The modern conception of rhetoric, which holds it to be characteristic of all discourse wherein one "symbol-using animal" influences another, as Kenneth Burke puts it, means that every expression becomes a candidate for analysis. This view does not guide the analysis of this study, since it would seem to make the project either too monumental to undertake or too arbitrary if undertaken. In general the approach used here is contextual; it attempts to view science and rhetoric as they were conceived and practiced by the participants in the debate. This does not mean that modern rhetorical theories are ignored. The insights of such astute observers as Burke and Chaim Perelman concerning the varieties of audience and the range of rhetorical message illuminate some of the mechanisms at work in the relationship between authors and readers that might otherwise have passed unnoticed.

The principal subject of rhetoric is taken to include that commonly agreed on in the classical period and honored in the Middle Ages and the Renaissance, that is, issues of public concern. The techniques and principles of rhetoric were indeed held to be applicable to any subject whatever, and these were adapted and incorporated in a number of other arts—most notably poetics, but also in painting, sculpture, architecture, and music. Despite these varied applications, however, the educated public was aware of the primary purpose of the art of rhetoric, and when it was employed to influence a sophisticated audience regarding conclusions in matters of physics or theology, both the author and the audience recognized its use. They would not have expected rhetorical proofs to be constitutive of knowledge in either of these sciences.

I am very fortunate in this undertaking in having so rich a trove of historical, scientific, and philosophical discussions of the issue to draw upon. It is only

---

works became more widely known. Galileo himself translated passages from Isocrates in his school days; see *Le Opere di Galileo Galilei* (henceforth referred to as *Opere*), ed. Antonio Favaro, 20 vols. in 21 (Florence: G. Barbèra, 1890–1909, rpt. 1968), 9:283–84.

because so much has been said on these aspects that a rhetorical analysis is feasible. I have especially relied on the outstanding work of a number of philosophers and historians of science.

My contentions regarding the nature of science in the period, and particularly of Galileo's science, has been inspired by the work of Professor William A. Wallace, who first introduced me to the nature of Galileo's physics and logical methodology in a Folger seminar a decade and a half ago. Since that time I have been examining the relation of rhetoric to scientific discourse, gaining insights periodically from his more recent published material. His latest two-volume work—an English translation and commentary on Galileo's logical treatises, plus a book-length introduction outlining their background and use in his scientific work—appears to offer solid corroboration for my sense of Galileo's scientific and rhetorical argumentation.[4] I do not pretend to understand all the intricacies of demonstrative methodology in Renaissance science, but I have aimed to convey the unprecedented manner in which rhetoric entered into the science of the day, as both disciplines were understood in the period.

4. William A. Wallace, *Galileo's Logic of Discovery and Proof: The Background, Content, and Use of his Appropriated Treatises on Aristotle's "Posterior Analytics,"* Boston Studies in the Philosophy of Science, no. 137 (Dordrecht and Boston: Kluwer Academic Publishers, 1992); idem, *Galileo's Logical Treatises: A Translation, with Notes and Commentary, of his Appropriated Latin Questions on Aristotle's "Posterior Analytics,"* Boston Studies in the Philosophy of Science, no. 138 (Dordrecht and Boston: Kluwer Academic Publishers, 1992).

CHAPTER ONE

The Expansion of Rhetoric into Science

When Copernicus introduced his new cosmology to the world in *De revolutionibus,* he knew that he would have a difficult task persuading people to give up a world picture that had served for more than a millennium. For this reason he includes in his introduction to that work an eloquent rhetorical apologia, replete with classical allusions and evocative figures, to convey his own passionate conviction of its truth. He uses rhetoric also later in the treatise to help remove the prejudices of his readers against what must have seemed a fantastic proposal—that the sun, not the earth, is in the center of the universe. For example, after presenting a summary of his reasons for overturning the Ptolemaic position, he declares:

At rest, however, in the middle of everything is the sun. For in this most beautiful temple, who would place this lamp in another or better position than that from which it can light up the whole thing at the same time? For, the sun is not inappropriately called by some people the lantern of the universe, its mind by others, and its ruler by others. [Hermes] the Thrice Greatest labels it a visible god, and Sophocles' *Electra,* the all seeing. Thus indeed, as though seated on a royal throne, the sun governs the family of planets revolving around it.[1]

His exercise of analogy here, glorifying and personifying the sun, is intended to induce his readers to shift their emotional allegiance from what he considered an erroneous explanation of the cosmos to a more accurate one. An elegant rhetorical appeal to the emotions such as this in a work devoted to science, although strange to us, certainly would not have surprised his audience. For Copernicus lived in a period when enjoyment of language urged expression in manifold ways and in a variety of literary and rhetorical genres. Writings devoted to scientific subjects were traditionally cast in the

1. Nicolaus Copernicus, *On the Revolutions,* ed. Jerzy Dobrzycki, trans. Edward Rosen (Baltimore: Johns Hopkins University Press, 1978), 22. I have used this English translation, made from the edition sponsored by the Committee for the History of Science and the Institute for the History of Science, Education, and Technology of the Polish Academy of Sciences, as offering the most accurate, if not the most graceful, version available in English. The text translates the manuscript of Copernicus, but Rosen indicates the places where this version is at variance with the published version. My rhetorical analysis I have restricted to the published version of the manuscript. Page references included in the text are to this edition.

forms of treatises or disputations. Their language was generally straightforward, and reasoning followed the teachings of Aristotle's *Analytics*. But the rediscovery of classical literature affected education in the schools and universities, inspiring its scholars and stimulating rhetorical flights even in traditionally spare works of science. All of the mathematicians and philosophers whose works are examined in this book were Renaissance humanists as well, and, in contradistinction to the popular notion of humanists, all had deep religious commitments.

As mathematicians, philosophers, and theologians contemplated the theory of the heavens advanced by Copernicus, some turned to rhetoric to assist them in advancing their evaluations of his thesis, for even though the issue was a scientific one it affected the heart as well as the head. If the earth could no longer be viewed as the center of the universe, not only were commonplace allusions and everyday reference points overturned, but scriptural passages could no longer be taken literally. Just as serious were the threats posed to reigning philosophical views by Galileo's discoveries that seemed to corroborate the Copernican thesis. His observations with the telescope called into question the existence of the heavenly spheres and the nature of the element of which the heavens were thought to be composed. As a result a new philosophy was born that threatened to unseat the old.

As the use of rhetoric increased, its domain was extended. Rhetoric was employed not only to frame discussions, to prepare the ground, to clarify, or to underscore the importance of the positions assumed on a question, but it was gradually introduced into the scientific proofs of the arguments. The growing role of rhetoric in these debates and in the religious controversies that Copernicus's new cosmology generated is the major subject of this book. Fortunately the participants in the Copernican dispute have left us an extensive record. We can witness first hand the gradual displacement of demonstration by dialectic and, ultimately, the employment of rhetoric as a critical means of gaining assent to a scientific theory.

The nature of Renaissance rhetoric is so varied that it is difficult to speak in general about it. This is certainly a reason no one has attempted a comprehensive history of the art for that period. One of the complicating factors is its changing relation to dialectic. Another is the differing emphases given to the traditional parts of the discipline. Nevertheless, it is possible to discern the major characteristics of rhetoric and of its sister arts, particularly dialectic.

The aims of rhetoric and dialectic, to begin, are different no matter what the age. Since the source of the difference was conveyed to scholars in the Middle Ages and the Renaissance through the teachings of the *Organon*, a review of Aristotle's conception of the various instruments of knowledge should help to clarify them for the modern reader. Very little of what was well known to scholars remains in the teaching of logic or of rhetoric today. The nature and aims of logic and its relation to the arts and sciences was once the subject of intensive study. Logic, which included the study of dialectic, was to continue as a discipline in many universities through the nineteenth century. Interest in the subject varied, but even during periods when the discipline was revamped, its parts reorganized and expanded or simplified, its basic elements remained the same.

For those unfamiliar with the history of these disciplines, a more detailed consideration of dialectic and rhetoric and their interaction as these were understood in the Renaissance follows. A brief overview of science and demonstration and the relation of these to dialectical reasoning precedes a summary of the art of rhetoric.

Dialectic and Demonstration

In the Renaissance as in the Middle Ages, the accepted method for proving a scientific theory was demonstration according to the canons provided by Aristotle. In the scholastic understanding, the demonstration of science results in certainty. Usually expressed in syllogistic form, the premises of a demonstration are based on sense experience or true principles and are concerned with the causes of phenomena. When the causes or true principles allow one to cast a proof in a valid form of the syllogism, one can frame a "necessary demonstration." Only in such a case can one speak of having attained the ideal: perfect knowledge, *scientia* in the Latin, *epistēmē* in the Greek. This concept of demonstration was still taught in the universities and *studia* during this period, primarily through texts and commentaries on Aristotle's *Prior* and *Posterior Analytics* and the *Physics*.[2] Students would have already learned the nature of terms and their incorporation in propositions from earlier studies, especially Aristotle's treatment in the *Categories* and *On Interpretation*. These texts, along with the *Analytics* and the *Topics*, comprise the *Organon* or the logical tools of Aristotle. With the help of the *Isagoge* or "Introduction" to the *Categories* written by Porphyry, and the numerous

2. For a discussion of methodology in the Renaissance see Neal Gilbert, *Renaissance Concepts of Method* (New York: Columbia University Press, 1960), and Charles Coulston Gillispie, *The Edge of Objectivity* (Princeton: Princeton University Press, 1960), chs. 1–3. Charles Schmitt discusses the prevalence of Aristotelian logic in Europe and its resurgence in England, a phenomenon which has been too much ignored in modern studies, in *John Case and Aristotelianism in Renaissance England* (Kingston: McGill-Queens University Press, 1983), 5–7, 13–41. Both medieval and Greek commentaries were studied, with the latter becoming increasingly important during the sixteenth and seventeenth centuries. The medieval Latin commentaries of Aquinas and Albert the Great continued to dominate the teaching of the Peripatetics in most institutions. They were supplemented by the recovery of the Greek commentaries, which were greatly concerned with the exact words of Aristotle.

commentaries written on these and the other works of the *Organon*, students in the high Middle Ages and the Renaissance would have begun their study of logic.

After mastering terms and propositions, students were introduced to valid and invalid forms of reasoning as these were described by Aristotle in the *Prior Analytics*, where the philosopher uses the structure of the syllogism to simplify the analysis. Then they studied the material content of the propositions of the syllogism as these were developed in the *Posterior Analytics* and illustrated in the various special sciences. In the *Physics*, for example, Aristotle is concerned with such things as the concept of nature, motion, the infinite, place, void, time, and change, and in the course of developing this material he formulates appropriate demonstrations. The interrelationship of induction and deduction is a part of the material developed here and in related treatises.

But if one were treating problems where causes could not be discerned with certainty or where basic principles were elusive, scholars of the day turned to dialectical reasoning. This exploratory process, whose genealogy includes the dialogic approach of Plato, was refined and explained by Aristotle in his treatise entitled the *Topics*. Dialectical reasoning, also termed "probable reasoning" because it treats matters where certainty is impossible, uses premises based on what is commonly thought to happen, or on what is regarded as true generally and for the most part. Thus, opinion, particularly that of informed or wise people, furnishes the premises of dialectic. Opinions on both sides of an issue are examined exhaustively until what appears to be true becomes clear. This kind of reasoning is also used to discover the first principles of a science, a point Aristotle mentions in the first chapter of the *Topics*.[3]

Dialectic reached the apex of its development in the disputations that guided the teaching and examination systems of the universities from the late Middle Ages through the nineteenth centuries. (In fact, some religious orders and ecclesiastical faculties still examine students in this fashion to the present day.) This erotetic logic, the logic of question and answer resting on informed opinion, furnishes the basic structure of the disputation, and is plainly visible in its more relaxed formulation in the Renaissance. Interest in dialectic actually increased in the sixteenth century after a period of decline in the early part of the Renaissance.[4]

3. Aristotle, *Topica*, trans. E. S. Forster, Loeb Classical Library (Cambridge, Mass.: Harvard University Press, 1976).

4. See the discussion by E. J. Ashworth, "Traditional Logic," especially 143–167, and that of Lisa Jardine on "Humanistic Logic," 191–92 in *The Cambridge History of Renaissance Philosophy* (1988). While interminable debates concerned with distinctions waned, scholars still employed the classical methodology, as the analyses in this book show. I have discussed the

Although the format of dialectical investigations changed during the Renaissance as these were published to reach a wider public, the writings of the controversialists still show evidence of the underlying structure, even when published as treatises, letters, or dialogues. This will become manifest in the chapters devoted to such works. Some authors continued to describe their treatises as disputations despite the rhetorical style in which they cloaked them, as in the case of Grassi's first treatise on the comets, discussed in the third chapter. Galileo, as the consummate Renaissance stylist he was, framed his questions and answers in a classical dialogue.

A textbook published in 1597, *Introductio ad logicam,* offers especially clear explanations and illustrations of the logical methodology employed during the period, and thus can conveniently serve for our overview of that subject. Its author, Ludovico Carbone, was intimately acquainted with the major academic approaches to logic, and also to philosophy, rhetoric, and theology, publishing works in all of these fields. His publications were well known in Italy. They are important for their systematic pedagogy and also for their preservation of the teaching of these subjects at one of the premier academic institutions of the day, the Collegio Romano. The Collegio, founded in 1551, was the chief *studium* of the Jesuit Order. Its program of studies was adopted in all of the Order's colleges, which were rapidly being established throughout Europe in the last half of the sixteenth century. Graduates soon spread these teachings far beyond the halls of the Jesuit colleges.

Little is known about Carbone other than that he studied at the Collegio Romano in its first decade and that he was a professor at the University of Perugia before his death in 1598.[5] His works on logic are particularly noteworthy because of their tangential connection with Galileo's early writings. As William A. Wallace has discovered, both the *Introductio* and another work published by Carbone in the same year, *Additamenta ad commen-*

---

erotectic structure of dialectic in relation to disputations of Galileo and others in "Dialectics and Rhetoric: Questions and Answers in the Copernican Revolution," *Argumentation* 5 (1990): 17–37. Neal Gilbert provides persuasive evidence that even the early humanists were not as prejudiced against dialectic as many scholars believe. They did not disparage dialectic in general, but rather the British "Terminist" variety. See Gilbert, "The Italian Humanists and Disputation," in A. Molho and J. A. Tedeschi, eds., *Renaissance Essays in Honor of Hans Baron* (Florence: Sansoni, 1971), 203–26. See also the works of Enrico Berti on ancient dialectics in *Logica Aristotelica e Dialectica* (Bologna: L. Cappelli, 1983) and "Ancient Greek Dialectic as Expression of Freedom of Thought and Speech," *Journal of the History of Ideas* 39 (1978): 347–70.

5. Marc Fumaroli describes Carbone's contribution to an understanding of Jesuit rhetoric in his *L'Age de L'Eloquence* (Paris: Droz, 1980), 182–86. See also my "The Rhetoric Course at the Collegio Romano in the Latter Half of the Sixteenth Century," *Rhetorica* 4 (Spring 1986): 137–51.

taria D. Francisci Toleti in Logicam Aristotelis (1597), were directly derived
from lectures given by Paulus Vallius at the Collegio Romano in 1587–88.
These, as the title states, were based on Franciscus Toletus's commentary
on Aristotle's logic. Toletus taught at the Collegio from 1559–62, and
Carbone had studied with him then. When Carbone began preparing lectures for his own students at Perugia, he evidently returned to the College to
avail himself of the latest teachings on logic, knowing that they would build
upon the foundation laid down years before in the lectures of Toletus. He
probably found what he sought in Vallius's lecture notes, deposited in the
library for the benefit of his students, as was the custom at the Collegio.
Many years after Carbone's work had appeared in print, Vallius published
his own two-volume treatise on logic and in the preface stated that a well-
known scholar had appropriated his lecture notes and published them. He
did not name the person, but his description leaves little doubt that his reference was to Carbone.[6]

Galileo, too, turned to the resources of the Collegio Romano to develop
a better grasp of logical methodology during his first teaching post at the
University of Pisa in the early 1590s. He evidently carefully copied and ab-
breviated lecture notes emanating from the College into a notebook that is
still preserved, MS 27. That manuscript, known to scholars as the "logical questions," is actually a commentary on Aristotle's *Posterior Analytics*.
Wallace has shown that these are the same lectures, the lectures of Paulus
Vallius, that Carbone appropriated in two books on logic.[7] Thus Carbone's

6. The details of Wallace's remarkable detective work are given in *Galileo and His Sources*
(Princeton: Princeton University Press, 1984), 10–20, wherein he further argues that the
teaching on demonstration in Vallius's lectures of 1587–88, and particularly that on the demonstrative regress, grounds most of Galileo's references to "necessary demonstration" in his
subsequent writings. In his most recent study, *Galileo's Logic of Discovery and Proof* (cf. n. 4 of the
Preface, above), Wallace analyzes in detail Galileo's arguments to show how they conform to the
Aristotelian canons as explained by Vallius. I mention this here since in what follows I do not
focus on demonstrative argument in Renaissance science, concentrating instead on its dialectical and rhetorical counterparts.

7. Wallace has offered convincing proof for the source and dating of the early writings of
Galileo, called the *Juvenilia* by Antonio Favaro, the nineteenth-century editor of Galileo's
works. These include a commentary on Aristotle's logical questions, MS 27, a commentary on
Aristotle's physical questions, MS 46, and *De motu*, MS 71, an early treatise on motion. Favaro
thought the logical questions were written by Galileo as a boy before he entered the University
of Pisa and thought only the last two manuscripts were worthy of inclusion in the National
Edition of his works (*Opere* 9:280–81). Wallace has shown that they were instead composed by
Galileo while preparing to teach at Pisa (*Galileo and His Sources,* 18–19). A contrary view is held
by Adriano Carugo and Alistair Crombie, who think that Galileo wrote these works much
later, in "The Jesuits and Galileo's Idea of Science and Nature," *Annali dell'Istituto e Museo di
Storia della Scienza di Firenze* 8, no. 2 (1983): 1–68. Wallace has described the source of
the logical questions (MS 27) in his introduction to and commentary on MS 27, a work

writings become particularly apposite for our study, since they reflect the views on dialectic and rhetoric that Galileo himself absorbed when writing out his logical questions.

Dialectical Reasoning

In his *Introductio*, Carbone describes the structure of dialectical reasoning.[8] The premises of the dialectical syllogism, he explains, are probable propositions commonly held to be true. (Rhetoric, as we shall see, resembles dialectic in this, for its premises are also founded on common opinion.) These opinions enjoy various levels of acceptability. Some are accepted by all ("that parents love their children"), some by well-informed people ("that the good in itself is preferable to the useful"), and others by experts in a field ("that the universe is one"). Expert opinions are not accepted by all experts, however, and when not accepted they are not classified as probables, for example, "that anything can come to be from anything else" (170–71). When one or more probable propositions are used in a syllogism, the syllogism is termed dialectical. As such, it may include a necessary proposition, but its conclusion, because of the problematic character of one of the premises, is nevertheless only probable.

Carbone is next concerned with the use of dialectic in solving a question or a problem. He says that the dialectical syllogism is used to investigate a proposition that is disputed; such statements are "said to be theorems, that is, statements worthy of consideration when not yet explored" (171). The dialectician takes up either side and proceeds to examine the issue. In such an interrogation a respondent may concede a proposition and phrase it as a question, for instance, "Is it true that one ought to seek after riches?" Put in this way the question becomes a dialectical proposition even though it is phrased as a question. It is an assumption for the respondent, one which he shares as a common opinion, and a challenge for the questioner. In this light, "Is it true that the earth stands still?" is a dialectical proposition, while "Does the earth move?" is a dialectical question. The first would be affirmed by most; but the second would be a doubtful query.

To help him explore a question or examine a proposition, the dialectician has recourse to the *topoi* or topics. These heuristic strategies are treated by

---

transcribed by William F. Edwards, *Tractatio de praecognitionibus et praecognitis* and *Tractatio de demonstratione* (Padua: Editrice Antenore, 1988). This manuscript has been translated in Wallace's *Galileo's Logical Treatises*, a companion volume to his *Galileo's Logic of Discovery and Proof*.

8. I have used Wallace's English translation of the *Introductio*. Page references following my quotations are to the manuscript.

Aristotle both in the *Topics* and in the *Rhetoric*.[9] Carbone simplifies and explains the purpose of topoi: "The argument or middle term here is commonly called a place, using the translation of the Greek *topos* or the Latin *locus* . . . although place designates the argument itself more than the seat of the argument or where the argument may be found. On this account a topic is generally defined as the seat of an argument or the place from which it can be obtained, for when topics are known arguments are easily discovered" (174). The reasoner in posing a question is actually asking whether a predicate can be applied to a subject. For example, one might ask, Is it true that the moon is composed of a crystalline fifth element? The inquirer is actually looking for the middle term that will permit these two to be joined: whatever might indicate the nature of the moon. Following Aristotle's treatment in the *Topics,* Carbone states that the middle term is what is furnished by the topoi, and that it is based on four major concepts or universals. Called the predicables by Aristotle, these are genus, species (both of which pertain to definition), property, and accident. The middles proposed to solve any problem pondered in a debate may be ultimately reduced to one of these four predicates. Porphyry added *differentia* to Aristotle's four predicables to clarify the invention of a definition by underscoring the consideration of "differences" in order to arrive at the species.

Aristotle notes among the major topoi related to definition: parts and wholes, relative things, opposites, and contraries; degrees of variation in differences; and numerical, specific, and generic sameness. Definition can be approached from a universal and from a particular standpoint, from what is prior and better known; the form and matter can also be examined, as well as the linguistic and grammatical details, inflections, or equivocity. Closely related to definition and sometimes helpful in arriving at the differentia that distinguishes something is the concept of property. In exploring the property of something, Carbone says, the arguer might have recourse to some of the same lines of reasoning as those used in definition. He might compare for "sameness" and "difference," and determine how a property belongs, and to what degree.

9. The history and interpenetration of the dialectical and rhetorical topical traditions are treated in Eleonore Stump's *Boethius's De Topicis Differentiis* (Ithaca: Cornell University Press, 1978) and her *Boethius's In Ciceronis Topica* (Ithaca: Cornell University Press, 1988); see also Niels J. Green-Pedersen's *The Tradition of the Topics in the Middle Ages* (Munich: Philosophia Verlag, 1984), and Otto Bird, "The Tradition of the Logical Topics: Aristotle to Ockham," *Journal of the History of Ideas* 23, no. 3 (1962): 307–23. Again see E. J. Ashworth on "Traditional Logic" and Lisa Jardine on "Humanistic Logic" in *The Cambridge History of Renaissance Philosophy.* Paolo Rossi finds that Renaissance uses of the topics limit choices in the face of multitudes of possibilities, "Mnemonical Loci and Natural Loci," in Marcello Pera and William R. Shea, eds., *Persuading Science: The Art of Scientific Rhetoric* (Canton, Mass.: Science History Publications/USA, Watson Publishing International, 1991), 77–88.

Both definition and property assume considerable importance in the arguments in the Copernican debate. In seeking to determine the nature of the moon's composition, the subject of the proposition posed above, Galileo had to consider how the moon's body as seen through the telescope compared to better-known crystalline bodies. Galileo found far different properties indicated by its appearance from those traditionally attributed to it, and he noted that it appeared to be more like the earth.

Accident is another predicable that figured prominently in scientific disputation, since it refers to something that might or might not belong to a subject. Whether the sunspots' shapes are accidental or not was one of the major issues debated by Scheiner and Galileo.

The topics of dialectic, then, have their origin in the predicables and enable one to find middle terms for the syllogisms of probable reasoning. As opposed to this, in the syllogism's most scientific form, necessary demonstration, the middle term is joined "necessarily" to the other terms on the basis of a causal connection that permits an inference to be drawn with certitude. In its dialectical form the syllogism cannot attain this rigor. Its middles are "something probable, invented to induce belief . . . so as to gain an assent to what is proved, although without absolute necessity" (172). In saying that they are "invented to induce belief" Carbone means by "invented" not that middles are made up, but that they follow from the process of topical invention just described, which is intrinsically related to ways of talking about how things are in reality. He points out that the predicables are an outgrowth of the "categories" described by Aristotle, namely, the categories of being, such as substance, quantity, quality, etc. Similarly, when he says that middles are invented so as to "induce belief," Carbone does not mean that these are chosen because they happen to convince an audience but rather that they are invented because in themselves they are convincing to the intellect. The dialectician uses the topoi to investigate the question from opposite sides so as to winnow the most probable solution.

In the animadversions at the end of the work, Carbone briefly summarizes the topoi as follows: "When something is proposed to be proved, we should examine the subject and the attribute of the question and for each we should look into the term, the definition, the denomination, the parts, the causes, the effects, the antecedents, the consequents, the adjuncts, the similars, the dissimilars, and the repugnants; from these, various arguments can be drawn, and these will offer us a vast supply for disputation" (206).

Rhetorical Reasoning

Like dialectical reasoning, rhetorical reasoning is also based on opinion, and it too employs topoi to invent middle terms. These however are chosen because they would seem to convince an audience. The proofs or arguments of rhetoric are concerned with contingent matters of everyday life and so can yield only what seems to be true. Again, rhetoric is like dialectic in attaining only to probabilities.

The differences between dialectic and rhetoric are significant. One of the most obvious is seen in the forms that discourse assumes in these arts. The question-answer logic that underlies dialectic takes a proposition-objection-refutation form in the scholastic genre of the disputation. This format was quite familiar to scholars in the period of our study. As such, it is different from the continuous flow of argument characteristic of a rhetorical piece. The uninterrupted discursive form of rhetoric is only a superficial difference, however, for both are concerned with supplying answers on both sides of a question.

The major difference lies in their aims. Dialectic seeks truth or what is probably true through the consideration of opposites. Rhetoric considers opposites too, but seeks what is persuasive or "persuasible" to an audience, as the Renaissance rhetorician Antonio Riccobono termed it in his commentary on Aristotle's *Rhetoric*.[10] Whether what is persuasible is as close to the truth as is possible to discern depends on the rhetor. The good rhetor will aim to persuade about what really seems to be the case. The sophist, on the other hand, seeks to persuade for the sake of persuasion itself, as Aristotle notes in the *Rhetoric*. The term is not a common one, yet Galileo uses the Italian *persuasibile* in the preface to the *Dialogue*. Perhaps he was acquainted with Riccobono's commentary. This would not be surprising, for he corresponded with Riccobono in 1588, before joining him on the faculty at the University of Padua. Riccobono taught there until shortly before his death in 1598.[11] Both dialectic and rhetoric, Aristotle observed, were considered to be applicable to any subject whatever, but whereas dialectic was normally applied to speculative matters, such as nature or metaphysics, rhetoric was thought to be concerned with three main areas—politics, justice, and virtue. Thus Aristotle offers advice for three kinds of rhetoric: political, judicial, and epideictic. The last variety, so useful for ceremonial occasions, was developed by the Greeks and Romans into a com-

10. Antonio Riccobono, *Aristotelis Ars Rhetoricae* (Venice, 1579), 209. Riccobono speaks of what is able to persuade, in Latin the *persuasibile*.

11. A letter from Riccobono of 11 March 1588, referring to a previous letter from Galileo is found in *Opere* 10:30. The passage from the *Dialogue* reads "ho giudicato palesare quelle probabilita' che lo renderebbero persuasibile, dato che la Terra si movesse," *Opere* 6:30.21–22. In Stillman Drake's translation of the *Dialogue* (Berkeley: University of California Press, 1962), the sentence reads "I have thought it good to reveal those probabilities which might render this plausible, given that the earth moves" (6). See the discussion of the *Dialogue* in chapter 9 below.

mon academic exercise and a form of public entertainment. Known as ceremonial, exhibitionary, or demonstrative rhetoric, it was used to special advantage by Renaissance authors in prefaces and dedicatory letters.

The distinctions found in Aristotle's teachings were familiar to many teachers of rhetoric in the period, especially in Italy, for the Greek text of the *Rhetoric* was newly studied by humanist scholars, becoming the subject of a number of commentaries.[12] Illustrious teachers such as Riccobono, Franciscus Robortello, and Carbone refer to the opinion of contemporary Italian scholars on Aristotle throughout their works. In addition, the writings of Cicero and Quintilian carried on the doctrine of rhetoric's concern with probable reasoning.

The methods used to accomplish their aims constitute another important difference between these sister arts. Dialectic uses argument or *logos* alone, while rhetoric employs two other means besides logos: *ethos,* the character of the speaker, and *pathos,* the emotions of the audience. Aristotle saw the last two appeals as specific to rhetoric. Ethos he thought to be especially compelling. Sincerity, good will, and knowledge of the subject are its source. It figured prominently in the discourse of Kepler and Galileo. Pathos was evoked also in the Copernican debates, both openly, through invective, laudation, allusion, and the exposition of examples, and implicitly through the audience's veneration or disapprobation of particular religious and philosophical tenets. In his commentary, Riccobono treats these appeals as the rhetorical equivalent of demonstration.

The logos of rhetoric is logical, but it is more informal in practice since its arguments are always developed with an audience in mind. Like dialectic, rhetoric employs induction and deduction, but in rhetoric induction takes the form of the example while deduction becomes the rhetorical syllogism, the enthymeme. Rhetorical induction does not follow the practice of enumerative induction practiced in dialectic. A few examples or even one representative example, vividly described, may be enough to make a point. The enthymeme, however, is for Aristotle the most important form of rhetorical reasoning. Its importance rests on the fact that it draws on the interests and values of the audience, thus enlisting their emotions as well as their minds. In enthymemic reasoning the orator need not articulate both of the premises of an argument or state the conclusion when these can be supplied by the audience. Aristotle noted that the audience particularly enjoys getting that unspoken element, becoming a participant in the argument.

12. F. Edward Cranz mentions ten Latin translations from the Greek, one Italian, and nine commentaries for the sixteenth century, *A Bibliography of Aristotle Editions: 1501–1600,* 2d ed. with addenda and revisions by Charles B. Schmitt. Bibliotheca Bibliographica Aureliana, vol. 38 (Baden-Baden: Valentin Koerner, 1984), 220–21.

While the nature of the enthymeme as Aristotle conceived it was not fully conveyed by Roman rhetoricians (partly because of ambiguities in the text and partly because of the imperfect transmission of peripatetic teachings), nevertheless the principles of enthymemic reasoning can clearly be seen at work in the writings examined here. For example, to argue that the Copernican system should not be accepted because it contradicts the teachings of the Scriptures is to use an enthymeme, the unstated major premise being that whatever contradicts the teachings of Scripture is untenable. The mere suggestion that biblical truth was challenged by the Copernican system was enough to strike fear into the hearts of the faithful. Just such fear moved the Grand Duchess of Tuscany, Madame Christina of Lorraine, to ask Galileo's friend Dom Benedetto Castelli for reassurance, and it further prompted Galileo's famous letter to Madame Christina, discussed at length in chapter 7.

The strategic use of the three appeals—logos, ethos, and pathos—constitute the essential artistry of rhetoric. For this reason they are also referred to as the "artistic proofs." Inartistic proofs are also employed by rhetoricians, but these are the givens: documents, laws, or authoritative pronouncements that can be brought to bear on an issue. Arguments from authority, the source of so much rancor in the controversy over cosmology, fall into this category.

The artistic and inartistic proofs, the nature of rhetorical argument, and the topoi are all treated within the first canon of the discipline of classical rhetoric: invention. The other four parts of the art are arrangement, style, memory, and delivery. Of these, style was to generate the greatest interest in the recovery of rhetoric during the Renaissance, although rhetorical invention was also to excite renewed interest, especially on the Continent.[13]

13. The best summary of classical rhetoric is by George A. Kennedy, *Classical Rhetoric and its Christian and Secular Tradition from Ancient to Modern Times* (Chapel Hill: University of North Carolina Press, 1980). An earlier study by Charles Sears Baldwin is smaller in scope but still a classic, *Ancient Rhetoric and Poetic* (New York: Macmillan, 1924, rpt. Gloucester, Mass.: Peter Smith, 1959). No survey of Renaissance rhetoric has yet been attempted on this scale. A number of studies of Italian rhetoric are germane to this study: The essays by Paul Oskar Kristeller, *Renaissance Thought and its Sources,* ed. Michael Mooney (New York: Columbia University Press, 1979), and that of Hanna Gray, "Renaissance Humanism: The Pursuit of Eoquence," *Journal of the History of Ideas* 24 (1963): 497–514; the longer works by Baldwin, *Renaissance Literary Theory and Practice: Classicism in the Rhetoric and Poetic of Italy, France, and England, 1400–1600,* ed. Donald Leman Clark (New York: Columbia University Press, 1939); Hans Baron, *The Crisis of the Early Italian Renaissance: Civic Humanism and Republican Liberty in an Age of Classicism and Tyranny,* 2 vols. (Princeton: Princeton University Press, 1955); Jerrold Seigel, *Rhetoric and Philosophy in Renaissance Humanism* (Princeton: Princeton University Press, 1968); Cesare Vasoli, *La dialettica e la retorica dell'Umanesimo* (Milano: Feltrinelli Editore, 1968); Nancy S. Struever, *The Language of History in the Renaissance: Rhetoric and Historical Consciousness in Florentine Humanism* (Princeton: Princeton University Press, 1970);

Invention

The dynamic methodology of invention, the topoi, are shared by rhetoric and dialectic.[14] Aristotle, in fact, refers readers of the *Rhetoric* to the *Topics* for a fuller understanding of these. Originally the topical invention described was not seen as a prescriptive methodology, for it was founded on Aristotle's study of psychology. His observations led him to conclude that the mind generally investigates a new thing by looking at what it resembles, by noting differences, and by attempting to classify it. The topoi were simply his extrapolation and systemization of these mental processes. In identifying and organizing them he thought to make these operations of the mind more efficient. As he conceived them, the topoi are prompts or stimuli to obtaining knowledge.

In the *Rhetoric*, Aristotle is less systematic in his treatment of the topics but he introduces a helpful distinction between the dynamic conceptual prompts applicable to any subject, which he calls common topics, *koinoi topoi*, and the material lore of a particular subject, which he calls the special topics, the *eidē*. The koinoi topoi, then, are the heuristic devices he treats in the *Topics*, where he classifies them formally under the predicables, though he does not do this in the *Rhetoric*. Aristotle also introduces in the latter work a kind of subgenre of the common topics, the *koinē*, which might be termed "very common" topics. These are basic ways of looking at a problem

---

and John W. O'Malley, *Praise and Blame in Renaissance Rome: Rhetoric, Doctrine and Reform in the Sacred Orators of the Papal Court, c. 1450–1521* (Durham: University of North Carolina Press, 1979). Like Baldwin, Bernard Weinberg also describes the relation of rhetoric to poetic in *A History of Literary Criticism in the Italian Renaissance* (Chicago: University of Chicago Press, 1961), ch. 1; essays by Kristeller, O'Malley, and others in James J. Murphy, ed., *Renaissance Eloquence* (Berkeley: University of California Press, 1983) are also illuminating. An excellent discussion of the relation of rhetoric to logic on the Continent as well as in England is in Wilbur Samuel Howell, *Logic and Rhetoric in England, 1500–1700* (Princeton: Princeton University Press, 1956). Brian Vickers makes a convincing case for the importance of the figures of speech as intrinsic to rational and pathetic appeals during the Renaissance in his *In Defence of Rhetoric* (Oxford: Oxford University Press, 1988), chs. 5 and 6. Maurice Finocchiaro has recently amplified his conception of rhetoric in "Varieties of Rhetoric in Science," *History of the Human Sciences* 3, no. 2 (1990): 177–93. His view is ahistorical, recognizing three kinds of rhetoric as operative in science: a stance not necessarily reflected in a scientist's work but used to project an image; eloquent communication; and persuasive argumentation. Finocchiaro does not distinguish beween rhetorical and dialectical argument and regards demonstration as logical and mathematical in character.

14. Aristotle *Rhetoric* 1.1.1354a. William M. A. Grimaldi's commentary on the first book of the *Rhetoric* is especially helpful in its discussion of the importance of this comparison, *Aristotle, Rhetoric I: A Commentary*, (New York: Fordham, Fordham University Press, 1980), 1–2. See also his valuable treatment of the topoi in *Studies in the Philosophy of Aristotle's Rhetoric* (Wiesbaden: Franz Steiner, 1972), 115–35.

by considering the following: whether something is possible or impossible; whether it is larger or smaller than something else, or more or less; and whether a thing has happened in the past or may be expected to happen in the future.

The second kind of topoi, the eidē, does not appear in the *Topics*. These are not like the common topics, patterns of thinking that may be applied to any subject; rather they are the matter or particular knowledge contained in the individual arts and sciences. When an inquirer needs more specific knowledge than the common topics can yield, he must seek these within the discipline itself. Carbone offers a good description of the eidē in the *Introductio* described above: "First one should learn the proper meanings of ambiguous words and the terminology of that art; then one should perceive the first principles on which the entire discipline depends. After that one should learn the subject matter in a general way, the several parts, causes, and properties; following that one should descend to particulars. One should do this in physics, metaphysics, and ethics, and in other arts, and in all of learning" (207).

In his discussion Carbone has drawn on Aristotle's treatment of the special topics as explained in the *Rhetoric*. It is in this context, also, that Aristotle makes an important distinction, one that is especially significant for the writings in the Copernican debate. He remarks that when one goes more deeply into a subject, one enters into that science and leaves the area of rhetoric and dialectic behind.[15] The distinction is central to Aristotle's conception of the nature of discourse in the arts and sciences. When one develops a scientific proof one must employ the principles of the science and the methodology either of dialectic or of demonstration. Rhetoric comes into play when an issue is presented to the public. It is used to persuade others to form a different or a deeper opinion or to persuade them to take a particular action. But, theoretically, what causes the practitioner of the science to come to a conclusion he regards as true and certain is not rhetorical reasoning but demonstration.

As a Renaissance scholar would perceive it, when a scientist turns rhetorician, he still rests his scientific conclusions on scientific evidence and reason, but he persuades his public to accept his views by rhetorical appeals. In doing so he may present himself as a serious, careful, well-respected scientist, using an ethical appeal; and he may impress upon his audience that they may well suffer harm if they do not act upon his opinion, calling pathos into play. Indeed he may or may not decide to provide the scientific proof for his conclusion in detail, depending upon its complexity and the interest or capacity of his audience.

15. Aristotle *Rhetoric* 1.2.1358a.

The special topics seem to have evoked little interest in rhetorical literature subsequent to Aristotle, for they were not developed by Cicero and later rhetoricians. Since for Cicero and Quintilian a broad education was essential preparation for an orator, the eidē, drawn as they were from the special knowledge of a subject, may have been considered by them superfluous.[16] The common topics, koinoi topoi, are picked up by Cicero and by Boethius in a somewhat altered form: in their presentations the "topics" or "places" became a means of expanding a subject more than a means of making predications.[17] With them the organizing principles were shifted from the predicables of definition, property, and accident to a new division of loci: that treating attributes of persons and actions, following the practice of Cicero's De inventione. Under "person" fall topics such as "name, nature, manner of life, fortune, habit, feeling, interests, purposes, achievements, accidents, speeches made."[18] In considering "action," Cicero explains that these attributes of the action are "partly coherent with the action itself, partly considered in connexion with the performance of it, partly adjunct to it, and partly consequent upon its performance."[19] Concerning performance of actions he turns to topics of place, opportunity, time, occasion, manner, and facilities. The topics of action sound familiar because they have retained a place in the judicial oratory of our own day. Those of person find wide application by our authors, particularly in letters of dedication. Copernicus, Galileo, Kepler, and others take advantage of the range of compliments the topics of person open up.

16. Carolyn Miller has investigated the plight of the special topics in "Aristotle's 'Special Topics' in Rhetorical Practice and Pedagogy," *Rhetoric Society Quarterly* 17 (Winter 1987): 61–70. See also Michael C. Leff, "The Topics of Argumentative Invention in Latin Rhetorical Theory from Cicero to Boethius," *Rhetorica* 1 (Spring 1983): 23–44.

17. See the discussion of the usage of the commonplace and the place of dialectical topoi in Sister Miriam Joseph Rauh's ground-breaking work, *Shakespeare's Use of the Arts of Language* (New York: Columbia University Press, 1947, 1949) abridged and reprinted as *Rhetoric in Shakespeare's Time* (New York: Harcourt 1962), 308–44; and the clear presentation of Sister Joan Marie Lechner, *Renaissance Concepts of the Commonplaces* (New York: Pageant, 1962), 76–113. Rauh shows that many of the figures of thought were simply topoi ramified in various ways. In fact, some Renaissance scholars such as Philipp Melanchthon classified figures under definition, causes, contraries, etc. (Rauh 36–40). For general background concerning the Continental influences on English logic and rhetoric see Howell, *Logic and Rhetoric in England, 1500–1700* (n. 13 above). Howell discusses also the effects of the reforms of Peter Ramus, a topic explored in depth by Walter J. Ong in *Ramus, Method, and the Decay of Dialogue* (Cambridge: Harvard University Press, 1958). Norman E. Nelson provides a persuasive critique of the reforms attributed to Ramus in "Peter Ramus and the Confusion of Logic, Rhetoric, and Poetic," *Contributions in Modern Philology* 2 (April 1947): 1–22.

18. Cicero *De inventione* 1.24.34. Trans. H. M. Hubbell (Cambridge, Mass.: Harvard University Press, 1949).

19. Ibid., 1.26.37–38.

In this form the topics become not so much modes of thinking as they do devices for the recall of material elements. Such an alteration meant that the Aristotelian technique lost some its dynamism as a heuristic and could become, as was to be the case in less imaginative hands, a method of storing and recalling commonplaces appropriate to a subject. It was this later conception of the rhetorical topoi that was taught in the schools of the Middle Ages and the Renaissance through the two major rhetorical texts: Cicero's *De inventione* and the Pseudo-Ciceronian *Ad Herennium*. And these were among the first volumes to be printed with the invention of the printing press. Along with the recovery of Quintilian's *Institutio oratoria,* the Ciceronian works continued to dominate the teaching of rhetoric, furnishing most of the basic principles and concepts for rhetorics written during the period.[20]

By the time the term passed into English in the sixteenth century the "commonplace" became associated with a repertoire of stock or "canned" responses. The creativity in the commonplace was then most often directed to literary uses and especially to finding expressive figures of speech. The essentially dynamic character of the topoi as a means of inventing new subject matter or discovering hitherto hidden significances in known things was not always grasped by teachers of rhetoric in the scholastic curriculum. In time the two different conceptions of the term were conflated by teachers, and the rhetorical topoi became a technique for filling in the blanks.

In dialectic, however, the topoi retained their dynamic character. We find them used continuously by the Copernicans and their opponents. The topoi as originally presented in Aristotle's *Topics* were recovered during the "twelfth century Renaissance," stimulating scholars to train these investigative tools upon puzzling issues of philosophy and theology. They found expression in the innumerable disputations and the *Summae* of the period, and in the Renaissance they guide the flow of ideas within the more elegant and discursive forms of the dialogue, letter, and treatise. As we have seen, by the sixteenth century new editions and commentaries on Aristotle's *Rhetoric* appeared, reviving knowledge of Aristotle's conception of the rhetorical topoi. Some scholars attempted to simplify the teaching of invention by combining rhetorical and dialectical topics. Carbone reflected the trend of the period when he published his *De oratoria, et dialectica inventione, vel de locis communibus* (1589). Peter Ramus carried simplification to extremes, giving

20. See Baldwin, *Renaissance Literary Theory and Practice,* 53–64; Kennedy, *Classical Rhetoric,* ch. 10.

21. The classic study of the significance of Ramus is Walter Ong's *Ramus*. Ong treats the earlier reforms of logic by Agricola, the development of Ramus's thought and the spread of his ideas in Northern Europe and Spain.

all of invention and arrangement to logic and allotting only style and delivery to rhetoric.[21]

Arrangement

The classical conception of the oration and the nature of its parts was adapted to all kinds of discourse, from letters, to sermons, to poetry. The protocols continued to be followed in the Renaissance, and, indeed, still lie behind the teaching of "organization" in composition classes today. As explained in the base text of rhetorical teaching of the Renaissance, *De inventione,* and repeated in almost every other rhetorical manual, there are six parts to the oration: exordium, narration, partition, confirmation, refutation, and peroration. The exordium is, of course, the introduction, and its purpose is to induce the audience to be "well-disposed, attentive, and receptive."[22] Narration, in this oratorical context, furnishes an account of the main facts or the background of the issue, and it can be dispensed with if these are well known. The partition sets forth the plan of attack the orator intends to follow in treating the issues in the case and is usually preceded or followed by the thesis to be supported. Following this, the confirmation first provides arguments for the orator's position, and then a refutation of the opponent's arguments ensues. The order of these can be reversed, depending on which side of the controversy presents its case first; or alternatively, the two parts can be interwoven, whichever the orator deems would best suit the audience's views. In the peroration the orator reviews his arguments, mentioning the shortcomings of the opponent's case and endeavoring to leave a favorable impression by arousing pity or sympathy in the audience.

In describing each of these parts Cicero offers advice for magnifying the orator's case. That these tips became second nature to scholars of the Renaissance is obvious from even a brief glance at their works. The methods he suggests for arousing audience attention in the exordium are perennially apropos: "We shall make our audience attentive if we show that the matters which we are about to discuss are important, novel, or incredible, or that they concern all humanity or those in the audience or some illustrious men or the immortal gods or the general interest of the state; also if we promise to prove our own case briefly and explain the point to be decided or the several points if there are to be more than one" (1.26.23).

Cicero's techniques for discrediting the opponent's argument are similarly apt: "Every argument is refuted in one of these ways: either one or more of its assumptions are not granted, or if the assumptions are granted it

22. Cicero *De inventione* 1.15.20.

is denied that a conclusion follows from them, or the form of argument is shown to be fallacious, or a strong argument is met by one equally strong or stronger" (1.42.79).

The peroration receives extended treatment with the explanation of efficacious approaches and commonplaces that might be used, such as *indignatio* and *conquestio* or lament (the last designed to evoke pity.) But this last tactic should not be prolonged, Cicero advises, citing the observation of Apollonius, "Nothing dries more quickly than tears" (1.56.109).

Style

Style was of great importance to the Sophists, but was given little attention by Aristotle. In the Roman era and again in the Renaissance, style became a dominant concern of teachers and scholars. Levels of style (grand, middle, and plain), principles of effective diction, and sentence construction were painstakingly addressed under the rubric of style. But the component that generated the most interest, although least familiar to us today, was figurative language. Since the number of figures introduced to students in the Renaissance was often more than two hundred, only those that appear most often or were prominent in the Copernican debate will be described here.

The best explanation of the stylistic devices practiced in the Renaissance is presented in the *Ad Herennium*. Long thought to be the work of Cicero but actually the work of an unknown author, this became the second major rhetorical text of the Middle Ages and the Renaissance. It was valued particularly because of its treatment of style, a matter not developed in Cicero's *De inventione*. Many pages of this *Rhetorica secunda* are devoted to descriptions and examples of the figures of speech, which Renaissance authors generally adopted. The basic division was between figures of speech and figures of thought, with the first category being subdivided into schemes and tropes. The schemes included some of the favorite devices of style in the Renaissance: those of word construction, involving changes or repetition of letters, spelling, and sounds, plus schemes of parallel construction of words, phrases, and clauses. Among these, antithesis and reasoning by contraries were especially favored by Galileo. The schemes of repetition were enjoyed by the controversialists because they are so useful in arousing emotion.

The tropes, so called because these figures change or turn the ordinary meaning, abound in the work of our authors. The most frequently used were metaphor, metonymy, and synecdoche, while litotes or understatement, personification, the rhetorical question, irony, and allegory were also quite common. Figures of thought included frankness, description, exemplification, address to different groups in the audience, emphasis, and dialogue. Again, Galileo excels in his imaginative application of these devices. That the figures were not simply added to embellish prose and poetry but to create and clarify meaning has been convincingly argued and illustrated by Rauh and Vickers.[23] So well-drilled in the figures were Renaissance writers that their use became second nature, resulting in an extraordinary facility in elucidating their subjects. Erasmus's *De duplici copia rerum et verborum* (1511) was a popular source of such copiousness.

Through a series of exercises, increasing in difficulty at each step, students were trained in the principles of style. This program, the *progymnasmata* of the Sophists, was recovered in the Renaissance by Rudolf Agricola, and his translation became a popular text throughout Europe. Translation, paraphrase, and imitation were among the steps to be mastered in turn, and these culminated in the writing of compositions or themes. We have an example of Galileo's own attempts to translate Isocrates from his early school years.[24]

Many of the principles of organization and style mentioned here were also incorporated in a number of other arts—poetry, painting, sculpture, music, architecture, and even theology. Thus, the disciplines of this period enjoyed a cross-disciplinary fertilization even though they remained, except for poetry, consciously separated from rhetoric.

The Teaching of Rhetoric

It would be impossible to provide an account of the varied emphases in the teaching of rhetoric to which the authors whose writings are encompassed in this study were exposed, but some general observations can be made. With the exception of Copernicus, most were educated in the sixteenth century; however, the Polish astronomer was subject to the influence of humanism at the University of Cracow and in his later studies at Ferrara and Bologna.[25]

In the fifteenth century the *studia humanitatis* of the Italian Renaissance began to dominate education in the schools and at a higher level in the universities, whence it spread throughout Europe. As a result the art of rhetoric was studied in a broader context, that is, in conjunction with grammar, poetics, moral philosophy, and history. The new academic discipline of poetics was closely allied to rhetoric, for in the Middle Ages the writing of poetry was often treated as a subgenre of rhetoric, while reading and interpretation of poetry and drama was treated in grammar. In the Renaissance poetics in-

23. Rauh, *Shakespeare*, describes the use of figures in England while Vickers, *Defence*, treats these in England and on the Continent.

24. *Opere* 9:283–84.

25. Paul W. Knoll, "The Arts Faculty at the University of Cracow at the End of the Fifteenth Century," in Robert S. Westman, ed., *The Copernican Achievement* (Berkeley: University of California Press, 1975), 152–55.

cluded the study of the newly recovered *Poetics* of Aristotle, Horace's *Ars poetica,* and the recovered literature of the Greeks and Romans. These works, often taught simultaneously with rhetoric, served to reinforce and demonstrate principles of style. Rhetoric was seen as the creative art behind the classical literature that so delighted scholars of the time. Eloquent expression was admired and cultivated by teachers at all levels of education. By the period of our interest, the late sixteenth and early seventeenth centuries, scholars like Galileo preferred to emulate the richness of classical expression through the vernacular, rather than attempt to mimic the language and style of Augustan Latin as had been done in the earlier Renaissance. The practice of rhetoric was also given new vitality by the revival of classical literature. The recovered orations of Cicero, which received much attention in rhetoric classes, stimulated a renewed interest in Ciceronian civic oratory. Not only epideictic or demonstrative oratory but political oratory was once more practiced in the forums and courts of Italian cities. The function of rhetoric was to make all discourse pleasurable, to delight and persuade at the same time. By delighting one persuaded, thought Sperone Speroni.[26]

The scholastic emphasis on logic, and on dialectic in particular, continued to be strong in the university arts faculties. Aristotle's *Organon* remained important throughout the period and into the nineteenth century.[27] In lower-level education, that in the schools, simplified revisions made by Agricola and his disciple Peter Ramus were felt most strongly in northern Europe and England during the late sixteenth and seventeenth centuries.

26. Speroni's views are described by Eugenio Garin, as are those of Stefano Guazo, who characterized the function of rhetoric in communication on any subject of concern as "civil conversation," in *Italian Humanism,* trans. Peter Munz (Oxford: Basil Blackwell, 1965), 158–62. Hans Baron's *The Crisis of the Early Italian Renaissance* notes a strong influence of rhetoric on politics. Jerrold Seigel describes the effect of scholastic and humanist training on discourse in *Rhetoric and Philosophy,* ch. 8. Recently Anthony Grafton and Lisa Jardine have raised doubts concerning the received notion that the ideals of the citizen-orator were effectively revived, *From Humanism to the Humanities: Education and the Liberal Arts in Fifteenth- and Sixteenth-Century Europe* (London: Duckworth, 1986), xii–xiii, 12–25.

27. Kristeller has pointed out the importance of Aristotle in the period in "Humanism and Scholasticism in the Italian Renaissance," *Renaissance Thought,* 85–105, and in other essays. The courses offered in the Jesuit College at Rome, the Collegio Romano, show the influence of Aristotle's *Rhetoric* and the *Topics;* see my "The Rhetoric Course at the Collegio Romano." In addition, the archives of the Universities of Pisa and Padua cite Aristotle's *Rhetoric* as a text in courses in the classics and studia humanitatis in the late sixteenth century (Archivio di Stato-Pisa, Rotulus almi studii; Archivio storico-Padova). Numerous commentaries on the *Rhetoric* during the period attest to its influence; see Charles Lohr, *Latin Aristotle Commentaries,* Vol. 2: *Renaissance Authors* (Florence: Leo S. Olschki, 1988). For a general view of rhetoric in Jesuit education see Aldo Scaglione, *The Liberal Arts and the Jesuit College System* (Amsterdam and Philadelphia: John Benjamins, 1986), 84–86; and Paul Grendler, *Schooling in Renaissance Italy* (Baltimore: Johns Hopkins University Press, 1989), 378–79.

The simplification of the teaching of rhetoric and dialectic introduced by the Reformers, aimed at reducing redundancies, actually had the effect of preserving the importance of logic and accentuating the subordination of rhetoric to it. In the reformed version, rhetoric became even more closely equated with style. The effect upon scientific discourse was to emphasize the separation of rhetoric from it and to magnify logic's importance as the instrument of the highest form of knowledge, science.

Yet rhetoric, in its classical five-part form, continued to be recognized as an autonomous discipline in the universities. Its principles were taught with varying emphases in Italy and throughout the rest of Europe during the sixteenth and seventeenth centuries. It was nourished by study of Cicero, the *Ad Herennium,* and Quintilian, by Latin translations and commentaries on Aristotle's *Rhetoric,* and by the new rhetorics of the period that synthesized the teachings of Aristotle, Cicero, and Quintilian.[28]

The Jesuit schools, which had multiplied throughout Europe, were very influential in preserving the classical conception of rhetoric, emphasizing as they did not only style but invention as the fountain of eloquence. Their interest in the fecundity of the topoi for both logic and rhetoric insured the survival of probable reasoning as the most effective tool for investigation of a subject. But their concern for Aristotelian logic also helped to preserve scholastic teaching regarding the difference between scientific and rhetorical proofs.

Such was the disciplinary experience of those who took part in the great controversy about the true nature of the cosmic system. They were well trained in the verbal arts, tediously rehearsed in the principles of dialectic and demonstration, and they thought eloquent expression and skill in rhetorical argument marked the superior, cultured courtier or scholar. In his day, Galileo was admired and reviled no more for what he discovered than for how he revealed it, embellished it, and defended it. The distinctions noted here between dialectic, rhetoric, and demonstration furnish the key to understanding the nature of the second revolution in science and how it was effected.

Current Views of Rhetoric and Dialectic

Recent discussions of rhetoric and science often do not consider the historical role of dialectic in science. Alan Gross's work *The Rhetoric of Science* mentions dialectic in passing at the end, seemingly as an afterthought to acknowledge that it was an approach recommended by Aristotle.[29] He

28. Kennedy, *Classical Rhetoric,* 195–219.
29. Alan G. Gross, *The Rhetoric of Science* (Cambridge: Harvard University Press, 1990).

would like, however, to redefine the provinces of these arts and make rhetoric include both logic and dialectic.

Marcello Pera, pursuing a similar vein of argument in an erudite essay in *Persuading Science: The Art of Scientific Rhetoric,* thinks dialectic's mission impractical.[30] "Not only are there no impersonal and universal rules of inquiry, but even if we take up a more modest position and consider local rules of method, it is easy to show that they cannot act as neutral yardsticks" (48). Nevertheless, the answer does not lie in discarding method. Pera would rather consider broadening dialectic to give it a rhetorical cast, making it a rational enterprise aimed at gaining the assent of an audience. He is well aware that Aristotle differentiated between "rhetoric and dialectics as the art of persuading and the art of arguing or confuting, respectively," but he feels compelled to depart from Aristotle's separation of the two disciplines (35). "Rhetoric is not the counterpart of dialectics," he writes, "rather it includes it. At least in science, persuading an audience or converting it includes contrasting rival opinions" (35). Again, for Pera science depends upon assent to theories. As he explains, "to be rational is to accept those theories, to work out those problems, to take those decisions that are supported by good reasons, 'good' in the sense that they won a victory in a concrete debate conducted according to a concrete configuration of the basis of scientific dialectics" (49). Pera sees "scientific rhetoric" as "the set of those persuasive, argumentative techniques scientists use in order to reach their conclusions" (35).

Philip Kitcher, in "Persuasion" in the same book, examines the need for terminology to describe what goes on in scientific persuasion. He coins the term "dead rhetoric" to describe devices of arrangement that indicate relationships of parts of a proof to each other that are so commonly used they have become invisible (8 and note 6).[31]

The Renaissance scholars whose works are examined here would understand these views but think them not precise enough. "Scientific rhetoric" would be for them a contradiction in terms. They would limit the aim of "good" scientific dialectic to the finding of what appears to be true, regardless of whether others assented or not. They would think that assent should indeed come in time and should be given by any rational person, but they would recognize the possibility that some may be blinded by prejudice or simply fail to see the point.

Galileo himself provides an example of the faith scholars placed in the

30. Marcello Pera, "The Role and Value of Rhetoric in Science," in Pera and Shea, eds., *Persuading Science,* 29–54.

31. Philip Kitcher, "Persuasion," in Pera and Shea, eds., *Persuading Science,* 3–27.

dialectical process in comments he made concerning the strength of his own arguments for the Copernican system.[32] He remarks that were the decision on Copernicus to be won by vote, he would acknowledge defeat and even permit each opposing vote to be counted as ten times to his one, if that decision were made "by persons who had perfectly heard, intimately penetrated, and subtly examined all the reasons and evidence of the two sides. . . . "[33]

The rational part of the argument would be expected to carry the decision. Should ethos or pathos attempt to tip the scale, correction or reproach would be an acknowledged right of the opponent. Our authors would have been acquainted with those additives as well as the ploys of deceitful reasoning Aristotle describes in *Sophistical Refutations*. Whether they always adhered to the ideals implicit in the notion of the separation of these fields is another matter. The "glue" used to bind parts in scientific proofs would probably be seen as more analogous to grammar than to rhetoric. That analogy has been used, of course, very effectively by both John Henry Newman and Kenneth Burke to examine the operation of elements of reasoning and of discourse.[34]

The appropriation of the term rhetoric to denote all varieties of persuasion has resulted in the current aggrandizement of the field, but at the same time such expansion has diminished the term's definitive power. "Persuasion" would seem a far more apt term to describe the conviction attendant upon a valid mathematical proof, a brilliant dialectical argument, or a rhetorical address. That broader term more clearly denotes all compelling discourse from demonstrations of science to declarations of love without implying that all are essentially "rhetoric." Such a classification would not preclude an examination of elements that *are* rhetorical, or, for that matter, literary, or grammatical, or philological.

32. These notes, called "Considerations on the Opinion of Copernicus" by Favaro (*Opere* 5:351–376), are discussed at length in chapter 7 below.

33. The translation is by Maurice Finocchiaro, *The Galileo Affair* (Berkeley: University of California Press, 1989), 73.

34. Newman's work analyzes the basis of religious conviction, *A Grammar of Assent* (Notre Dame, Ind.: University of Notre Dame Press, 1979), and Burke describes the role of symbols in persuading humans to act, *A Grammar of Motives* (Berkeley: University of California Press, 1945).

PART ONE

THE CELESTIAL
REVOLUTION

CHAPTER TWO

Copernicus's Revolutionary Thesis

To argue that a significant change took place in scientific discourse during the century following the publication of Copernicus's *De revolutionibus orbium coelestium*, a convincing proof text is needed. The best place to begin would appear to be Copernicus's monumental work itself, for it was offered as a persuasive defense of a new cosmology. At the time of the publication of his great work, Copernicus was a respected and well-known astronomer. He presented his case in the traditional manner, using what he conceived to be methodology expected by his readers. An examination of the kinds of proofs he offered should serve to point up the differences alleged in later works that debated his thesis.

In the first eleven chapters of the first book of *De revolutionibus*, Copernicus summarizes what he considers to be the major arguments for the new system. Our analysis of this part of the work will not be exhaustive nor will it provide new mathematical insights; rather it will demonstrate in brief the character of the proofs offered in this early stage of the revolution in discourse. The place of mathematical demonstrations in his discussion, the use of topoi as a basis for the dialectical arguments, and the role given to rhetoric all clearly show the Aristotelian nature of his conception of the discourse appropriate to astronomy. The focus of his arguments in the preface and the first book shows that Copernicus feared his thesis would be resisted, but not so much on theological grounds as on philosophical. He merely notes that scriptural difficulties might be raised by insignificant troublemakers. The shift to a strong rhetorical character so obvious in the later Copernican debate occurs as the heliocentric thesis becomes more widely known and its arbiters are forced to face not only philosophical objections but theological problems engendered by the new system.

The Ptolemaic Precedent

De revolutionibus orbium coelestium was closely patterned after the text that mapped out the reigning Aristotelian-Ptolemaic cosmology, the *Almagest* of Ptolemy. In fact it followed so nearly the mathematical program of that work that Alexandre Koyré could say: "Copernicus discovered practically nothing. By shifting the centre of all motion from the Earth to the sun, or more exactly, to the centre of the Earth's orbit he inverted the system of the Universe, but not the mathematical structure of astronomical knowledge; and from this point of view he was the most important follower of Ptolemy."[1] Koyré grants that Copernicus used more modern mathematical tools than Ptolemy—trigonometry rather than the more primitive Greek astronomy—but it was not in his computations and observations that the later astronomer outstripped the earlier. It was his vision of the nature of the universe, its geometrical form, that enabled Copernicus to posit a system superior in explanatory power to Ptolemy's.

Nonetheless, Copernicus was not "modern," says Koyré.[2] The universe of this Renaissance astronomer is founded on assumptions that would hardly be acceptable today. The arguments he presents are premised not on the old Aristotelian astro-biology but on a new astro-geometry.[3] Whether we accept Koyré's Platonizing of Copernicus or whether we grant the astronomer more realist insights regarding the nature of the universe, his argumentative strategies are instructive for what they demonstrate about the way in which scientists of the period established what are today called theories.

Since so much of what Copernicus says in arguing for the heliocentric system is related to the reigning thesis of Ptolemy, a brief look at the *Almagest* seems appropriate. We know little about Claudius Ptolemaeus, but ancient sources and his writings indicate that he studied and later wrote his mathematical treatises in or near Alexandria during the first part of the second century A.D. He lived until about 178, and we can assume from his writings that he had the benefit of a first-rate Hellenistic education. The *Almagest,* especially its first book, is remarkably clear and pleasant reading even for the nonspecialist, and it enunciates at the outset the metaphysical and physical principles on which the work is based.[4] Rhetoric does not play an appreciable role in the book. The text originally formed part of what was

1. Koyré, *Astronomical Revolution,* 24.

2. Ibid., 65–66.

3. Other historians of science see Copernicus in a different light. See the arguments by Edward Rosen in *Three Copernican Treatises* (New York: Dover, 1959), 22–33. Rosen discusses the opposing views and the position of Pierre Duhem, who criticized Copernicus for his realist views (33).

4. I have quoted from the translation of G. J. Toomey, *Ptolemy's "Almagest"* (London: Duckworth, 1984), referring also to the translation by R. Catesby Taliaferro in *Great Books of the Western World,* vol. 16 (Chicago: Encyclopaedia Britannica, 1952), 526–28. Taliaferro explains the origin of the title *Almagest* in the Arabic appellation of "the greatest" *al megistē* (the Arabic prefix *al* added to the Greek *megistē*). The work had originally been entitled *The Mathematical Composition* (ix). Olaf Pedersen provides a balanced and erudite discussion of the *Almagest* in his *A Survey of the Almagest* (Odense: Odense University Press, 1974).

termed "The Mathematical Collection." Eventually the astronomical text became known by its sobriquet: "The Greatest," *megistē* in the Greek prefixed by the Arabic article *al*. My comments will be focused only on the first of the thirteen books of the *Almagest,* where Ptolemy lays out the general assumptions of the work.

In the preface Ptolemy analyzes the sciences in general with regard to their subject matter and aims. He separates the theoretical sciences from the practical, noting the progressive character of the former and the repeated application demanded by the latter. For this reason, Ptolemy says, he trained himself always "to strive for a noble and disciplined disposition" even in small matters, while devoting his intellectual endeavors to theories "which are so many and beautiful," especially those termed "mathematical." Following Aristotle, he divides the theoretical sciences into three kinds: the physical, the mathematical, and the theological, ranging from the visible through the invisible realm. All things are composed of substance, form, and motion, he says, but none of these three are visible in themselves. They can be the objects of thought only. Consideration of "the first cause of the first motion of the universe" leads one to the study of "an invisible and motionless deity," which is the subject of theology. Knowledge in this type of science comes only through conjecture and imagination. Physical science, on the other hand, studies the quality of matter and motion of corruptible things in the sublunar regions. Intermediate between these sciences is the study of form and local motion in regard to figure, number, magnitude, position, time, and such, which is the domain of mathematical science. Its subject matter falls between the other two "because it can be conceived of both with and without the aid of the senses" and because "it is an attribute of all existing beings without exception, both mortal and immortal." It changes with mortal things and preserves the unchanging form of celestial ethereal nature (35–36).

The reason for this philosophical discussion, and for our interest in it, becomes apparent in the subsequent exposition. In speaking of the matter studied by theology and physics, Ptolemy says, we have here only "guess work." This is so for theology "because of its completely invisible and ungraspable nature," and for physics "because of the unstable and unclear nature of matter." He adds, "for there is no hope philosophers will ever be agreed about them." Only in the area of mathematics can "sure and unshakable knowledge" be provided. Arithmetic and geometry use "indisputable methods" (36). Here Ptolemy departs from Aristotle, who believed that demonstration was likewise possible in the other theoretical sciences. Ptolemy says it is the certainty gained in mathematics that has led him to take up its study. It is the only science that deals with "things that are eternal and unchanging." By the very nature of its study it can shed light on the other two theoretical sciences through its investigations of the motion of the heavenly bodies and of the kinds of motions of material bodies. Not only is the study of mathematics productive of the highest intellectual activity, but Ptolemy regards it as morally edifying as well: "With regard to virtuous conduct in practical actions and character, this science, above all things, could make men see clearly; from the constancy, order, symmetry and calm which are associated with the divine, it makes its followers lovers of this divine beauty, accustoming them and reforming their natures, as it were, to a similar spiritual state" (37). The manner in which Ptolemy frames the contribution of mathematics is the major rhetorical element of the text. Astronomy is judged to be the highest endeavor of man; its calling demands a priestlike dedication and affords a kind of sanctifying "grace."

Ptolemy ends this introductory discussion of the philosophical foundations of his work by saying that his intent is simply to supplement the work of the ancients with "as much advancement as has been made possible by the additional time between those people and ourselves" (37).

Ptolemy's Method of Argument

In the second chapter of the book, Ptolemy describes the order he will follow in treating the matter: first a general overview of the relation of the earth to the heavens and then the particulars. In covering particulars he plans to describe the movements of the sun and moon, and, finally, the fixed stars and the planets. Copernicus was to model the first book of his work on that part of the *Almagest*. Ptolemy provides the following summation of the topics to be covered: " . . . the heaven is sensibly spherical in shape when taken as a whole; in position, it lies in the middle of the heavens, very much like its centre; in size and distance it has the ratio of a point to the sphere of the fixed stars, and it has no motion from place to place" (38).

In chapter 3 he examines the first postulate: that the heavens move like a sphere, explaining how the earliest observers arrived at this opinion. He conjectures that it probably derives from what the ancients thought was reasonable on observing the movements of the heavenly bodies. They saw that the sun and moon and other bodies always moved in parallel circles from their rising to their setting and that these bodies would again appear in the same orderly pattern after a period of being invisible. They also noted that the stars revolve in circles around the pole star, some making smaller, some larger circles depending upon their distance from the center. The stars that move out of sight for a time were seen to move in regular patterns, those farthest from the center disappearing for longer periods, those closer for shorter periods.

Ptolemy then takes up opinions opposed to this principle and refutes each. For example, the view that the stars may be moving in a straight line to infinity he answers with a question: how can one explain the fact that each star is seen daily moving from the same place? How could they turn around and come back after rushing to infinity?

After refuting another equally implausible theory, that the stars are lit by the earth on rising and extinguished by it at their setting, he offers a number of other arguments for the spherical movement of the heavens, among them that instruments for measuring time agree only with the spherical hypothesis, that circular movement offers the most facile and least impeded path. His conclusions here are based on observations and inference; many that follow are informed by Aristotelian principles.

Regarding the last consideration of this chapter, the movement of the ethereal spheres, Ptolemy reasons from the formal and material causes of Aristotelian physics—that the ether has the finest and most homogenous parts of any body, and, since it must have a form adequate to embrace these parts, it could be only a sphere—to the conclusion that the ether also "moves in a circular and uniform fashion" (40). The spheres, as he explains in another work, are interlocking concentric spheres; on the surface of each of the inner spheres the planets are apparently embedded. The outermost sphere or first sphere is that of the fixed stars and it embraces all the others; the next largest sphere contains Saturn, followed by those carrying Jupiter, Mars, the Sun, Venus, Mercury, and the Moon.[5]

In treating the shape of the earth in the fourth chapter, Ptolemy founds his arguments entirely on sight observation and mathematics. Those who think that Columbus was among the first to show that the earth is round should review Ptolemy's astronomy. He offers a number of proofs attesting to the conclusion, some from the sighting of eclipses of the moon, which are recorded as occurring at different times depending on the location of the observer. Observers in the East see it at a later time than those in the West, he notes. He adds that these facts would not be true for any other shape of the earth. For example, if the earth were concave, people in the West would see the stars first, and if it were flat, all would observe the stars to rise and set at the same time. He goes on to refute other possibilities, ending with the proof most familiar to moderns, "if we sail towards mountains or elevated

5. The cosmology described was known as the Ptolemaic system throughout the Middle Ages and the Renaissance but the details of the system were not explained in the *Almagest*. Bernard R. Goldstein has published a translation of the long unknown and rare Arabic version of Ptolemy's *Planetary Hypotheses*, which does explain the nesting spheres: *The Arabic Version of Ptolemy's "Planetary Hypotheses,"* Transactions of the American Philosophical Society, vol. 57, pt. 4, 1967.

places from and to any direction whatever, they are observed to increase gradually in size as if arising from the sea itself in which they had previously been submerged; that is due to the curvature of the surface of the water" (41).

With regard to the position of the earth in relation to the celestial bodies, he argues from appearance, adding proofs based on Aristotle's biology. The arguments Ptolemy premised on Aristotelian principles of form and its relation to substance he must have thought to be weak demonstrations. These he would not term "necessary" or "certain" because they treat matters of physics, not those of mathematics. For Ptolemy, as noted, held that only mathematics can yield certain or irrefutable proofs. Arguments from appearances are sometimes demonstrative in a weaker sense also. At times his proofs are simply dialectical or merely probable. The weaker varieties he distinguishes by such prefatory or concluding statements as we can "suppose" or "imagine."

In the case of arguments for the earth's sphericity, Ptolemy uses demonstrative reasoning of the *quia* variety—knowledge of the fact that comes through recognition of an effect. To be termed a necessary demonstration, or a demonstration *propter quid,* he would need to offer causes for the appearances, knowledge of the reasoned fact, and this he does not attempt. Aristotle, Copernicus, and Galileo also argue in the same way when they speak of the curved shadow cast by the earth on the moon from the sun's light as an effect of and evidence for the earth's shape. (From the standpoint of the causes of the *shadow,* however, the argument is propter quid.) Olaf Pedersen notes that it is surprising that Ptolemy did not offer this proof of Aristotle's for sphericity of the earth.[6] Ptolemy's reasoning takes the form of a "weak" necessary demonstration when he speaks of the sphericity of heavenly bodies, noting that the cause of this shape is their nature: an ether composed of homogeneous parts.

Concerning the stability of the earth and its position in the heavens—the subject of the fifth through the seventh chapter—Ptolemy argues from sense experience, observation of the heavens, and principles he has already established. He grants that it would be far simpler to conjecture that the earth moves than that the heavens move around it. However, he says, since falling objects always fall at right angles to the earth's surface, one has to assume that it is in the center of the universe. Likewise, light bodies, such as fire, fly upwards from the earth. Up and down are relative directions with respect to the earth, he says. The heavier, coarser bodies find a resting place

6. Pedersen notes that it is best that the spheres were omitted in the *Almagest,* for Ptolemy would have otherwise sacrificed clarity (46). Wallace treats clearly the character of scientific reasoning, *propter quid* and *quia,* in his *Causality and Scientific Explanation,* 2 vols. (Ann Arbor: University of Michigan Press, 1972, 1974), 1:10–24.

on or in the earth, while it remains unmoved by the impact of these falling bodies since it is so much larger. In chapter 7 he projects what might transpire from the earth's movement: "If the earth had a single motion in common with other heavy objects, it is obvious that it would be carried down faster than all of them because of its greater size: living things and individual heavy objects would be left behind riding on the air, and the earth itself would very soon have fallen completely out of the heavens. But such things are utterly ridiculous to think of" (44).

Ptolemy's major difficulty in developing accurate measurements of the movements of the spheres was what appeared to be the departure of some of the planets from the orderly circular movement about a common center ordained for them by their sphericity. In the eighth chapter, to account for these aberrations he proposes the concepts of epicycles and deferents developed by Apollonius and of the eccentrics devised by Hipparchus in the second century B.C. The epicycle helps to explain the apparent retrograde motion of a planet by positing that the planet itself makes a smaller circle while embedded in the sphere. The center of that circle describes a uniform motion along a larger circle, called the deferent, which is concentric with the sphere.

This complicated movement serves to explain why Mars seems to move backwards, but it does not explain why Mars seems to move faster in part of its orbit or why the Sun seems to spend longer periods in the orbit's summer quadrant. For these problems Ptolemy employs another adjustment, the eccentric. He simply envisions the center of their orbits as offset some distance from the earth, thus creating an "ex-centered" or off-centered orbit.

The eccentric works well enough for the sun in explaining its appearing to spend a greater time in the summer quadrant while it preserves a uniform motion around the earth. It does not work quite so well in explaining Mars' movements, however, which necessitate the introduction of another variable, the equant. This is a point of uniform angular motion related to the center of the epicycle that allows the motion to appear to be irregular from the standpoint of the epicycles' movement along the deferent, but regular in reference to the equant. Thus, the motion preserves in one sense the principle of uniform motion, although with regard to the deferent and the rotation around the earth the motion is irregular. That point was to bother Copernicus so much that it led him to consider another explanation.[7]

With this Ptolemy has come to the end of his general treatment of the universe. In a brief ninth chapter he outlines the rest of book 1, wherein

7. The complicated system is described in Thomas S. Kuhn, *The Copernican Revolution* (Cambridge, Mass.: Harvard University Press, 1957), 64–72; and by Owen Gingerich in "Ptolemy, Copernicus, and Kepler," *The Great Ideas Today* 1983 (Chicago: Encyclopaedia Britannica, 1983), 141–51.

particulars are given concerning principles of measurement employed in the foregoing matter and the geometric theorems necessary for what is to follow. In the succeeding twelve books Ptolemy elaborates upon what he has introduced earlier regarding the celestial bodies.

In Ptolemy's presentation we find little rhetoric, save in the ethos he wishes to project in describing the lofty aims of astronomy and in the moral character that emerges from his conception of the effect of such study on the astronomer. In advancing his conclusions he exposes premises carefully and clearly; in refuting opposing positions, he does not denigrate the persons advancing them. Nor does he play on the interests or emotions of the audience, unless his claims that his conclusions are based on recognized principles can be said to appeal to the audience's desire for regularity. His discourse is the purest scientific discourse of any we shall examine here. Clearly he thinks as did Aristotle that when one enters into the subject of the science one leaves aside the realm of rhetoric.

The Copernican Solution

De revolutionibus orbium coelestium libri VI, or, in English, *Six Books on the Revolutions of the Heavenly Spheres*,[8] was written in Latin and in some respects preserved much of Ptolemy's cosmology. Copernicus seems to have retained the concept of ethereal spheres and he used eccentrics and epicycles to explain some celestial movements; nevertheless the thesis of the work, its heliocentric doctrine, was too inconceivable for most of his contemporaries. Copernicus realized that what he was proposing was a revolutionary hypothesis and one that would probably bring down on his head the wrath of many in the learned world. His acknowledgement of that eventuality in the opening words of the preface, a dedicatory letter addressed to Pope Paul III, is fashioned in the style characteristic of him: at once assured yet deferential, elegant yet clear. In the preface, and throughout the letter, he reveals the richness of the learning he had gained through his own private study and through his formal education, first, in his native Poland at the University of Cracow and, later, in Italy at Bologna, Padua, and Ferrara.

The Polish astronomer was born in 1473 at Torun, and after his father's early death he was accorded the protection of his uncle Lucas Watzenrode, Bishop of Warmia. An ecclesiastical career was open to him following his studies at Cracow. In 1497, while continuing his education in Italy, he was elected canon of the Cathedral of Frauenburg (Frombork), a post he held

8. The English translation used in this analysis is Edward Rosen's. See note 1 to chapter 1. The Latin text cited is the facsimilie edition of the manuscript, Mikołaj Kopernik, *Dzieła Wszystkie*, vol. 1 (Warsaw and Krakow: Panstwowe Wydawnictwo Naukowe, 1972).

until his death in 1543. He was not a priest, as Galileo was to claim in his *Letter to Madame Christina of Lorraine, Grand Duchess of Tuscany,* but he did gain a doctorate in canon law. This was in addition to his degrees in the liberal arts and in medicine. Thus, when he took up his post again on his return from Italy, he was a humanist scholar of the first rank, with knowledge of Greek, Latin, the classics, mathematics, and astronomy. He continued his interest in astronomy, constructing an observatory in 1513, from which he could use astronomical instruments to make measurements of the heavenly bodies. His expertise was widely recognized and attracted scholars eager to learn astronomy from him.[9]

Prefatory Matter

Since Copernicus entrusted to his pupil, Georg Joachim Rheticus, the supervision of the printing of the book at Nuremberg, and because the author lay in his final illness as the publication was completed, scholars agree that he did not contribute much to the design of the frontmatter. The title page and foreword, described below, were not his, but obviously he himself composed the dedicatory preface addressed to Pope Paul III after years of concern about the reception his book might receive. As a canon of the Cathedral of Frauenberg, Copernicus thought it suitable as well as prudent to seek an ecclesiastical patron, and evidently envisioned an audience composed primarily of Roman Catholic scholars: mathematicians, astronomers, natural philosophers, and learned ecclesiastics.

Copernicus probably also made the decision to place first in the prefatory materials a letter to him composed in 1536 by the Dominican cardinal of Capua, Nicholas Schönberg. In the letter the cardinal urges Copernicus "to communicate this discovery of yours to scholars" and to send him "at the earliest possible moment" his writings on the subject along with "the tables and whatever else you have that is relevant." The cardinal's request demonstrates the interest of the Church in new reckonings that would permit preparation of an accurate calendar, which the Lateran Council under Pope Leo X had called for. Copernicus mentions this fact at the end of his preface, adding that the Bishop of Fossombrone had also suggested that the astronomer devote his attention to that task.

The numerous references Copernicus makes in the dedicatory letter to the interest of the ecclesiastical hierarchy in his research almost create the impression that the Church supports his new cosmology. More importantly,

9. Details of Copernicus's life are drawn from Edward Rosen's article on Copernicus in the *Dictionary of Scientific Biography*, vol. 3 (1971): 401–11; and Charles Glenn Wallis's biograpical note accompanying his translation of *De revolutionibus, Great Books of the Western World*, vol. 16 (1952): 499–500.

in addressing the pope, Copernicus tenders a compelling appeal to the Church to continue its support and protect him from any imputations of impiety.

We now know that this appeal was not enough to spare the work a negative reading by the pope's theologian, a Dominican and Master of the Sacred Palace, Bartolomeo Spina, just a year after it was written. Another Dominican theologian who was also an astronomer, Giovanni Maria Tolosani, in a critique of *De revolutionibus* records that Spina read the book and expressed a desire to see it condemned, but he was taken ill before he could do so.[10] Tolosani adds that he hopes his own work will accomplish the same purpose. But the book did not seem to draw significant criticism from the hierarchy again until the discoveries of Galileo with the telescope in 1610 had cast into doubt some of the basic principles of Aristotelian-Ptolemaic cosmology. Even then no official action was taken until 1616, when the book's thesis was condemned.

Dedication to the Pope

Since the content and the rhetorical elements of the dedicatory preface reveal a great deal about the concerns and intent of Copernicus in writing and permitting the publication of *De revolutionibus,* these merit an examination.

The pope whom Copernicus addressed was an erudite man, a person of culture and a prelate concerned with the preservation of the faith during this critical time for the Church. Thus, with confidence in the rectitude of his conclusions, but also with some apprehension regarding the reception that would await them, Copernicus begins the dedication of his great work to Pope Paul III:

I can readily imagine, Holy Father, that as soon as some people hear that in this volume, I ascribe certain motions to the terrestrial globe they will shout that I must be immediately repudiated together with this belief. For I am not so enamored of my own opinions that I disregard what others may think of them. I am aware that a philosopher's ideas are not subject to the judgement of ordinary persons, because it is his endeavor to seek the truth in all things, to the extent permitted to human reason by God. Yet I hold that completely erroneous views should be shunned.

10. Eugenio Garin discovered this early challenge to the work. See his *Rinascite e rivoluzioni: Movimenti culturali dal XIV al XVIII secolo* (Bari, 1976), 255–95, cited by Robert Westman, "The Copernicans and the Churches," in D. C. Lindberg and R. Numbers, eds., *God and Nature* (Berkeley: University of California Press, 1986), 87–89. Westman mentions that the Copernican theory was not discussed by the Council of Trent, but that Tolosani composed his critique in the spirit of the Council, condemning *De revolutionibus* for its transgressions against Aristotelian natural philosophy and logic. Further details on Tolosani are given below in ch. 6.

In this frank *captatio benevolentiae* Copernicus hopes to disarm his readers and capture their sympathy. Here he follows the formalities of dedicatory letters, though he omits the usual effusive compliments. A subclass of the formal epistle, the dedicatory letter adapts the format of the classical oration, adding to it a salutation, which can be lengthy, deferential, or simply nominative. The exordium becomes the captatio benevolentiae in the letter genre. It is designed to capture the reader's goodwill and attention before the narration describes the circumstances prompting the letter. Unlike the oration, a formal letter does not usually include a confirmation as such to present arguments supporting a proposition, but it transforms this part into a petition, which embodies the raison d'etre of the epistle. It seeks something: patronage, protection, or notice, or all three. Copernicus's petition is obvious, given the letter's addressee: to place the work under the protection of the pope. But when Copernicus does so explicitly in the last part of the letter, his candor harmonizes with the opening tone, a candor that rises almost to impertinent boldness in the conclusion. Generally Renaissance letters end in perfunctory fashion, and in the dedicatory variety the conclusion may simply provide a bridge to the text of the work itself.

The letter genre will figure prominently in this study, for besides the ubiquity of dedicatory letters, the published private letter was exploited by many of the disputants in the Copernican controversy. Rhetoric courses in schools and universities not only acquainted students from boyhood with the three kinds of orations of classical rhetoric—judicial, political, and ceremonial—but they also drilled students in the art of letter writing. Paul Grendler notes that the medieval *ars dictaminis* was replaced in the Renaissance by the study of familiar letters, using Cicero's newly discovered epistles as models. These were far more informal, their organization dictated by occasion and topic. Nevertheless, he remarks, students were still made cognizant of the various parts of letters and of the relationship of writer to recipient, displayed in the text of the letter in this period rather than in the elaborate grammatical equations of the medieval salutation.[11] The courtesies remained important when seeking patronage.

After acknowledging the fantastic character of the hypothesis he proposes, Copernicus takes up its relation to his profession as a philosopher in what on the surface seems to state a gratuitous truism—that the task of the philosopher is the pursuit of truth. The implications of this statement for

---

11. Grendler, *Schooling in Renaissance Italy,* 217, 230–233. Cicero's *De inventione* and the Pseudo–Ciceronian *Ad Herennium* were still major texts for pre-university schooling, while Aristotle's *Rhetoric* was studied in the universities along with other works of Cicero and the writings of Quintilian. See the account of rhetorical education in the period in George Kennedy's *Classical Rhetoric.* The reforms of Peter Ramus had not yet been effected.

the nature of Copernicus's work are significant. The question of whether Copernicus regarded his thesis as true or not, or whether he thought it simply a better explanation than Ptolemy's because it served to save the appearances more efficaciously, was to become an important issue. If one were to take only the evidence of Copernicus's statements in the preface and the text of *De revolutionibus,* one would have to conclude that he thought the thesis of his book referred to cosmological realities.

Modern Copernican scholars have advanced even more diverse interpretations regarding the import Copernicus attached to the system. Robert Westman summarizes three major views before offering yet another, his own rhetorical reading.[12] He says that some, like Alexandre Koyré and Thomas Kuhn, think Copernicus presented simply an "esthetic" argument, a more pleasing, satisfactory, and harmonious picture of the heavens. Others, such as Imre Lakatos and Elie Zahar, find such a "subjectivist" view erroneous. They think that Copernicus believed the theory was superior to the previous one because it could predict celestial events far more accurately. A third view, that of Noel Swerdlow and Otto Neugebauer, holds that Copernicus did think his theory was a "true" representation of the cosmic system. For his part Westman believes that the key to what Copernicus thought lies in an image, drawn from Horace and presented in the preface, from which the astronomer argues for the essential unity and symmetry of the universe. Not through an esthetic appeal of the beauty of the system, but through a rhetorical appeal to "decorum," does Copernicus seek to persuade his audience. We shall return to the point when we arrive at the image in our examination of the preface.

The person who first made the veridical status of the system a subject of conjecture was Andreas Osiander. This Lutheran theologian completed the oversight of the publication of *De revolutionibus* at Nuremberg after Rheticus accepted a teaching position in another city and found he could not be present to see the book through the final stages of production. The task entrusted to him permitted Osiander to insert an unsigned foreword to

12. Robert Westman, "La préface de Copernic au pape: esthétique humaniste et réforme de l'église," *History and Technology* 4 (1987): 365–66. An interesting sidelight is provided by Alexandre Koyré (97–98), who observes that Arthur Koestler is one who takes issue with most views of Copernicus, saying that the astronomer really thought the system he described was hypothetical and so agreed with Osiander, citing Koestler's *The Sleepwalkers,* 567. Confusion about Copernicus's meaning continued in the next century; Cardinal Robert Bellarmine made the point strongly to Galileo and Foscarini that Copernicus was speaking hypothetically. See the discussion of the hypothesis in Amos Funkenstein, "The Dialectical Preparation for Scientific Revolutions," in Robert S. Westman, ed., *The Copernican Achievement* (Berkeley: University of California Press, 1975), 165–203; and the response to Funkenstein by Maurice A. Finocchiaro, "Commentary: Dialectical Aspects of the Copernican Revolution: Conceptual Elucidations and Historiographical Problems," in the same volume, 204–12.

the treatise preceding all of the materials we have discussed. The foreword had the effect of substantially altering the force of Copernicus's argument. Whether the astronomer ever saw the foreword is unknown, since the printed copy of his book did not reach him until the day of his death, May 24, 1543.[13] We do know that he had correspondence with Osiander and did not agree with the Lutheran that he should consider his hypothesis simply a convenient means of calculating the movements of the heavens.

In a letter to Copernicus, Osiander wrote:

I have always felt about hypotheses that they are not articles of faith but the basis of computation; so that even if they are false it does not matter, provided that they re-produce exactly the phenomena of the motions. For if we follow Ptolemy's hypoth-eses, who will inform us whether the sun's nonuniform motion occurs on account of an epicycle or on account of the eccentricity, since either arrangement can explain the phenomena? It would therefore appear to be desirable for you to touch upon this matter somewhat in your introduction. For in this way you would mollify the pe-ripatetics and theologians, whose opposition you fear.[14]

The reply from Copernicus is no longer extant, but Kepler, who saw it, said that the astronomer had no intention of complying with these instructions but had decided to maintain his own opinion "even though the science should be damaged."[15] Copernicus's letter to Pope Paul III contains a num-ber of statements that evince strong realist convictions; however, the fore-word provided by Osiander was to deceive many in his day. Rheticus, who knew of Copernicus's convictions on that score, was able after its publica-tion to get Osiander to admit that he had authored the foreword to the work, but the Nuremberg printer, Johannes Petreius, refused to add a cor-rection to the text. Thus, some readers came to the opinion that Copernicus was hypocritical, giving conflicting statements about his views.[16]

Far from being equivocal, Copernicus was intent upon maintaining that his hypotheses truly accounted for the phenomena. If we can take his state-ments in the preface at face value, he was neither an instrumentalist nor a fictionalist. Here is the text of his preface to Pope Paul III immediately fol-lowing the opening passage quoted above:

Those who know that the consensus of many centuries has sanctioned the concep-tion that the earth remains at rest in the middle of the heaven as its center would, I reflected, regard it as an insane pronouncement if I made the opposite assertion that

13. Edward Rosen describes the circumstances in the notes accompanying his translation of *De revolutionibus*, 334–35.

14. Ibid.

15. Ibid., 335.

16. Koyré (34–38) discusses the difficulties and the efforts of Tiedemann Giese to correct what he saw as a despicable act of Osiander.

the earth moves. Therefore I debated with myself for a long time whether to publish the volume which I wrote to prove the earth's motion or rather to follow the example of the Pythagoreans and certain others, who used to transmit philosophy's secrets only to kinsmen and friends, not in writing but by word of mouth, as is shown by Lysis' letter to Hipparchus. (3)

His conviction is set forth plainly here. But no wonder readers were unsure of Copernicus's honesty, for preceding this letter to Pope Paul, on the very first page of the work, Osiander had declared that an astronomer "cannot in any way attain to the true causes," but "will adopt whatever suppositions enable the motions to be computed correctly from the principles of geometry for the future as for the past." A few sentences later, to make sure there is no doubt of his meaning, Osiander adds, "And if any causes are devised by the imagination, as indeed very many are, they are not put forward to convince anyone that they are true, but merely to provide a reliable basis for computation." Osiander's insertion may have been prompted by his own views on the nature of astronomy, but, as he had mentioned in his letter to Copernicus, he was also greatly concerned about the furor he was sure the book would arouse among theologians and philosophers. The opening sentence of his preface mentions his fears that "the novel hypotheses," which have already been widely reported, may have given "serious offense" to those scholars who think the liberal arts should not be "thrown into confusion."[17] Presumably for Osiander, as for Ptolemy, certainty could not be attained in astronomy, and those who claimed more would upset the academicians.

The conception of the mathematician-astronomer's task described in the foreword was also subtly introduced in the publisher's blurb on the title page of De revolutionibus and in the quotation that follows it:

Diligent reader, in this work, which has just been created and published, you have the motions of the fixed stars and planets, as these motions have been reconstituted on the basis of ancient as well as recent observations, and have moreover been embel-

17. Koyré thinks that Osiander was only trying to protect the book against the ire of theologians and peripatetic philosophers and that was why he had written to Copernicus and also to Rheticus before the book's publication to suggest a "phenomenalistic theory of knowledge" (35). See the discussion of Osiander's motives in Koyré, 34–38, and n. 8, p. 97; and in Rosen's notes to De revolutionibus, 333–334. A defense of Osiander as a conciliator in the face of anticipated religious opposition appears in Bruce Wrightsman's "Andreas Osiander's Contribution to the Copernican Achievement," Copernican Achievement, 213–43. Wrightsman thinks Osiander omitted his name from the preface because his anti-Catholic stance was well known and he did not want to create further prejudice against the work.

lished by new and marvelous hypotheses. You also have most convenient tables, from which you will be able to compute those motions with the utmost ease for any time whatever. Therefore buy, read, and enjoy.

Let no one untrained in geometry enter here

Selection of the motto that reputedly graced the portal of Plato's Academy was probably the work of Osiander, Edward Rosen surmises, while the advertisement was the work of the printer, who had written similar ones for other works he had published.[18] The question of whether one can achieve a true demonstration in mathematics is obliquely raised here. This was an issue debated through the centuries. The Platonist position was that mathematical forms underlie what we see in nature, and that we can only be sure of the computations themselves and not in their conformity to what we see. For these scholars the physics of Aristotle lends support to the contention that mathematics cannot really tell us about nature with certainty, since natural science is arrived at through a knowledge of physical causes. The ambiguity of some of the texts of Aristotle permitted this interpretation.

On the other hand, another Aristotelian school, and one to which the Jesuit scientists who were to challenge Galileo belonged, based its views on assiduous study of Aristotle's *Posterior Analytics, Physics, De caelo,* and the *Meteorology* along with Greek, Latin, and Arabic commentaries on these works. They believed that much stronger evidence in Aristotle supports the validity of a mixed science where geometric proofs could be applied to nature, yielding causes and ultimately a science with necessary demonstrations.[19] Aristotle had conjectured physical causes for the twinkling of stars, for example, that presumed observation and mathematical computation. The nearness of the planets explains why they do not twinkle. Measuring the distances from the observer at which lanterns shine without a twinkle and when these are seen to twinkle yields the information that distant lights twinkle. It can be argued strongly that Copernicus belonged to this school, for this was the position taught at Cracow during Copernicus's studies of mathematics and physics there.[20] He would not have thus been afraid of throwing "the Liberal Arts into confusion" by asserting that demonstrations could be attained in physics.

18. Rosen, notes to *De revolutionibus,* 333.
19. Wallace provides an illuminating discussion of these schools in *Causality* 1:10–24; see also his *Galileo's Logic of Discovery and Proof,* sec. 3.4.
20. Paul W. Knoll describes the content of studies at the University of Cracow in Copernicus's time in "The Arts Faculty at the University of Cracow." Knoll points out that "the Jagiellonian University was indeed an important center for astronomical studies" (149). The Aristotelian approach to a mathematical physics was well entrenched there (148–149).

Instrumentalists, such as Osiander, would argue that the astronomer was simply making mathematical computations and was too far removed from the possibility of ascertaining causes; he can only claim that his hypothesis seems to explain phenomena. The Greek commentator on Aristotle of the sixth century, Simplicius, whose position will figure largely in Galileo's *Dialogue Concerning the Two Chief World Systems,* seems to have been the first to have mentioned "saving the appearances." He says that Plato suggested that mathematicians might raise hypotheses "to save the appearances presented by the planets."[21]

After referring to the revolutionary character of his thesis and his qualms about revealing it, Copernicus offers his explanation for the practice of the Pythagoreans who imparted their thoughts only to their disciples. This is a significant passage, for it shows his understanding of the danger Osiander had foreseen. It also foreshadows some of the critical reaction in Galileo's day. Copernicus suggests that the Pythagoreans were secretive but not because of elitist notions concerning their philosophy. Rather this was because they feared ridicule from those too lazy to undertake a deep study of it, unless they thought it might be lucrative; or they were reticent because they feared their ideas might be taken up by some who, because of their "dullness of mind," could make no contribution anyway. The indirect ad hominem reference to his critics he will enlarge upon later.

Alluding to the advice of Horace to let a work lie buried for nine years before its publication, Copernicus says he finally yielded to the importunities of friends after four times nine years. The friends he names are Cardinal Schönberg of Capua and Tiedemann Giese, Bishop of Culm; the latter, he notes, is a "close student of sacred letters as well as other literature." These men urged him not to be afraid to make his writings available to students of astronomy, arguing that the stranger his teaching seemed to be the more honor he would gain from it when "the fog of absurdity" was dispelled by his "most luminous proofs." The a fortiori argument is an ingenious ploy and was surely meant to challenge his audience to read on.

Conspicuously absent from the list of friends was his pupil Georg Joachim Rheticus, undoubtedly the one who was most responsible for the publication of *De revolutionibus.* It was Rheticus who had published in 1540, with Copernicus's permission, an account of his teacher's new cosmology, *Narratio prima.* The account, written in the form of a letter to Rheticus's former mentor, John Schöner, was to have been followed by a second fuller account. The enthusiastic reception accorded the *Narratio prima* was probably one of the major factors both in gaining the support of Copernicus's friends and in convincing the astronomer himself to permit

21. Wallace, *Causality,* 1:21.

the publication of *De revolutionibus*. This happy outcome is the reason a *Narratio secunda* was not necessary. Given these circumstances, why did Copernicus not explicitly recognize the contribution of Rheticus? Was it a deliberate slight, or as Edward Rosen suggests, was it because Rheticus was a Protestant, teaching at Wittenberg, a stronghold of Lutheranism, when Copernicus wrote his prefatory letter—a letter which seeks to place the work under the protection of the pope and which mentions the support of a cardinal and a bishop for the work of its author? Or was the reason, as Bishop Giese consolingly suggests in a letter to Rheticus, not Copernicus's lack of appreciation for what Rheticus had done for him, but "a certain apathy and indifference (he was inattentive to everything which was nonscientific) especially when he began to grow weak."[22] We shall never know, but Rheticus must have appreciated the inclusion of his *Narratio* in the second edition of *De revolutionibus*, published in 1566.

An interesting facet of the involvement of Rheticus and Osiander in the production of the *De revolutionibus* is the fact that their religious orientations did not affect their enthusiasm for the Copernican solution. Both were Lutherans and both espoused the heliocentric explanation, even though Osiander preferred a qualified reading of the nature of hypotheses. This reading was the position that Cardinal Bellarmine was to take in his discussion of the Copernican system in his letter to Foscarini, discussed below. For his part, Luther was to call Copernicus a "fool" and to denounce the Copernican thesis.[23] Not until much later was the suppression of the hypothesis identified solely with Catholicism.

In the long narration of the prefatory letter Copernicus presents his reasons for considering solutions to the movements of the heavenly bodies other than the traditional ones, and refutes the opposing position, showing the lack of agreement in computations and methodology of the proponents. Their uncertainties about the motions of the sun and the moon cause inconsistencies in their calculations of the length of the year, which they derive by means of different principles. Some use homocentric spheres but fail to show that their calculations are in accord with the phenomena. Others conjecture epicycles and eccentrics and are able to predict the appearances fairly accurately, but they do not give good reasons for introducing these or explain how they can be correlated with the first principles of motion. Nor can they show that these parts are integral to the overall structure of the universe.

22. Rosen translated and printed the letter in *De revolutionibus*, 339–40.

23. Koyré, 74. Dorothy Stimson recounts the negative reception of the Copernican thesis by Francis Bacon, the Louvain professors, Sir Thomas Browne, George Herbert, and Richard Burton, and also Descartes' Brahean view, *The Gradual Acceptance of the Copernican Theory of the Universe* (New York: Baker and Taylor, 1917), 70–90.

To sum up the result of such calculations, Copernicus employs an illuminating and telling analogy, inspired probably by the familiar opening
lines of Horace's *Ars Poetica*. Horace draws the would-be poet's attention to
the principles of harmony and proportion: "If a painter chose to join a human head to the neck of a horse, and to spread feathers of many a hue over
limbs picked up now here now there, so that what at the top is a lovely
woman ends below in a black and ugly fish, could you my friends, if favoured
with a private view, refrain from laughing?"[24] The reigning cosmological
system was a similar creation for Copernicus: "Their experience was just like
some one taking from various places hands, feet, a head, and other pieces,
very well depicted, it may be, but not for the representation of a single person; since these fragments would not belong to one another at all, a monster
rather than a man would be put together from them" (4.25–29). Thus, previous theories have presented astronomers with a fantastic conglomerate of
disparate parts. The Horatian analogy could be expected to strike a familiar
chord, as Westman notes, and to urge acceptance of a system whose parts are
in harmony with the whole. The dialectical topos of the unity of part to
whole also works beneath the surface to undergird the poetic image. Here,
dialectic, poetic, and rhetoric blend to magnify the need for reform.[25]

Herbert Butterfield points out that what troubled Copernicus was not
just the wheels within wheels of the epicycles but also the introduction of
the equant, which called for a uniform angular movement around a point
that was not central. This he thought inappropriate, given the principle of
circular motion Aristotle had enunciated for the heavenly orbs.[26]

Copernicus continues, finding the monster to have resulted from its cre-
ators' overfondness for the "method" itself, by which he seems to have
meant the symmetry of their calculations. But they have neglected to examine the principles behind the method and have thus entertained false
hypotheses.

Probably to dispel the notion that he was given to novelties, Copernicus
describes the extensive study he made of all scholarly opinions proposed
throughout the ages concerning the movements of the spheres. (Doxographical argument was a powerful tactic still, if the authorities cited were revered.) Copernicus relates that Cicero stated that Nicetas thought the earth
moved, and Plutarch mentioned that Philolaus the Pythagorean thought

24. The English translation of the first lines of *Ars Poetica* is by H. Rushton Fairclough, in
Horace: Satires, Epistles and Ars Poetica, Loeb Classical Library (Cambridge: Harvard University Press, 1978).

25. Westman (373) sees the interaction of poetics and rhetoric here. I have added the dialectical topos of whole/part, which would seem to be a habitual argumentative tactic.

26. Herbert Butterfield, *The Origins of Modern Science*, rev. ed. (New York: Macmillan,
1967), 36–37.

the earth and the moon revolved around the sun and that Heraclides and Ecphantus also believed in a heliocentric movement of the earth. On reflection he realized that such an hypothesis (his proposition) was the only one that permitted all the motions to fit well together: " . . . not only do their phenomena follow therefrom but also the order and size of all the planets and spheres, and heaven itself is so linked together that in no portion of it can anything be shifted without disrupting the remaining parts and the universe as a whole" (5.16–19). The picture has become more pleasing: the parts make up a unified whole—the monster is dispersed and a harmonious figure emerges.

To sketch out this new picture, Copernicus next describes the organization of the book. He will first give a general description of the universe, including the movements of the earth and the spheres. In the remainder of the work he will correlate "the motions of the other planets and of all the spheres with the movement of the earth" and strive to show "to what extent the motions and appearances of the other planets and spheres can be saved if they are correlated with the earth's motions" (5.23–26).

If "acute and learned astronomers" will only study seriously what he sets forth here, he says directly following this partition, then he thinks that they cannot but agree with him. In this passage Copernicus makes a strong appeal to those whose opinion he values most and yet most fears.

The astronomer then places the book at the disposal of the general public (of scholars) as well, saying he does not seek to escape the judgment of anyone. He does so through a petition in which he asks the pope to protect the work. "For even in this very remote corner of the earth where I live you are considered the highest authority by virtue of the loftiness of your office and your love for all literature and astronomy too." But Copernicus tempers his compliment with candor and humor: "Hence by your prestige and judgement you can easily suppress calumnious attacks although, as the proverb has it, there is no remedy for a backbite" (5.32–36).[27]

In the preface Copernicus makes only a passing reference to the possibility that his hypothesis might contradict scriptural passages, and he seems unwilling to countenance any disapproval on that score. Uncharacteristically, he uses strong language: "Perhaps there will be babblers who claim to be judges of astronomy although completely ignorant of the subject and, badly distorting some passage of Scripture to their purpose, will dare to find fault with my undertaking and censure it. I disregard them even to the extent of despising their criticism as unfounded" (5.37–40). Copernicus does not even consider the possibility that theologians will raise

---

27. Rosen mentions that this proverb probably came to Copernicus from Erasmus via Giese's correspondence with the great humanist, *De revolutionibus*, 342.

serious questions about the cosmology of the work. Instead he ends with the statement that "astronomy is written for astronomers," and remarks that these scholars will no doubt think that he has made a contribution to the Church by aiding in the reform of the calendar.

Not surprisingly, the first passage in *De revolutionibus* to be censored by the Congregation of the Index was that beginning with "Perhaps" and continuing through the maxim about astronomy.[28] The passage was viewed as an affront to theologians concerned that the new hypotheses might threaten the very foundations of their discipline, then regarded as the queen of the sciences.

Copernicus's Line of Argument

Book 1 of *De revolutionibus* begins with a graceful introduction reminiscent in tone and content of Ptolemy's introduction to the *Almagest*. In addition, the opening chapters follow closely the pattern of organization of those of the earlier book. The similarity is explained by Rheticus in the *Narratio prima*. He describes how Copernicus was persuaded by Bishop Giese of Culm to undertake a serious study of the heavens with an eye to reforming the calendar, which was so greatly needed, and to seek a "correct theory and explanation of the motions [of the heavens]."[29] When Copernicus completed his investigations he first decided to publish the results in tabular form, as had been done in the Alfonsine Tables, which revised the computations of Ptolemy but gave no accompanying explanations. Rheticus says that Copernicus hoped in this way to avoid controversy but still provide enough information to enable scholars to infer the principles that lay behind the computations. Giese, however, persuaded the astronomer that his contribution to scholars would be much greater if he modeled his work on that of Ptolemy and gave a full explanation of his new hypothesis. Rheticus refers to Copernicus's cosmology as a revision of "the hypotheses," and calls it the "new hypothesis" as compared with the hypotheses of the ancient astronomers and the "common" or "vulgar" hypotheses of his day.[30]

The Role of Hypotheses

Before returning to the text of *De revolutionibus,* it would be well to reflect on the nature of the term "hypothesis" itself and on its relation to Copernicus's scientific method. As it was used by Copernicus and others of his

28. Ibid., 343.
29. Edward Rosen's translation of Rheticus's *Narratio prima* in *Three Copernican Treatises,* 192.
30. Ibid., 31.

time, the term does not have quite the same meaning that it does for us to-day. For us it is a conjecture one posits and seeks experimental evidence to prove.[31] For those trained in the scholastic method, a hypothesis meant an assumption that could be made, given a preceding thesis, the Greek *hupothesis* meaning "to stand under." In Latin the correspondent term for *hupothesis* was *suppositio*, as in "let us assume" or "grant the supposition" that such is the case.

In considering Ptolemy's understanding of certainty, we noted two varieties of demonstrative reasoning: *propter quid*, reasoning from cause to effect, or a priori demonstration; and *quia*, reasoning from effect to cause, or a posteriori demonstration. In that discussion we were mainly concerned with the process of reasoning from cause to effect. But thesis and hypothesis were also used in a posteriori reasoning, reasoning from effect to cause. Again, this reasoning is not simply conjectural; rather, it takes the form: on the supposition of this effect, such must be the cause. As noted earlier, Copernicus and Galileo argued from effect to cause in speaking of the observed phases of the moon (seen by them as effects). The cause, they reasoned, is the moon's spherical shape, which, when illuminated by the sun and seen from the earth, presents the appearance of phases.[32] Here the hypothesis would treat the cause, and the thesis, its connection with the effect. For Ptolemy, arguments pertaining to physics, where causes could not be known for certain, could not produce perfect demonstrations. Only dialectical arguments yielding probable truths could be framed. By Copernicus's day many scientists accepted the possibility of combining mathematics with physics to create a mixed science in which demonstration would be possible. Osiander and his school did not, while Copernicus seems to take the validity of a mixed science for granted, as he and his pupil Rheticus made clear.

In the *Narratio prima* Rheticus sheds light upon the subject. He speaks in several places of the approach used by Copernicus in arriving at his hypotheses. In astronomy as in physics "one proceeds as much as possible from effects to principles," and Rheticus conjectures that if Aristotle "could hear the reasons for the new hypotheses, [he] would doubtless honestly acknowledge what he proved in these discussions, and what he assumed as unproved principle." Rheticus adds that Copernicus "decided that he must assume

---

31. In the present day the term hypothesis is used for what medieval and Renaissance scholars refered to as the antecedent in a conditional syllogism: If p, then q. Here p is the antecedent, which is posited hypothetically, and q is the consequent that follows from it.

32. For a detailed analysis of this argument and an explanation of how, as a demonstration, it concludes necessarily, see Wallace's *Galileo's Logic of Discovery and Proof*, ch. 5, sec. 5.1. In subsequent sections of chapter 5 Wallace explains how the argument provides a paradigm in terms of which Galileo's subsequent demonstrations of mountains on the moon, the satellites of Jupiter, and the phases of Venus can readily be understood.

such hypotheses as would contain causes capable of confirming the truth of the observations of previous centuries, and such as would themselves cause, we may hope, all future astronomical predictions of the phenomena to be found true." These remarks leave no doubt that Copernicus thought his goal was knowledge of the reality behind what he observed in the heavens. Rheticus continues to explain Copernicus's method, recounting his veneration for the ancients, which entailed his use of their observations and a consideration of their hypotheses. He notes that it was Copernicus's own observations that led ultimately to his choice of heliocentrism as the only true explanation.[33] After this recital, Rheticus reassures his former teacher, Schöner, to whom he addressed the *Narratio* (and, no doubt, the greater audience he anticipated for the work), that Copernicus wanted nothing more than to be a follower of Ptolemy and the ancients, that he had "no lust for novelty," but that he was compelled to come to different conclusions because of "the phenomena" and "mathematics."[34]

Here it is difficult to know what Copernicus thought about the validity of the *proofs* he could offer in support of his hypotheses, for he himself does not say. As it stands, the explanation he offers is not complete. The hypothesis is proposed and it satisfies Aristotle's criteria by providing a statement of fact or a definition, but to be fully demonstrative the thesis requires further support, a series of physical demonstrations to prove each of the movements of the earth.[35] These are not provided by Copernicus; thus, although the hypotheses he posits may satisfy the appearances, as Osiander pointed out, it does not prove that the earth is moving and that it moves around the sun. These require physical proof which he did not have. Until Foucault's experiments with the pendulum demonstrated the earth's rotation, and Bessel

33. *Narratio prima*, particularly 163–66, and 138–39, 140–43.

34. Ibid., 186–187.

35. Aristotle explains the difference between a thesis and a hypothesis in a demonstration in the *Posterior Analytics* 1.2.72a. I have used G. R. G. Mure's translation in *Great Books of the Western World*, vol. 8 (Chicago: Encyclopaedia Britannica, 1952), 98. The nature of hypothesis for Copernicus is discussed by Edward Rosen in *De revolutionibus*, 344. He mentions that when Copernicus means something hypothetical he uses the term *coniectura*. The debate over the "reality" of the two systems is also touched on later in his notes. Rosen (352–53) quotes Poincaré's unequivocal assertion that Ptolemy's system is denied by science and the just as unequivocal assertion by Fred Hoyle that neither Ptolemy's nor Copernicus's theories are physically right or wrong. Rom Harré discusses the fictionalist and realist views of hypothesis in relation to Copernicus in *The Philosophies of Science* (London: Oxford University Press, 1972), 80–89. Pierre Duhem provides an illuminating discussion of the evolution of what hypothesis meant to whom and when in his *To Save the Phenomena* (Chicago: University Chicago Press, 1969). He describes realist views and fictionalist views from their origins to modern times. His quotation from Agostino Nifo expresses well the scholastic view (48–49). See also Wallace's discussion of Galileo's extensive use of suppositions in his own demonstrations, *Galileo's Logic of Discovery and Proof*, chs. 5 and 6.

could show through stellar parallax that the earth was moving around the sun (when the telescope was perfected enough to reveal this), physical proof was not available. Aristotle himself had mentioned that parallax could determine this question, but it is not apparent to the naked eye nor was it observable even with Galileo's telescope; consequently, one could not offer this proof. Since parallax was not observed, arguments could be made that other explanations were just as feasible. As Pierre Duhem points out in speaking of Nifo's critique of Ptolemy, these arguments may be sufficient to explain the appearances but they are not necessary proofs.[36]

As we have seen, Copernicus does believe that his hypothesis is an explanation of the reality of the heavens, but whether he believes he has sufficiently demonstrated this is not clear. Although he does not think of the system as merely probable, most of his arguments are dialectical, used to provide the ground for the mathematical computations he gives in the text and tables. The computations may contain demonstrations, but the general hypothesis is what must be proved. Perhaps he did not ask himself whether he had proved his hypothesis, or perhaps, seeing that what Aristotle demanded as proofs was unattainable here, he thought the answer self-evident. The fault he finds with the hypotheses of others—their overfondness for demonstration, the "method" itself, which distracts them from the principles—could be interpreted as his impatience with the formalities of scholastic logic. Such might be the case despite his veneration of the ancients.[37]

The reason for our lengthy consideration of hypothesis and proof in this context is of course because of the bearing these have upon the rhetorical content of Copernicus's writing. If he did know that something would seem lacking in his proof in the eyes of others, he might have been more inclined to make up for it by rhetorical appeals. Since his knowing or not knowing is unclear, perhaps an answer will be forthcoming through an analysis of his use of rhetoric.

A Place for Rhetoric

The introduction of book 1 of De revolutionibus provides Copernicus with an appropriate occasion for rhetorical expression, and he chooses to take advantage of it. He fashions an exordium in the classical vein, as a good humanist would be expected to do.

Having decided to pattern his work after Ptolemy's, Copernicus treats

36. Duhem, 53.
37. Rosen (*Three Copernican Treatises*, 28–29) notes that Copernicus interchanges the terms hypothesis, assumption, and principle, and all of these he uses to refer to what he posits as propositions. But this seems in line with Aristotelian logic, where a hypothesis or supposition can be an assumption and a principle as well.

some of the same points in his own introduction. First he comments upon the value of a study of the heavens in words that will in turn be echoed by Galileo:

Among the many various literary and artistic pursuits which invigorate men's minds, the strongest affection and utmost zeal should, I think, promote the studies concerned with the most beautiful objects, most deserving to be known. This is the nature of the discipline which deals with the universe's divine revolutions, the asters' [i. e., stars'] motions, sizes, distances, risings and settings, as well as the causes of the other phenomena in the sky, and which, in short, explains its whole appearance. (7.4–10)

In the development of this encomium to astronomy, Copernicus uses rhetorical topoi especially germane to his purpose. He begins with an extended definition in which he compares astronomy to other studies, examines the etymology of the focus of the art (i.e, the heavens), speaks of its nature, its aim, the esteem accorded it, and finally places it in relation to other disciplines: "Unquestionably the summit of the liberal arts and most worthy of a free man, it is supported by almost all the branches of mathematics." Like Ptolemy he notes its aid in perfecting moral virtue, and he, too, extols the intellectual pleasure it offers. Those who knew Ptolemy's work, and among them the critics Copernicus most feared, would have found these parallels reassuring, especially when he echoes the ancient astronomer by noting that such study inevitably leads one to "admiration for the Maker of everything, in whom are all happiness and every good." But he shows his knowledge of virtue to be superior to pagan understanding by citing the scriptural reference that follows: "For would not the godly Psalmist [92:4] in vain declare that he was made glad through the work of the Lord and rejoiced in the works of His hands, were we not drawn to the contemplation of the highest good by this means, as though by a chariot?" (7.26–29).

Copernicus next turns to secular authorities who have acclaimed the value of astronomy. Plato, he notes, has dwelt on the benefit of such study to the commonwealth in the *Republic,* where the philosopher states that it permits a calendar to be devised, which in turn encourages public festivals and sacrifices. Moreover, Plato believed that astronomy is a necessity for the scholar, for "it is highly unlikely that anyone lacking the requisite knowledge of the sun, moon, and other heavenly bodies can become and be called godlike" (7.37–38). The elevation of the astronomer almost to divinity is a theme that Galileo was to carry to a dangerous point in the *Dialogue.*

Astronomy, thus, is a divine science, but, Copernicus points out, it has had its "perplexities." Its "principles and assumptions, called 'hypotheses' by the Greeks" have been a source of disagreement. The measurements of the movements of celestial bodies, which are dependent upon earlier observations made over centuries, Ptolemy was able to bring "almost to perfection." The problem is that "very many things . . . do not agree with the conclusions which ought to follow from his system, and besides certain other motions have been discovered which were not yet known to him" (8.8–10).

Copernicus reinforces his point with the testimony of Plutarch, whom he says complained that astronomers still have not been able to compute the movements of the celestial bodies satisfactorily.

Noting that these difficulties should not be an excuse for indolence, Copernicus brings his introduction to an end, saying that he will try to find a resolution of them. In doing so he intends to treat some things quite differently from his predecessors, but "I shall do so thanks to them, for it was they who first opened the road to the investigation of these very questions."

In beginning the development of his hypotheses in the first chapter, Copernicus does not change the tone of his discourse from that of the introduction. In fact, throughout the book he retains the same clear, almost conversational, style. I say almost conversational, for Copernicus maintains an attention to the formal structure of the argument throughout and declines to deliver asides or wander from the subject.

The Heliocentric Thesis

Just as Ptolemy did, Copernicus moves from general matters in book 1 to particulars in succeeding books. The brief first chapter is devoted to the shape of the universe. The universe is spherical, says Copernicus, citing the principle that a form must be suitable to its matter: a sphere is the most perfect form, one that cannot be increased or diminished, and it is the form best suited for enclosing the largest amount of matter, as "wholes strive to be circumscribed by this boundary, as is apparent in drops of water" (8.31–32). The sphere is thus best suited to celestial bodies.

From this Aristotelian argument based on the topos of property, Copernicus moves in the next chapter to proofs of this postulate, mentioning the familiar evidence of objects on land as observed from ships and vice versa. Rosen points out that many of the proofs in these first chapters follow the text of Pliny's *Natural History*.[38]

In a departure from Ptolemy, Copernicus considers in chapter 3 "How earth forms a single sphere with water." He seems to have taken up the subject because some Peripatetics have argued from principles of transmutation of the elements that there would have to be ten times more water than land. Copernicus argues dialectically that the earth has less water than land, first because the earth would be flooded otherwise and both elements are gov-

---

38. Rosen, *De revolutionibus*, 345.

erned by gravity, and, from final cause, because in this way earth's creatures are preserved. He then develops geometrical and logical proofs to show that the opposite view cannot be maintained. In this context he mentions the geography of the known world, which is composed of islands surrounded by water, and he notes the recent discovery of more islands, among these, America.

He concludes that the land and water "together press upon a single center of gravity" and that earth being heavier supports water in pockets on its surface, so that there is more land than water even though more water is visible. At the end of this chapter he observes that the earth along with its waters reveals its shape to be "a perfect circle" through the shadow it casts during eclipses of the moon. This he says effectively refutes the contentions of some ancients who have proposed various other shapes: that the earth is flat, hollow, or like a drum, or a bowl, or a cylinder.

Circular Motion

After these preliminary considerations, Copernicus begins in chapter 4 the difficult task of introducing his own hypotheses. Turning to the motion of the heavenly bodies, he at first follows Ptolemy, who considers the same subject in the third chapter of the *Almagest*. But instead of offering observational evidence, as Ptolemy does at the outset as a foundation for his explanation of celestial movements, Copernicus begins by citing some principles of Aristotelian cosmology: that the movement of celestial bodies is regular, circular, and eternal. He explains that circular movement is appropriate to a spherical body because it has no beginning and no end and that the spheres turn in many such orbits: "The most conspicuous of all is the daily rotation . . . that is, the interval of a day and a night. The entire universe, with the exception of the earth, is conceived as whirling from east to west in this rotation. It is recognized as the common measure of all motions, since we even compute time itself chiefly by the number of days" (10.35–39).

What seems to be a lack of regularity in some of these movements as seen from the earth only appears to observers to be so. This must be true, he contends, because the heavenly bodies are "objects constituted in the best order" (10.20–21). Copernicus then starts to counter the reigning explanations for the apparent irregularities in the movement of the planets by using the same principle of order or regularity, one which was commonly assumed in all the natural sciences.

The cause "may be either that their circles have poles different [from the earth's] or that the earth is not at the center of the circles on which they revolve." This latter possibility Ptolemy had projected with eccentrics, epicycles, and the equant, but his own solution is more radical still, as he suggests at the end of the chapter: "Hence I deem it above all necessary that we should carefully scrutinize the relation of the earth to the heavens lest, in our desire to examine the loftiest objects, we remain ignorant of things nearest to us, and by the same error attribute to the celestial bodies what belongs to the earth" (11.30–34). By speaking in the first person plural and placing the blame for our common ignorance on *our* laudable desire to examine things of the highest order, he avoids the appearance of condescension, and makes it easier for the reader to accept what follows. A less astute understanding of the audience might have led him to rail at the ignorance or stupidity, or both, of those ancients and moderns who have maintained erroneous opinions. Perhaps the reason for his adroit handling of the proposition he will present may lie not so much in a conscious desire to manipulate the audience as an empathic response to those whom he respects, who are staunch Ptolemaicists and Peripatetics. In either case, the results would be the same.

In looking at the text of Copernicus's arguments, the preface excepted, one cannot help but be struck by his conciliatory approach. The insights of the modern psychologist Carl Rogers regarding the dynamics of conflict resolution helps to explain the underlying structure of his approach.[39] The first and most important move in changing another's opinion is to show sympathetically that the opposing view is understood. Copernicus sympathetically introduces the evidence before he attempts to controvert differing opinions.

If you hope to persuade the opponent to your side, Rogers says, you next need to explain your point of view objectively and then show how it would benefit the other were he to accept it. Copernicus seems intuitively to have approached the issue in this manner. The advantage he offers to the reader is a better understanding of the real nature of the heavens and a more efficient

---

39. Carl Rogers thinks the major difficulty in convincing others of a different point of view is the obstacles to its acceptance in the opponent's mind, the main one of which is the loss of face that would result in giving up a position the opponent has supported. To overcome that fear, a speaker or author should state the other person's point of view with understanding and empathy, showing respect for the opposing opinion; then after outlining the shared ground in the two views, the author explains the context of the alternative view demonstrating how the other might benefit in adopting it. If done well, such an approach, says Rogers, has the effect of calming the atmosphere and increasing the possibility of change. Rogerian argument was introduced to rhetoricians through a discussion in Richard E. Young, Alton L. Becker, and Kenneth L. Pike, *Rhetoric: Discovery and Change* (New York: Harcourt, Brace and World, 1970), 274–90. Copernicus also may be said to have used a tactic Kenneth Burke thinks essential to successful rhetoric. He sees the aim of rhetoric as "identification." That is what the author attempts; the act of identifying he calls "consubstantiation." From the Burkean point of view then, in this case, unlike the passage in the preface where Copernicus calls his opponents "babblers," the astronomer has attempted to identify with his audience, and in this way he carries his readers with him to the next points. See *A Rhetoric of Motives* (Berkeley: University California Press, 1969), 20–23, passim.

means of predicting celestial events. Seen in this light, the manner in which he presented his argument may help to explain why the Church was slow in condemning the hypothesis, and why it did not criticize him directly. Copernicus's conciliatory approach reflects his training in dialectics, where objections were fully aired; yet the agonistic air of debate pervades that kind of discourse, whereas the text of *De revolutionibus* conveys a spirit of congenial, even fraternal reform. More commonly in the debates of the day dialectical techniques were liberally laced with rhetorical tactics of vituperation, which embittered both sides.

In the next chapter, Copernicus considers whether circular motion suits the earth and its position in relation to the heavens. He addresses the problem from the topos of property, i.e., what is proper to an entity. He says, "we must in my opinion see whether also in this case the form entails the motion" or we will not be able to understand what takes place in the heavens. Most people would agree with the Ptolemaic solution and think "the contrary view to be inconceivable or downright silly," he acknowledges. "Nevertheless, if we examine the matter more carefully, we shall see that this problem has not yet been solved" (11.40–44).

The Earth's Motion

Next Copernicus discusses how one can determine which is moving, the object or the observer, in cases where one or the other are moving or where both are moving and possibly at different speeds. Our observation point limits our perception: it is from the earth that "the celestial ballet is beheld in its repeated performances before our eyes." If, on the other hand, rotation were to be assigned to the earth, the objects outside it would appear to be moving in the opposite direction. This motion of the earth would explain the apparent movements of the sun, moon, and the other heavenly bodies. Having raised this possibility, he distances himself from the provocative question: "it is not at first blush clear" why motion is attributed to the stars that encompass everything, to the "framework," that is, to "the enclosing" rather than "the enclosed." Moving again to authorities who have previously suggested the heliocentric solution, he names "Heraclides and Ecphantus, the Pythagoreans, and . . . Hicetas [Nicetas] of Syracuse, according to Cicero" (12.15–16).

After having introduced the idea of the earth's movement and offered the testimony of authorities who have proposed it, the astronomer addresses the problem that follows from it: the position of the earth in relation to the planets, the sun, and the moon. First he notes that almost all agree that the earth is in the center of the universe. But he explains that even someone who would say that the earth is not in the center would admit that its distance from that center is very "insignificant" in relation to its distance from the stars and that the planets' positions in relation to it are much more notice-able. The seemingly erratic movements of the planets have led to the sup-position that they move around a center some distance from the earth. The fact that they are sometimes nearer and sometimes farther from the earth "necessarily proves that the center of the earth is not the center of their circles" (12.29–30). With the phrase "necessarily proves" [*necessario arguit*], Copernicus indicates that this is a necessary demonstration, and this point will be used to advantage in building the rest of his argument. He adds that what is not clear is whether the earth or the planets are moving closer or farther away.

At the end of this very important chapter, having carefully prepared the reader, Copernicus can say confidently, "It will occasion no surprise if, in addition to the daily rotation, some other motion is assigned to the earth." The hypothesis at which he hints is that besides its diurnal rotation the earth also moves with other motions as one of the celestial bodies. Again citing an authority, he says that such was the view of Philolaus the Pythagorean, "no ordinary astronomer" as we know from Plato's own respect for him (12.30–35).

The Earth's Position

"The immensity of the heaven compared to the size of the earth" is the sub-ject of chapter 6. It is important because of its bearing on the question of the earth's position in the universe. Ptolemy had shown through geometrical measurements that the earth is at the center of lines drawn from the ecliptic path of the constellations. Copernicus argues that the earth's being at the center of measurements from the ecliptic is not relevant. Since the distance is so great between the earth and the stars, drawing bisecting lines from the horizons can only show that it appears that the earth is at the center. A line drawn from a place on the surface of the earth and one drawn from its center to the same point in the stars would appear to be parallel, and in fact become one and the same line, because of the great distance of the stars from the earth. The distance between the two lines would be as nothing compared to the length of the line itself. By this combination of geometrical and optical evidence and probable reasoning Copernicus is able to counter the explana-tion of Ptolemy so that his own view has equal force. He then makes the telling point that these computations serve to underscore the vastness of the heavens and thus show how difficult it is to conceive that they revolve and not the earth.

In the next two chapters, 7 and 8, Copernicus treats in detail the reasons the ancients thought the earth was motionless in the center of the heavens. The main argument rests upon Aristotle's concepts of the heaviness and lightness of the elements. Earth as the heaviest element would by its own weight fall toward the center of the earth. All of the particles of earth in the cosmos tend to this center, falling at right angles until they are stopped by the earth's surface. There they all come to rest; thus earth itself must be at rest because of its weight.

Ancient principles of motion also furnish the foundation of the Ptolemaic view of the universe. For Aristotle, earth and water move naturally downward toward the center of the earth, while air and fire have an upward movement away from it. The four elements can also move horizontally if propelled by force, but only celestial bodies move naturally in a circular motion, and this around the center. Ptolemy argued that the earth could not move diurnally because it would be against its nature and such a movement would entail a forced or violent motion of incredible speed. That kind of movement would certainly burst the earth asunder, flinging its particles outward. Objects could not then fall at right angles to the surface since that surface would have moved onward. In addition clouds would be observed to float towards the West. These are all topical arguments drawn from the nature of the thing, its genus and what is proper to it.

Natural Motion and Gravity

The refutation of this position in chapter 8 is also based on Aristotelian principles. Copernicus argues that the motion of the earth could be a natural and not a violent motion. He rehearses Aristotle's views: natural motion is that motion by which objects on earth, composed of the earthly or watery elements, move downward when they are dislodged from a higher place. In falling they are seeking their proper place, the center of the universe. The elements of fire and air move naturally upward because they seek their proper places in the regions above the earth. Celestial bodies, composed of the fifth element, move naturally in circles. Violent motion is different from natural motion, having as its cause a force external to an object. The effects of natural and violent motion, he explains, also would be different: natural motion preserves the unity and order of the entity, whereas violent motion does not. "Ptolemy has no cause, then, to fear that the earth and everything earthly will be disrupted by a rotation created through nature's handiwork, which is quite different from what art or human intelligence can accomplish" (15.31–33). If motion were natural to the body, then all of nature's works would move naturally with it. This argument founded on principles of nature is just as probable as Ptolemy's, but no more so. Copernicus's next argument, however, does have greater force than Ptolemy's. He asks why Ptolemy is not more concerned (a fortiori) about the even swifter motion that would have to take place in the immense heavens to impel its objects around the earth each day? In addition, Ptolemy's reasoning about the earth's motion, if applied to the heavens, would require the heavens to expand infinitely. But a fundamental axiom of physics denies movement to the infinite, he explains, and it is thought that beyond the heavens there can be nothing. So "there is nowhere the heavens can go" (16.1).

Leaving the further development of this subject to the philosophers, Copernicus returns to an initial point, the form of the earth. In this regard he presents a powerful argument from analogy:

> Why then do we still hesitate to grant it the motion appropriate by nature to its form rather than attribute a movement to the entire universe, whose limit is unknown and unknowable? Why should we not admit, with regard to the daily rotation, that the appearance is in the heavens and the reality in the earth: This situation closely resembles what Vergil's Aeneas says: "Forth from the harbor we sail, and the land and the cities slip backward" [*Aeneid* 3.72]. For when a ship is floating calmly along, the sailors see its motion mirrored in everything outside, while on the other hand they suppose that they are stationary, together with everything on board. In the same way, the motion of the earth can unquestionably produce the impression that the entire universe is rotating. (16.10–20)

This chapter especially disturbed the Sacred Congregation charged with expurgating the work after the Church's decision of 1616 to suppress advocacy of the Copernican system, a decision addressed fully below in parts 2 and 3. The censors emended the passage to make Copernicus's statements about form conjectural. They also removed the clause in the quotation above regarding "reality" and "appearance," but left the analogy to Vergil since it speaks only of resemblance.[40]

In answer to Ptolemy's argument regarding the movement of the clouds, Copernicus counters that the clouds, rain, and atmosphere would move with the same motion as the earth, either because, being composed of the same elements as things on the earth, they would partake of the same natural motion, or because in their closeness to the surface they would be moved along easily with it.

Copernicus then discusses Aristotle's treatment of motion, showing that there is a complexity in nature not explained by the simple motions he attributed to matter, and he concludes that Aristotle must have thought of these purely as logical concepts. Edward Rosen explains that Copernicus had to give up some of Aristotle's principles of motion to ascribe motion to the earth. For Aristotle had said that parts have the same motion as the whole to which they belong; thus, since parts of the earth move in straight lines towards the earth, the earth could not then have a circular motion. Copernicus assigns circular motion to the earth as a whole, but grants that parts of the earth still move toward the center in straight lines.[41]

40. Rosen, *De revolutionibus*, 352.
41. Ibid., 354.

Copernicus cleverly invokes Aristotelian principles in the next passage, however, when he says that immobility is "deemed nobler and more divine than change and instability, which are therefore better suited to the earth than to the universe" (17.28–30). He uses a further argument based on what is suitable or proper when he again mentions that it is "absurd" to propose that the framework—the fixed stars—would move, rather than what they frame—the earth.

Acknowledging that these refutations have only dialectical force, Copernicus concludes: "You see, then, that all these arguments make it more likely that the earth moves than that it is at rest." The Sacred Congregation ordered this passage stricken from the chapter, obviously because they could not grant that the arguments rendered this conclusion more probable.[42]

The tactic of presenting forceful dialectical arguments to counter the received cosmology was to be employed to great advantage by Galileo in the *Dialogue Concerning the Two Chief World Systems* when advocacy of the Copernican system over the Ptolemaic was forbidden. The nautical analogies Copernicus introduces in his refutations are repeated and expanded by Galileo.

In chapter 9, Copernicus takes up the question of whether another motion can be attributed to the earth. Since he has already argued convincingly that motion could be assigned to the earth and has posed the possibility of diurnal rotation, in this chapter he turns to another—the annual revolution of the earth around the sun. He will later address a third—the tilt of the earth that explains the seasons.

As in the previous chapter, the Sacred Congregation sought to change the wording so that Copernicus's statements implying true conclusions for his proofs indicate only hypothetical results. The opening sentence originally read "Accordingly, since nothing prevents the earth from moving, I suggest that we should now consider also whether several motions suit it, so that it can be regarded as one of the planets" (17.44–46). The censors ordered the first part of the sentence emended to read: "Accordingly, since I have assumed that the earth moves"; they then retained the middle clause and eliminated the last one altogether.[43]

The related problem of whether the earth is the center of gravity of the universe gains Copernicus's attention. He develops an important principle in this regard: "Gravity is nothing but a certain natural desire, which the divine providence of the Creator of all things has implanted in parts, to gather as a unity and a whole by combining in the form of a globe. This impulse is present, we may suppose, also in the sun, the moon, and the other

42. Ibid.
43. Ibid., 354–55.

brilliant planets, so that through its operation they remain in that spherical shape which they display" (18.6–10). Thus the other celestial bodies have centers of gravity also. He implies that they are no different in composition from the earth, which would eliminate the concept of a different celestial element, the quintessence.

The Sun, Planets, and Stars

Copernicus ends chapter 9 by asserting that the sun is at the center of the universe, adding that "all these facts are disclosed to us by the principle governing the order in which the planets follow one another, and by the harmony of the entire universe, if only we look at the matter, as the saying goes, with both eyes" (18.21–23). Surprisingly, this portion of the chapter was not censored, even though Copernicus speaks of his conclusions as "facts." The principle he mentions was first articulated in the preface as the major reason for adopting a different cosmology, and it prepares the way for his consideration in the next chapter of the order of the heavenly spheres. The allusion to looking at the universe with both eyes is a reference to a proposition in Euclid's *Optics*, newly translated by Giorgio Valla, but there the author's meaning is purely physiological. Copernicus obviously means to look at the evidence with the inner eye, the eye of the intellect, as well as the physical eye.[44] The reference serves to emphasize Copernicus's contention that sense evidence is not enough, but must be subjected to reason and to astronomical computations.

Turning to a description of the positions of the planets and the sun, Copernicus places the orbits for the planets then known in the same order ascribed to them today. He may have assumed that the planets are embedded in spheres, but again he does not distinguish the stuff of the spheres as the fifth essence, the ether. Of the earth's sphere and its relation to the moon and the sun he says: "Hence I feel no shame in asserting that this whole region engirdled by the moon, and the center of the earth, traverse this grand circle amid the rest of the planets in an annual revolution around the sun. Near the sun is the center of the universe. Moreover, since the sun remains stationary, whatever appears as a motion of the sun is really due rather to the motion of the earth" (20.35–39). Given their views, we should not wonder why the Sacred Congregation ordered "asserting" changed to "assuming."[45] Copernicus here says that the sun is near the center rather than *at* the center because he had to assume eccentrics for the planets to explain their paths.

At the outer edge of the planetary orbits Copernicus envisioned the immovable sphere of the fixed stars. He says this proposal seems preferable

44. Ibid., 355.
45. Ibid., 358.

to a multitude of spheres required by a geocentric universe: "On the contrary, we should rather heed the wisdom of nature. Just as it especially avoids producing anything superfluous or useless, so it frequently prefers to endow a single thing with many effects" (20.45–47).[46] Before proceeding, Copernicus again concedes the incredible nature of what he proposes: "All these statements are difficult and almost inconceivable, being of course opposed to the beliefs of many people. Yet, as we proceed, with God's help I shall make them clearer than sunlight, at any rate to those who are not unacquainted with the science of astronomy" (21.1–4).

The next principle he introduces assigns the positions of the planets in relation to each other: "the size of the spheres [in which the planet is embedded] is measured by the length of the time" of their revolution; the larger the sphere, the longer it takes to complete its journey and the further it will be away from the sun.

In this passage he echoes the dependency of the proofs on the science of astronomy, announced in the preface. These mathematical proofs he has not yet furnished in detail, but they will be introduced in chapters to follow. Next he describes the order of the spheres beginning with the outermost; he pauses when he reaches the sun to remark on the appropriateness of its being the center of all of this activity. The principle is rhetorical, arguing from *aptus* (or *prepon* in the Greek), what is fitting, rather than from the dialectical topos of property, what belongs as an attribute.[47] It is here that Copernicus delivers the impassioned encomium to the sun quoted in the first chapter of this study. The image of a majestic sun beneficently illuminating the universe softens the impact of his annunciation that the sun and not the earth is the center of the celestial orbits. Copernicus again turns to authorities; this time they are diverse figures, certain to appeal to a variety of interests: Hermes, the mysterious author of hermetic wisdom, Sophocles, whose dramatic works delighted Renaissance scholars (even though Copernicus wrongly cites *Electra* instead of *Oedipus at Colonus*), and Aristotle.

At the end of the chapter, Copernicus offers an interesting argument from final causality in explaining the immense distance of the stars from Saturn (the furthermost planet visible with the naked eye). He mentions that the stars twinkle and by this sign show their greater distance, a distance which is necessary, "for there had to be a very great difference between what

46. Ibid., 359. The famous observation originated in Galen.

47. The ubiquity of the principles of *prepon* (propriety or fitness) and *kairos* (timeliness) is described by James L. Kinneavy in his "*Kairos*: A Neglected Concept of Classical Rhetoric," in Jean Dietz Moss, ed., *Rhetoric and Praxis* (Washington: The Catholic University of America Press, 1986), 82. Cicero carried on these principles from the Greek tradition, whence they permeated medieval and Renaissance thought.

moves and what does not move." He concludes, "So vast, without any question, is the divine handiwork of the most excellent Almighty." This last sentence of the chapter was deleted by the Sacred Congregation, no doubt because they felt uncomfortable with the unequivocal statement of the universe's immensity, which might even be construed as infinite.[48]

Copernicus completes his general survey of the heliocentric system in the eleventh chapter of book 1 by examining the third motion he attributes to the earth. The title of the chapter is noteworthy because it claims to offer a demonstration: "De triplici motus telluris demonstratio," or "A demonstration of the threefold motion of the earth." The Sacred Congregation seems to have been frightened by the wording and ordered its rewording to read "De hypothesi triplicis motus terrae, eiusque demonstratione" or "On the hypothesis of the threefold motion of the earth and its demonstration." The point of the emendation is at first reading difficult to discern since what was gained by the insertion of "hypothesis" seems taken away by saying that it has been demonstrated, in effect proved.[49] But adding hypothesis weakens the effect of the term demonstration and reminds the reader of its problematic character.

Owen Gingerich, in commenting on the text of the emendations, notes that only one translation has rendered the term "demonstratio" correctly, namely, as "explication," when translating the revised title into English: "On the Hypothesis of the Three-fold Motion of the Earth and its Explication." This translation would remove the "absurdity" of Copernicus's seeming to offer a certain proof of an hypothesis, says Gingerich.[50] He sees the chapter as providing an "explanation," not a "geometrical theorem."

Given the terminology of the day, however, both translations appear to say the same thing. The Sacred Congregation used the last part of the first sentence of the published text in chapter 11 to restate and clarify Copernicus's meaning. He says there, in the English translation, that he wishes to give "a summary of this [Earth's] motion, insofar as the phenomena are explained by it as a principle" (22.40–41). The Latin text reads "in summa exponemus quatenus apparentia per ipsum tamque hypothesim demonstrentur" (10). In other words, he proposes to explain how the appear-

48. The traditional universe was much tinier than that proposed by Copernicus. For early estimates of its size, see Albert van Helden, *Measuring the Universe: Cosmic Dimensions from Aristarchus to Halley* (Chicago and London: University of Chicago Press, 1985).

49. The translation is my own. Rosen's text reads "Proof of the Earth's Triple Motion"; while this adequately reflects the sense of the title, it leaves out the important distinction of the terms "demonstration" and "hypothesis." The change decreed by the Sacred Congregation is listed in *Opere* 19:401.

50. "The Censorship of Copernicus' *De revolutionibus*," *Annali dell'Istituto e Museo di Storia della Scienza di Firenze* 6, no. 2 (1981): 56. Gingerich cites J. F. Dobson and S. Brodetsky; *Nicolaus Copernicus, De Revolutionibus, Preface and Book 1.*

ances may be demonstrated if one assumes the hypothesis of the threefold movement of the earth. Such a demonstration, termed *ex suppositione* as we have noted above, would be appropriate to a mixed science. This weaker form of demonstration serves to explain through geometrical illustrations how appearances are saved by the hypothesis.

The kind of demonstration indicated in that first sentence of the chapter, taken together with Osiander's preface, may have influenced Cardinal Bellarmine's views on Copernicus's work. In his letter to Foscarini (discussed at greater length below), Bellarmine states that he thinks that Copernicus simply reasoned "ex suppositione" about the earth's motion and that no "true demonstration" has yet been presented.[51] The latter, as we have noted, would offer physical evidence of the causes. But Galileo, who like Bellarmine read *De revolutionibus* before censorship was imposed, thought that Copernicus offered demonstrations of the motion of the earth. In his "Considerations" he declares "I should propose nothing but the reading of Copernicus's own book; from it and from the strength of his demonstrations one could clearly see how true or false are the two ideas we are discussing."[52] The two ideas referred to are the views that the earth's movements should be held to be sufficiently demonstrated, "poter esser dimonstrata," or to be thought of as suppositional, "ex suppositione" (*Opere* 5:351). For Galileo, Copernicus's mathematical demonstrations, sense observations, and dialectical arguments were strong enough to compel belief. He states that Copernicus's opinions were confirmed by the observations of astute natural philosophers such as Pythagoras, Heraclides, and Aristarchus, among the ancients, as well as Gilbert and Kepler among the many moderns (352).

The eleventh chapter contains, in fact, a geometrical proof that the movements ascribed to the earth are consistent with the observed phenomena. The third motion, Copernicus explains, is necessary to account for the equinoxes, and he provides diagrams to accompany this proof. Rosen suggests that Copernicus included the third motion because he assumed that celestial bodies were rigidly tied to an orbit around a center, as if the earth revolved on a spoke around a hub.[53]

In the remainder of the great work Copernicus provides detailed proofs

51. Roberto Bellarmino to Paolo Antonio Foscarini, 12 April 1615, *Opere* 12:171.9–10.

52. "Considerations on the Copernican Opinion," are unpublished notes written evidently in response to Bellarmine's letter, where Bellarmine commended both Foscarini and Galileo for speaking ex suppositione as, he thought, did Copernicus, *Opere* 5:351–70.

53. The explanation is imperfect and the proof erroneous in the view of modern commentators; see Noel Swerdlow, "The Derivation and First Draft of Copernicus's Planetary Theory: A Translation of the Commentariolus with Commentary," *Proceedings of the American Philosophical Society* 117, no. 6 (1973): 445–50.

and tables to show that what he has claimed concerning the motions of the earth is borne out by measurements of the positions of the planets in relation to each other. He was able to eliminate most of the troublesome epicycles and, particularly, the equant required by Ptolemaic astronomy. The result was a simpler and more unified whole, even though it was to require revision to banish the last epicycles through the ellipses posited by Kepler.

Rhetoric and Dialectics

In considering the whole of the first book of *De revolutionibus,* one can see that rhetoric indeed played a significant role in the arguments Copernicus presented. But it is a role of lesser importance than that given it in the writings of Galileo. Copernicus employs rhetoric chiefly to prepare the reader for what is to follow. He does not permit it to participate in the arguments. These, as we have seen, are chiefly dialectical, founded on mathematical and geometrical premises. On the other hand, he does not attempt to offer physical proofs for his hypotheses, as Galileo will try to do.

Alexandre Koyré remarks that the absence of physical proof in Copernicus's treatise should not astonish us. Rather, we should admire "the power and boldness of this mind" that dared to surmount the barriers of received opinion, a feat greatly admired by Galileo.[54]

In answer to the question posed earlier in the chapter, as to whether Copernicus *knew* he did not have adequate proof to satisfy the canons of the day, it seems obvious that he did not think his proofs were inadequate, given the inability to obtain physical demonstration. Thus he does not use rhetoric to make up for the lack. Instead he appears to have put aside the problem and to have argued in the manner of Ptolemy, using dialectical arguments, observation, and geometrical proofs to show that his hypotheses are supported by his and others' calculations. The geometric proofs themselves contain demonstrations, but only demonstrations of the internal truths of the computations. These are not combined with physical evidence of the crucial points: the motions of the earth, as Kepler was to point out. Since Copernicus's arguments are merely dialectically superior to those offered by the reigning view, he may claim only that his hypotheses are more probable.

On the other hand, to argue well dialectically is to argue persuasively, convincingly. In a sense it can be said that Copernicus's explanation does "save the appearances" better than Ptolemy's. The picture he paints of the universe may be in effect only a picture, a more esthetic creation, but Copernicus's remarks on the subject show that he thought he had depicted the real world. Thus, Robert Westman's insight regarding the value of the

54. Koyré (54) summarizes the contribution of Copernicus.

Horatian image of monstrous composition in setting forth Copernicus's intent is significant if we see that image as a rhetorical confirmation that magnifies the point. It demonstrates his cultural literacy but need not be seen as illustrative of a tropic view of the universe.[55] Copernicus thought that the celestial bodies should exhibit the harmony and unity ascribed to them by Aristotelian principles, but these principles had been unfortunately violated by Ptolemy and others. The conformation demanded by heliocentrism contributed to a better Aristotelian order. An harmonious, circular, and unified order was what Aristotle thought was out there, even if he had not the fortune to uncover its center, and Copernicus believed him.

55. The tropic approach was proposed by Giambattista Vico in his *Scienza nuova seconda,* 1738. See Hayden V. White, "The Tropics of History: The Deep Structure of the *New Science*" in Giorgio Tagliacozza and Donald P. Verene, eds., *Giambattista Vico's Science of Humanity* (Baltimore: Johns Hopkins University Press, 1973), 65–85. White maintains with Vico that man has viewed the world through progressive use of the master tropes, moving from metaphor, through metonymy and synecdoche, to irony. Michael Mahoney analyzes these in *Vico in the Tradition of Rhetoric* (Princeton: Princeton University Press, 1985), 79, 229–32.

CHAPTER THREE

Evidence from the Heavens:
Galileo and Kepler

The Copernican cosmology did not immediately cause the consternation its author feared. Although *De revolutionibus* may not have been "an all time worst seller," as Arthur Koestler declared, it does not seem to have generated heated discussion, nor to have gained a large following for the new system during the latter half of the sixteenth century, either because people feared the consequences of espousing a potentially heretical position or because they had no firm conviction of its truth.[1]

The first Continental scholar of note to favor the system in print was Giordano Bruno. He praised the vision of Copernicus in perceiving the nature of the universe in a dialogue entitled *The Ash Wednesday Supper,* published in 1584. Bruno's dialogue provides a fictionalized account of his celebrated debate with Oxford scholars in the previous year. This passionate and bizarre espousal of the Copernican view is described in a later chapter. Besides his earlier defense, in 1591 Bruno published *De immenso,* an encomium of Copernicus with excerpts from *De revolutionibus.* In this work he extols the astronomer for his contribution but laments his thinking of the system only in a mathematical way and not recognizing its occult significance. For Bruno, Copernicus's discovery that the planets revolved around the sun provided cosmic corroboration for the infinitist and animist view of the universe.[2]

---

1. Owen Gingerich quotes Koestler and discounts the view expressed in his *The Sleepwalkers.* Gingerich has examined over 250 copies of the first edition and found many annotations in them. He thinks the silence was due to fear of new ideas. See Gingerich, "From Copernicus to Kepler: Heliocentrism as Model and Reality," Symposium on Copernicus, *Proceedings of the American Philosophical Society* 117, no. 6 (Dec. 1973): 520; hereafter noted as "Heliocentrism."

2. Rosen states that Bruno was the first to foster the theory on the Continent; see his *Kepler's Conversation with Galileo's Sidereal Messenger* (New York and London: Johnson Reprint Corporation, 1965), p. 141, n. 358. Frances Yates argues that Bruno's thought is founded on hermeticism in *Giordano Bruno and the Hermetic Tradition* (Chicago: University of Chicago Press, 1964), 153–56, passim. Robert Westman discounts the Hermetic influence on Bruno, seeing his acceptance of Copernicanism as a consequence of his belief in the principle of suffi-

Kepler's Celestial Physics

A young Johannes Kepler was next to praise Copernicus's achievement, claiming in his first book, the *Mysterium cosmographicum* of 1596, that this system at last furnished the key to understanding the geometrical structure God had ordained for the world. Kepler was not satisfied with his description there, however, and he strove for almost a quarter century to perfect the dimensions of this grand design, an effort that culminated in the *Harmonice mundi* of 1619. He believed that the system he had at last uncovered would blend all aspects of the physical world, the arts, and the sciences. Despite his preoccupation with divine mathematics, it is surprising that Kepler was the first astronomer to demand hard physical evidence for the cosmologies then being debated.[3]

This unique melding of mysticism, mathematics, and empirical science was characteristic of most of Kepler's published work. Despite the mystical significance Kepler ascribes to his cosmology, his arguments are founded on the Aristotelian dialectical and demonstrative principles we have seen in *De revolutionibus*. The point is amply illustrated in the discussion that follows. Rhetoric also finds a place in Kepler's writings, for he was well versed in its principles and devices. But a natural rhetoric appears to be at work in his writings as well. It is apparent in the ethos that seems to emanate spontaneously from his character, which he then uses to unusual advantage. Born in 1571 in Weil der Stadt, Germany, of indigent parents, Kepler was dependent upon his unusual gifts of intellect for the success he attained in life. Able to win the attention of his teachers from an early age, he eventually gained scholarships for study at the University of Tübingen. He intended initially to become a Lutheran clergyman and spent more than two years in theological studies at the university after receiving his master's degree. But the brilliance he had displayed in earlier studies of mathematics and astronomy while he was a pupil of Michael Maestlin led university officials to request that he fill a vacant post in mathematics at Graz before he finished his theological degree. He reluctantly agreed to go, but he taught there for a number of years before joining Tycho Brahe as his assistant in Prague in 1600.

During the second year at Graz no students enrolled in Kepler's classes in

cient reason; see his "Magical Reform and Astronomical Reform: The Yates Thesis Reconsidered," in Westman and J. E. McGuire's *Hermeticism and the Scientific Revolution* (Los Angeles: The William Andrews Clark Memorial Library, 1977), 24, 27.

3. See the entry on Kepler by Owen Gingerich in the *Dictionary of Scientific Biography* 7 (1973): 289–312; 290–91; referred to hereafter as "Kepler."

astronomy, prompting the administrators to ask Kepler to teach rhetoric and Virgil in addition to arithmetic.[4] He was well prepared for this new duty, having been introduced to rhetoric at a time of a revived interest in the subject resulting from the pedagogical reforms of Rudolf Agricola, Philipp Melanchthon, and Peter Ramus. Kepler's knowledge of rhetoric was used to advantage in his ingenious defense of Tycho and his endorsement of Galileo's *Sidereus nuncius*.

The Cosmic Mystery

While at Graz, Kepler formulated the conceptual design of the universe described in the *Mysterium cosmographicum*. In that work he mentions that he became interested in Copernicus's hypotheses when he heard the lectures of Michael Maestlin. The realignment of the planets in Copernicus's scheme inspired him to search for mathematical regularities in their relationships to each other.

When Kepler sought publication of the *Mysterium cosmographicum*, the prospective publisher asked that he seek an evaluation from authorities at the University of Tübingen. They approved it, but suggested he delete a passage in which he tried to reconcile heliocentrism with Scripture. They also asked that he include a full explanation of the Copernican system. The deleted portion was later to become part of his *Astronomia nova* (1609), and it was also translated into English by Thomas Salusbury for inclusion in his *Mathematical Collections*. The treatment of the Copernican hypothesis Kepler added was competently rendered, pleased the authorities, and thus became an important part of the revised text.

In Kepler's first discussion of the geometrical structure of the universe he envisions five three-dimensional polyhedra nested between the planets' orbits. The preface conveys the depth of his conviction regarding the significance of this design and at the same time provides an illustration of the direct ingenuous style characteristic of the man:

Dear Reader, it is my intention in this small treatise to show that the almighty and infinitely merciful God, when he created our moving world and determined the order of the celestial bodies, took as the basis for his construction the five regular bodies which have enjoyed such great distinction from the time of Pythagoras and Plato down to our own days; and that he co-ordinated in accordance with their properties the number and proportion of the celestial bodies as well as the relationships between the various celestial motions.[5]

4. Biographical details are taken from Gingerich, "Kepler."
5. The passage is translated in Koyré, *Astronomical Revolution*, 128. The entire work has since been translated by A. M. Duncan with a commentary by Eric Aiton in Johannes Kepler, *Mysterium Cosmographicum* (New York: Abaris Books, 1981).

The sense of immediacy his writing conveys is illustrated in the passage that follows. After noting that the solution came to him one day as he was teaching astronomy, he says:

Behold, reader, the invention and whole substance of this little book! In memory of the event, I am writing down for you the sentence in the words from that moment of conception: The earth's orbit is the measure of all things; circumscribe around it a dodecahedron, and the circle containing this will be Mars; circumscribe around Mars a tetrahedron, and the circle containing this will be Jupiter; circumscribe around Jupiter a cube, and the circle containing this will be Saturn. Now inscribe within the earth an icosahedron, and the circle contained in it will be Venus; inscribe within Venus an octahedron, and the circle contained in it will be Mercury. You now have the reason for the number of planets.[6]

Concerning the place of this work in the history of science, Owen Gingerich remarks that even though the thesis was erroneous, Kepler "established himself as the first, and until Descartes the only, scientist to demand physical explanations for celestial phenomena."[7]

Kepler sent copies to Tycho Brahe, Galileo, and other scholars. Tycho replied with a lengthy critique, Galileo with a brief acknowledgement of his positive reaction. In his letter, dated August 4, 1597, Galileo also mentions that he thought the Copernican explanation preferable to the Ptolemaic, especially because it supported some ideas of his own: ". . . as from that position I have discovered the causes of many physical effects which are perhaps inexplicable on the common hypothesis. I have written many reasons and refutations of contrary arguments which up to now I have preferred not to publish, intimidated by the fortune of our teacher Copernicus, who though he will be of immortal fame to some, is yet by an infinite number (for such is the multitude of fools) laughed at and rejected."[8] The letter is remarkable, first, because it shows that Galileo feared at this time to openly espouse Copernicanism, but it reveals that he has seriously entertained its validity, having moved beyond the position recorded in his early notebooks.[9] His reference to "causes" and "physical effects" seem to indicate that he is already working out his theory that the motion of the earth causes the tides, the argument that was to figure prominently in the *Dialogue Concerning the Two Chief World Systems*. Kepler wrote back, urging Galileo to take a public stand on the question. But Galileo did not respond to the letter.

The work brought Kepler fame and at last his disparate ambitions seemed

6. Gingerich translates this passage in "Kepler" (290).

7. Ibid., 292.

8. Stillman Drake's translation, *Galileo at Work* (Chicago: University of Chicago Press, 1978), 41.

9. A discussion of Galileo's notes on Aristotle's *De caelo* is in my "Dialectics and Rhetoric."

to come together. To his former teacher Maestlin he wrote: "I wanted to become a theologian; for a long time I was restless. Now, however, behold how through my effort God is being celebrated in astronomy."[10]

The realist nature of the Copernican thesis was an issue that Kepler thought extremely important, and it figures centrally in his writings. In the *Mysterium cosmographicum* he explores the subject in depth, displaying extensive familiarity with Aristotelian principles of logic. In arguing for the Polish astronomer's belief in its verisimilitude, he says that he could not have thought his thesis would be false and yet yield accurate conclusions in the same way as the calculations of the ancients agreed with the appearances:

Furthermore, Copernicus contested none of the hypotheses which agree with observation and account satisfactorily for the appearances; if anything, he accepted and explained them. [At first sight] it seems that he changed many things in the accepted hypotheses, but in fact it is not so. Indeed, it may be that the same conclusion results from two pre-suppositions, different in species, because the two [pre-suppositions] fall in the same genus, and it is in virtue of the genus [of cause and not of specific nature] that the result in question is produced. Thus, Ptolemy had shown [the reason for] the rising and setting of celestial bodies, but [he did not prove it] with respect to the nearest middle term, namely, with respect to the fact of the Earth's immobility at the centre [of the Universe]. Neither did Copernicus make this demonstration with respect to the middle term [which in his system corresponds to the Earth's immobility in Ptolemy's], namely, the fact that the Earth completes one revolution about the centre of the Universe whilst remaining at a certain distance from it.[11]

Kepler goes on to show how Ptolemy rested his argument on a false middle term when he assumed that what "happened because of a genus, happened because of the species." Despite the faulty logic, Ptolemy could demonstrate regarding the appearances. Copernicus, on the other hand, argued using principles that led to the "constant cause," which ancient astronomers did not know. Kepler mentions that even the great Danish astronomer Tycho Brahe had to admit that Copernicus's hypothesis did explain the phenomena.

Tycho had developed an alternate theory with which Kepler did not agree, although he marveled at the accuracy of the Dane's observations. Tycho thought that the earth remains at rest while the planets revolve around the sun and the whole ensemble orbits the earth. His astronomical

10. Gingerich quotes from *Kepleri opera omnia*, ed. C. Frisch, 13:40, in "Kepler," 291.

11. Koyré, 131; the brackets and interpolations are Koyré's. In the *Mysterium cosmographicum* Kepler praises Maestlin for introducing him to the method of argument in philosophy. Westman cites the passage in "Kepler, Maestlin and the Copernican Hypothesis" in Jerzy Dobrzycki, ed., *The Reception of the Copernican Heliocentric Theory* (Dordrecht: Reidel, 1972), 8.

readings and predictions were remarkably consistent with that explanation and convinced many scientists of the day who found the Ptolemaic system inadequate. Tycho's system had an added advantage: it did not conflict with Scriptural statements about the earth and the heavens.

Koyré appraises Kepler's evaluation of the Tychonian system:

> Having thus shown that the foundation for truth can be nothing but the truth, and having at the same time very skilfully shown the structure of the real logical foundations (purified by formal analysis) of Ptolemy's and Tycho Brahe's arguments (which analysis moreover reveals their complete agreement with Copernicus), Kepler thereafter felt justified in passing from 'astronomy', i.e., the purely computational study of celestial phenomena, to 'physics and cosmography', i.e., the study of reality.[12]

To clarify Kepler's methodology further, Koyré quotes the astronomer: "For I have no hesitation in asserting that everything, that Copernicus has demonstrated a posteriori and on the basis of observations interpreted geometrically, may be demonstrated a priori without any subtlety of logic."[13] Kepler refers here to the need to find real causes, the process of demonstration by means of the *regressus*, the term frequently used to refer to it. In this method, one proceeds from the effect (a posteriori) to discover the cause, as Copernicus did. Once the cause is grasped, it can become a middle term, which one then needs to examine to see whether it can sustain the conclusion, the effect (a priori).[14] In other words Copernicus reasoned from the appearances to the cause, heliocentrism, and then after reflection used this cause as the basis for his computations regarding celestial movements.

The New Astronomy

In the *Astronomia nova* or *Commentarius de stella martis* (Commentary on Mars), as it was also called, published in 1609, Kepler continued in this vein. He describes the methodology that led to the discovery of the laws of planetary motion. On the back of the title page of this work he revealed to the world the real author of the foreword to *De revolutionibus,* and declared that Osiander had falsely described the nature of Copernicus's hypotheses. In the next paragraph he remarked that Peter Ramus, professor of philosophy and rhetoric, had said he would relinquish his chair to anyone who could develop an "astronomy without hypotheses." Kepler adds that if Ramus were alive, he, Kepler, would claim the chair.[15]

12. Koyré, 133.
13. Ibid., 135.
14. This method was also termed "resolution and composition." For a full explanation of the method see the discussion in Wallace, *Causality and Scientific Explanation,* 1:140–49.
15. Gingerich, "Heliocentrism," 521–22. The entire *Astronomia nova* has yet to be rendered into English, but portions of it have been translated by William Donahue and Owen

In the introduction to this long and difficult work Kepler, ever conscious of the reader, appeals to those who disagree with him to examine the "principles of the demonstrations" he employs in the book. These principles or *suppositiones* will remain firm if no one refutes them, and "the demonstration built on them will not topple." He then sets forth the methodology he will follow in the text: "I will also do what is customary to physicists by mixing the necessary with the probable so as to draw from the mixture a plausible conclusion. Since in this work I have mixed celestial physics with astronomy, no one ought to be surprised that I have employed a certain amount of conjecture. For this is the nature of physics, of medicine, and of all natural sciences, which employ other axioms besides the most certain evidence of the eyes" (310). These words make clear Kepler's understanding of the variety of proofs open to a mixed science.

Noting the three major explanations offered for the movement of the celestial bodies, he explains that his main purpose in this work is "to correct the astronomical theory," especially that relating to the motion of Mars, so that the tables are in accord with the appearances. By this time he has worked out one of the difficulties in the Copernican hypothesis—the orbits of the planets. Having been put to work on the problem of Mars by Tycho Brahe, he finally found that the reason the planet's orbit could not be properly worked out was because it was not circular, as had been assumed, but elliptical. Once this was discovered the problem of computing the orbits of the planets accurately was solved.

Kepler outlines the contents of the work as follows:

In the meantime, in presenting this and cheerfully following my course, I shall also review Aristotle's metaphysics, or rather his celestial physics, and I shall inquire into the natural causes of motions. From such a consideration, very clear arguments will arise to the effect that only Copernicus's opinion about the universe (with a few changes) is true and the other two are demonstrably false, and so on. . . . I have tried many tentative methods, built partly from ways well trodden by the ancients and partly in imitation of them or on their example; no method would succeed except that which enters from the actual physical causes of the motions, which I establish in this work. (310–12)

Kepler next turns to consider the cause of the motions of the celestial bodies. Since this is the crucial point on which the Copernican question can be demonstrated according to the rules of science of the day, the manner in which Kepler treats it makes a continued examination of this text worthwhile.

Although Kepler describes the process he followed in working out his

Gingerich, working from a preliminary draft prepared by Ann Wegner. These appear in *The Great Ideas Today 1983* (Chicago: Encyclopaedia Britannica, 1983), 306–41; subsequent page references in the chapter are to this translation.

explanation in exhaustive detail, he does not always do so with cogency. Fortunately, his introduction to the work does offer a clear and well-organized overview. After describing the parts of the work that support the first step—his projection concerning the center of the orbit of Mars—he turns to the nonuniform way in which a celestial body moves. Whether either the earth or the sun is thought to be moving, the planet moves faster when it approaches the other body at rest, and it moves slower when it is at some distance from it. He first shows that Ptolemy is wrong in requiring a theory of the sun for each planet and that Tycho is right in finding that one theory could provide the foundation for these calculations. But Copernicus, he says, is more right than either. Tycho, like Ptolemy, had to suppose that each planet moved "not only by its own motion but by the actual motion of the Sun, mixing the two into one and from this mixture effecting a coiled course." Because of this intertwined mechanism Tycho had to discard the idea that the planets were embedded in solid spheres. Copernicus relieves the planets from this extra motion, which accrued simply from the way Tycho looked at the problem.

He thinks that Tycho, in dismissing the spheres but retaining the intelligences that move the planets, would require these spirits to do too many things: "to look after the origins, centers and periods of both motions." If the earth were thought to move, on the other hand, "many of these things can be accomplished not by animate faculties, but by physical ones—undoubtedly magnetic" (314).

Even though the cause posited here by Kepler—magnetism—was not that ultimately accepted by astronomers, Owen Gingerich points out that his reasoning that physical causes are at the basis of celestial physics led directly to Newton's theory of gravitation and to other advances of modern science.[16]

Kepler outlines the demonstrations that he believes are persuasive, although not necessary:

Assuming the Earth does in fact move, it is demonstrated that it adopts the rules governing its swiftness and slowness from a measure of its approach or recession from the Sun. But the same thing holds for the other planets as well, so that they are urged on or retarded according to this approach or recession from the Sun. The demonstration of these things is up to this point geometrical.

From this very reliable demonstration may now be drawn a physical conjecture, that the source of the motions of the five planets is in the Sun itself. Furthermore, it is very likely that the source of the Earth's motion is at the same place as that of the other five planets, namely, in the Sun. For this reason it is therefore very probable that the Earth moves, since there is an apparent plausible cause for its motions. (314)

16. Gingerich, "Ptolemy, Copernicus, and Kepler," 175–80.

The cause is merely conjecture, which means the argument as a whole has only dialectical or probable force. But it is founded upon the sense evidence of naked-eye observations and the laborious calculations he has made of Mars's elliptical orbit.

A comparison of arguments for the movement of the earth and for the sun follows. Kepler notes that Brahe and Copernicus find the source of motion in the sun, but for Brahe that source must move and carry the planets with it around the earth. Kepler argues dialectically that it makes more sense for the sun to be at rest and the planets, including the earth, to move around it than for the earth, which is much smaller than the sun, to be at rest and the sun and the planets be carried around it. Like Copernicus, he also adds a rhetorical proof from *prepon* when he suggests that the sun is appropriately at the center because of its "dignity" or its "brilliance."

Kepler mentions that an erroneous view of gravity is at the base of some of the objections to the new cosmology, and he outlines three axioms related to gravity:

Every physical substance, insofar as it is physical, tends to remain at rest at every position in which it is placed in isolation, outside the sphere of influence of a related body.

Gravity is a physical effect operating mutually between related bodies to unite or join them (which, in the order of things, the magnetic capability [*facultas*] also is) with the result that the Earth draws a stone much more than the stone seeks the Earth.

Heavy bodies (especially if we place the Earth at the center of the universe) are not borne toward the center of the universe as such but toward the center of a related spherical body, such as the Earth. So wherever the Earth is located or wherever it is carried by its animate capability, heavy bodies are always drawn toward it. (315–16)

Basing his reasoning upon these suppositions, he conjectures that the moon and the earth are attracted to each other and that if each were not sustained in their individual orbits they would be drawn together in proportion to their sizes. The waters on the earth would also be pulled toward the moon and would rise up and "flow around the body of the Moon." In this connection he correctly attributes the source of the tides to the pull of the moon on the waters. "This is imperceptible in inland waters, but is noticeable wherever there are very wide expanses of ocean," he says (316).

The effect of the moon on the tides was something Kepler thought Galileo should take into account when the Florentine spoke obliquely about the "causes of the natural effects" [*naturalium effectuum causae*] of the earth's motions as proof for the Copernican hypothesis in his letter of 1597. Kepler guessed that Galileo referred to the tides as the effects of the earth's motion, and he spoke of his surmise in a letter to a friend, noting that

Galileo should consider the moon as the cause rather than the movements of the earth.[17] By the time Galileo wrote the *Dialogue* Kepler's opinion about the moon's effects, expressed in the *Commentary on Mars,* was well known, but Galileo magisterially dismisses the German astronomer's notion that the moon could cause the tides as giving credence to occult forces.[18]

Countering arguments raised by Ptolemy, Kepler says that if it is granted that the moon has such an effect upon the earth, the earth's influence is all the more likely to extend far beyond its body. Therefore, earthly matter would not be able to escape the tug of the earth. Projectiles or flying objects would be propelled along with the earth by this "magnetic force" (317–18).

The a fortiori argument is followed by an argument from *proprium* in relation to the very fast speed with which the earth would have to move to sail through the universe. Kepler argues dialectically that indeed it is more fitting that the earth move whereas such movement of the heavens would be "inconsistent" and "unnatural."

Finally, Kepler takes up the apparent contradictions in Scripture that he had originally included in his *Mysterium cosmographicum.* Since this interesting defense is best seen in the context of the wider effort to reconcile the system with Scripture, it is treated in part 2 of our study along with other important exegetical works.

In the conclusion or peroration of his introduction, Kepler returns to the main points of his argument, reaffirming his aim "not to treat astronomy on the basis of fictitious hypotheses, but according to physical causes" (322).

Kepler has provided some ingenious explanations, proposing what he believes are the basic physical causes for the heaven's apparent motions. That these explanations did not quite provide the motive power, for example, of the sun's and the planet's rotation on their axes, was increasingly obvious to him, and he tried in later works to develop a better explanation, eventually positing meditative souls as the source of movement.[19] His efforts were ultimately to issue in the pattern described in *Harmonice mundi,* which he thinks must inform all of the changes we witness in the universe. But even though Kepler was sure he had succeeded in discovering the grand design, he was able to convince very few. Nevertheless, Newton was to build upon the three descriptive laws Kepler devised.[20]

We have seen that Kepler was very careful to distinguish between what he claimed was certain and what he conjectured on the basis of dialectical argu-

17. Kepler to Giangiorgio Herwart von Hohenburg, 26 March 1598, *Opere* 10:72, no. 61.

18. *Dialogue,* 462. The point is discussed in chap. 9.

19. Koyré, 283–325, especially 323.

20. The paradox of Kepler's rejection by contemporaries and his influence on Newton is noted by Koyré, 362–64.

ments. He longed for physical proofs to support his conjectures and those of Copernicus, but he never permitted himself to fabricate them, nor to use rhetoric to exaggerate his findings. Dialectical argument sustained his harmonic solution.

Galileo's Heavenly Message

Despite the bold defense of heliocentrism in Kepler's treatises, his work did not succeed in convincing the public. The novelty and strangeness of his ideas inspired caution. On the other hand, his honest enthusiasm, deference to his readers, and observance of preferred logical methodology probably dampened passionate opposition. He had many friends, and those with whom he differed on these subjects did not feel the gauntlet had been thrown at their feet.[21] It was to be otherwise with the most famous proponent of Copernicanism. His writings were widely read and provoked intense controversy. He made many enemies.

Copernicanism did not stimulate much spirited discussion at all until the appearance of Galileo's *Sidereus nuncius* in 1610, the year after the publication of Kepler's *Astronomia nova*. In his momentous little book Galileo recounts his discoveries with the telescope, claiming that they furnish remarkable support for the thesis of Copernicus. The evidence he describes could not be ignored; it was the talk of the educated world. Moreover, the aggressive manner in which he argued for the Copernican system in his lectures and in his informal debates in gatherings of the political, social, and ecclesiastical elite brought the controversy over heliocentrism to a crisis. When Galileo wrote to the German astronomer that he favored the Copernican thesis in 1597, he seems to have thought it solved the riddle of tidal fluctuations. His primary interest during his early teaching career at Pisa and Padua was in mathematics and physics, and it was curiosity about the variations and periods of tides that evidently led him to consider the motion of the earth as an explanation. His syllabus of 1606 shows that as late as 1606 he was still teaching traditional astronomy.[22]

Basing his opinion on Galileo's letters and writings, Stillman Drake believes that contrary to the opinion of most historians of science it was not until the appearance of a new star in October 1604 that Galileo became much concerned with astronomy.[23] Observations of the star indicated that

21. Gingerich describes Kepler's popularity, *DSB*, 307.
22. Wallace has discussed Galileo's teaching on Ptolemy and Copernicus in "Galileo's Early Arguments for Geocentrism and his Later Rejection of Them," in Paolo Galluzzi, ed., *Novità Celesti e Crisi del Sapere* (Florence: Istituto e Museo di Storia della Scienza, 1983), 31–40.
23. Stillman Drake, *Galileo against the Philosophers* (Los Angeles: Zeitlin & Ver Brugge, 1976) xi, 1–32.

it must be beyond the sphere of the moon. If these sightings and measurements were true, then Aristotle's view that novelties could not appear in the heavens because they were eternal and unchanging would have to be false. The phenomenon generated much excitement and provided material for provocative lectures by Galileo at the University of Padua, which more than a thousand students attended. Drake thinks that Galileo was hopeful that the star might offer evidence for the system, but when it began to diminish, he was forced to dismiss that evidence. During the period 1605–9, Drake argues, Galileo was disenchanted with the thesis and had little interest in it. Not until his observations with the telescope did he become convinced of the truth of Copernicanism and publicly declare his support for the system. Certainly a rhetorical analysis of Galileo's published work on this question supports Drake's analysis.

Rhetorical Overtures

As the next most important work in inaugurating the new science after *De revolutionibus*, the *Sidereus nuncius* is pivotal for our consideration of the use of rhetoric in promoting heliocentrism, for it marks the beginning of a shift in argumentative strategies from total reliance on demonstration and dialectic in advancing scientific claims to a partial reliance on rhetorical arguments to bear some of the burden of proof.

This brief work was hastily written in order to reveal to the world as soon as possible the marvels of the cosmos disclosed by the new instrument. Since the primary aim of Galileo was to communicate his discoveries, he devotes much of the text to description. He also makes a number of conjectures and dialectical deductions concerning the import of the discoveries and indulges in very little rhetorical argument in the body of the work, save at one critical point. The introduction is another matter. There rhetorical prose was expected, if not required.

Galileo's training and experience energized by an inordinate natural talent infused his writings with a rhetorical power, unexcelled in the annals of science. His rhetorical education was begun as a lad when he attended the monastery school at Vallombrosa, and it continued during his studies at the University of Pisa from 1581 until 1585. During the time he was enrolled at the university the required humanities courses included study of Aristotle's *Rhetoric, Ethics,* and *Poetics,* the orations and other works of Cicero, along with readings from works of Pindar, Sophocles, Virgil, and Isocrates.[24] The *Opere* includes a translation of Isocrates Galileo probably made during his

24. The *Rotulus* of Almi Studii Pisani records the titles of works to be read during the academic term, Archivo di Stato Pisa, MS G77, fols. 187v–206v.

stay at Vallombrosa.[25] That he maintained an interest in literature and rhetoric throughout his life may be inferred from the quality of his style and from allusions contained in his letters and published writings, but direct evidence is also extant. Favaro has preserved two lectures he gave at the Florentine Academy on the dimensions and location of hell in Dante's *Inferno* and a long satirical poem "Against Wearing the Toga," the academic garb required at the University of Pisa. Six sonnets attributed to the Father of Modern Science are also included among his works, along with a prose consideration of Tasso's *Jerusalem Liberated* and a poetic response to Ariosto's *Orlando Furioso*.[26] These were composed while he was teaching mathematics at Pisa from 1589 to 1592.

That Galileo was aware moreover of the difference between rhetorical, dialectical, and demonstrative discourse can be assumed, given his careful copying of Jesuit notes on logical argument, and his later references to those techniques. Disputation or prolonged dialectical debate was still the method of examination for degrees at Pisa and Padua, while public exhibitions of model disputations continued to be presented as edifying entertainment. From all contemporary accounts Galileo excelled in this skill, as shall become evident.

When he began his observations with the telescope in 1609 Galileo was teaching at the University of Padua, where he had attained the chair of mathematics at the end of 1592, but he had long desired a position at the court of the Medici in Florence so that he could continue his researches and write undisturbed by teaching obligations. The approaches to this objective that he had previously begun as early as 1601 had not yet borne fruit, and so it is not surprising that his book of startling revelations should bear a lavish dedication to the young duke Cosimo de' Medici, whom Galileo had tutored in mathematics some years before. Nothing serves more effectively to introduce his rhetorical skills than this dedication.

Galileo begins the dedicatory letter with a consideration of the monuments human ingenuity has created to honor virtue, from statues to pyramids.[27] Then he names more enduring things, such as literary works, and

25. The translation of Isocrates is in *Opere* 9:283–84.

26. Galileo's family had extensive cultural and literary contacts. His father was a lutenist and musicologist, well acquainted with classical languages and mathematics. The lectures and the poem may be found in *Opere* 9:31–57, 212–23. The writings on Tasso and Ariosto are in the same volume, 60–148, 149–54. Ludovico Geymonat discusses the early works of Galileo in *Galileo Galilei: A Biography and Inquiry into his Philosophy of Science,* trans. Stillman Drake (New York: McGraw-Hill, 1965), 9–16.

27. Albert van Helden's recent translation of the work, *Sidereus nuncius or The Sidereal Messenger* (Chicago: University of Chicago Press, 1989), attends closely to the optical phenomena

next turns to the most incorruptible of all: the christening of stars and constellations in honor of heroes and gods, like Jupiter, Mars, Mercury, and
Hercules. He notes that this practice has since passed out of favor, especially
after Augustus Caesar, thinking to honor Julius, named a new star Julian
only to find that it was a comet that disappeared soon afterwards. Galileo
reveals a profound ingenuity when he says that he can offer far more enduring honor to his prince by giving the newly discovered satellites of Jupiter
the duke's family name. He adds that God himself has seemed to indicate the
suitability of so naming them, for just as these four satellites cling closely to
Jupiter so the virtues of "clemency, kindness of heart, gentleness of manner,
splendor of royal blood, nobility in public affairs, and excellency of authority and rule have all fixed their abode and habitation in Your Highness" (24–
25). The ascription is particularly appropriate, he adds, since the young
duke was born under the influence of Jupiter, who is reputed to be the seat
of these very virtues. In a flight of eloquence he amplifies the point: "Jupiter,
I say at the instant of your Highness's birth, having already emerged from
the turbid mists of the horizon, and occupied the midst of the heavens, illuminating the eastern sky from his own royal house, looked out from that
exalted throne upon your auspicious birth and poured forth all his splendor
and majesty in order that your tender body and your mind (already adorned
by God with the most noble ornaments) might imbibe with their first
breath that universal influence and power" (25).

In the next line, through a rhetorical question, he neatly exploits the distinctions between dialectic and demonstration: "But why should I employ
mere probable arguments, when I can demonstrate it almost through necessary reason?" (*Opere* 3, pt. 1:59–61). Galileo goes on to relate his own experience of the reality of Cosimo's virtues when "by divine will" he became his
servant and instructed him in mathematics.[28] He claims that it was these
virtues that "inflamed" his soul so that he has worked day and night to bring
glory to him and thereby show his gratitude. He then reminds the duke that
he is his subject "not only by choice but by birth and lineage." Near the end
of the encomium, he employs the rhetorical ploy of contraries in a manner
characteristic of him: "And if I am first to have investigated them, who can
justly blame me if I likewise name them, calling them the Medicean Stars, in
the hope that this name will bring as much honor to them as the names of

---

and to the significance of Galileo's observations. I have generally cited the translation of Stillman Drake in *Discoveries and Opinions of Galileo* (Garden City, N.Y.: Doubleday, 1957) because
of Drake's more graceful rendering and sensitivity to the rhetorical overtones. Favaro includes a
facsimile of the Latin edition in *Opere* 3, pt, 1, 59–96, with which I have compared the translations. When two numbers are referenced in my discussion, the first numbers refer to Drake's
translation, the second to the facsimile. My own translation simply cites *Opere*.

28. Galileo taught Cosimo in the summer of 1605 when the Prince was fifteen.

other heroes have bestowed on other stars?" (25). Naming the stars thus honors them more than it honors the Medici.

In closing the letter Galileo mentions Cosimo's illustrious ancestors and, returning to the point of the sentence just quoted, mentions that although the glory of his progenitors is enshrined in history, only Cosimo's virtue justifies the naming of the stars. He ends with the anticipation that Cosimo will no doubt surpass all others in his deeds and thus be forced to vie only with himself.

In this piece of ceremonial rhetoric Galileo has used to advantage most of the topoi of amplification usually exploited in the genre: the category of external circumstances, including the topoi of fortune, descent, education, wealth, power, and titles, and the topos of character, comprising qualities such as those discerned in Cosimo. Only one of the usual topics is omitted: physical attributes, probably because of the fragile constitution of the young man. Moreover, Galileo has framed these to ascend at each reference to a more impressive attribute, following the scheme of climax. And the simile subtly informs the whole passage.

In the concluding passage Galileo takes pains to insure that the auguries he has divined are not misinterpreted: "Accept then, most clement Prince, this gentle glory reserved by the stars for you. May you long enjoy those blessings which are sent down to you not so much from the stars as from God, their Maker and Governor" (26).

The letter has all of the traditional components—*salutatio, captatio, narratio, conclusio*—except one: a *petitio*. But superlative rhetorician that he was, he did not need that, for the court, aware of his earlier requests for preferment, would understand that a petition was implicit in the tribute itself. And it worked. Four months after the publication of *Sidereus nuncius,* Galileo received the desired appointment as chief mathematician and philosopher to Cosimo, Grand Duke of Tuscany.[29]

The title of Galileo's book has engendered dispute ever since its printing. The problem is that the word *nuncius* can mean either "message" or "messenger"; for this reason I have referred to the work by its Latin title. Kepler took the word to be messenger and so entitled his response to the work, written only a few days after reading it, "Conversation with the Sidereal Messenger." Edward Rosen notes that this misunderstanding led to the capital Galileo's enemies made of the title, one of whom ridiculed him for appointing himself an ambassador from heaven.[30] According to Stillman Drake and Albert van Helden, who have both translated the work, Galileo intended "message" and so in English the title should be "The Sidereal Mes-

29. Drake, *Discoveries,* 65.
30. Rosen, *Kepler's Conversation,* xv and 51, n. 2.

sage." However, Drake entitles his English version "The Starry Messenger" because "starry" is more familiar than "sidereal" and "messenger" because this, he believes, has been the most common understanding of the title.[31]

The first page bears the heading "Astronomical Message," and beneath it epitomizes the "message" as that "which contains and explains recent observations made with the aid of a new spyglass [*perspicilli*] concerning the surface of the moon the Milky Way, nebulous stars, and innumerable fixed stars, as well as four planets never before seen, and now named The Medicean Stars."

An exordium follows in which Galileo attempts to convey the splendor and novelty of his discoveries. His valuation is charged with the emotion he felt upon first seeing what had been hidden from human eyes throughout history. In this, one of the first modern "scientific communications" concerning empirical discoveries, we find a mixture of ethical, pathetic, and logical elements, as Galileo records his reactions along with the data themselves.

Galileo mentions the multiplication by the telescope of the host of fixed stars, and then describes something even more remarkable: "It is a very beautiful thing, and most gratifying to the sight, to behold the body of the moon, distant from us, almost sixty earthly radii, as if it were no farther away than two such measures—so that its diameter appears almost thirty times larger, its surface nearly nine hundred times, and its volume twenty-seven thousand times as large as when viewed with the naked eye" (27–28). The real composition of the Milky Way, which always had been termed "nebulous," is next disclosed, but the most remarkable discovery, he declares, is "four wandering stars not known or observed by any man before us" (28).

Novelties in the Heavens

After the brief introduction, Galileo relates in the narratio the circumstances by which he came to make the telescope. He describes how he constructed it and tells prospective observers how to measure distances between objects seen.

While Galileo expresses himself poetically at times in these descriptions and reflections, he relates what he has seen with precision, and frames his conclusions dialectically or demonstratively.

After describing the face of the moon, he hazards some daring conclusions about its spots and surface: "From observations of these spots repeated many times I have been led to the opinion and conviction that the

31. Van Helden, xi; Drake, 19. Drake (19) notes that Galileo himself observed that *nuncius* had two different meanings and said that he meant only "message."

surface of the moon is not smooth, uniform, and precisely spherical as a great number of philosophers believe it (and the other heavenly bodies) to be, but is uneven, rough, and full of cavities and prominences, being not unlike the face of the earth, relieved by chains of mountains and deep valleys" (31; 62.28–63.3). The "uneven, rough and very wavy line" at the edge of the illuminated part of the moon would not be seen if a spherical solid were illuminated in this way. Galileo illustrates the phenomena by a drawing of the moon at quarter phase in which he dramatically magnifies the contours at the edge, no doubt to emphasize the point. Using poetic language to argue from similarities, he describes the significance of this effect: "There is a similar sight on earth about sunrise, when we behold the valleys not yet flooded with light though the mountains surrounding them are already ablaze with glowing splendor on the side opposite the sun. And just as the shadows in the hollows on earth diminish in size as the sun rises higher, so these spots on the moon lose their blackness as the illuminated region grows larger and larger" (32; 63.18–64.3). While "flooded with light" [*lumine perfusas*] and "ablaze with glowing splendor" [*splendore fulgentes*] are not the sort of simple, concrete expressions we expect to find in a discussion of evidence, their replacement with "neutral" language would not alter the demonstration he has offered. They do serve to make the observation come to life by setting it before the eyes, as both Aristotle and Cicero advised in discussing the attributes of good style. This passage illustrates well the unusual blend of science and poetry in his nature, his capacity for acute penetration and sensitive eloquence, employed with great effect in many genres.

In another part of this section on the moon, however, Galileo moves the discourse into the realm of rhetoric as he tries to prepare his audience to accept his conclusions, much in the manner of Copernicus. He concludes that what permits us to see the faint image of the rest of the moon when it is in its quarter phases is the light of the sun, thrown back upon the moon by the earth. Others have supposed the light to come from the moon itself or from Venus or even from the sun. After showing through dialectical argument based on sense experience that the light could not come from any of these sources, he moves from his conclusion to pose a series of questions and from these draws a conclusion expressed in a rhetorical mode:

Now, since the secondary light does not inherently belong to the moon, and is not received from any star or from the sun, and since in the whole universe there is no other body left but the earth, what must we conclude? What is to be proposed? Surely we must assert that the lunar body (or any other dark and sunless orb) is illuminated by the earth. Yet what is so remarkable about this? The earth, in fair and grateful exchange, pays back to the moon an illumination similar to that which it receives from her throughout nearly all the darkest gloom of night. (44; 74.2–10)

Galileo's whimsical conceit helps to make the whole idea far more accept-able.

A passage following closely on the above foreshadows the rhetorical arguments of the *Letter to Christina*, written five years later, and the *Dialogue Concerning the Two Chief World Systems* of 1632. He extrapolates concerning the earth's reflection of sunlight:

Let these few remarks suffice us here concerning this matter, which will be more fully treated in our *System of the World*. In that book, by a multitude of arguments and experiences, the solar reflection from the earth will be shown to be quite real—against those who argue that the earth must be excluded from the dancing whirl of stars for the specific reason that it is devoid of motion and of light. We shall prove the earth to be a wandering body surpassing the moon in splendor, and not the sink of all dull refuse of the universe; this we shall support by an infinitude of arguments drawn from nature. (45)

Here Galileo exploits the hubris of Renaissance man who gloried not only in his own powers but in the marvelous riches of the earth. This orb could now lay claim to a place among the planets, the wandering stars. Far from being inferior to the moon, it can claim more than equal status. The lines are designed to stir the emotions of readers to accept the implications he has outlined, and also to encourage them to anticipate his forthcoming book, where all will be made clear through "arguments and experience."

A closer look at the original Latin of this passage clarifies its significance for our study of the increasing prominence of rhetorical argument in Galileo's writings. His *et rationibus et experimentis,* while literally translatable as Drake has rendered it above, is more pointedly dialectical and demonstrative in connotation in Latin, implying as it does a distinction between dialectical arguments (*rationes*) and physical proof (*experimenta*). The use of *et* and *et* emphasize this, so that "both by rational argument and by physical proof" is closer to Galileo's thought. Similarly, the final clause of the sentence quoted above, "this we shall support by an infinitude of arguments drawn from nature" is expressed more precisely in the Latin, "esse demonstrationibus et naturalibus quoque rationibus sexcentis confirmabimus." Although the *et* combined with the *quoque* is somewhat improper usage, the passage again implies a twofold type of justification, as in an alternative translation: "this we shall confirm by natural demonstrations and also by innumerable logical arguments" (75.7–14). Thus, Galileo promises he will prove heliocentrism by demonstration, an obligation he is not actually able to honor in the work that he eventually entitled *Dialogue Concerning the Two Chief World Systems;* the new title implies an indeterminacy. The question of whether he believed he had attained a demonstration in the argument from the tides presented in the *Dialogue* was, and still is, roundly disputed. But evidently when he wrote *Sidereus nuncius* he thought demonstration was possible.

Galileo next treats the fixed stars, describing how different they appear from the planets. The planets are round, "like little moons and flooded all over with light," while the fixed stars "have rather the aspect of blazes whose rays vibrate about them" (47). He notes the varying degrees of brilliance the stars display and discloses that many, many more appear with the telescope than are visible with the eye, so many "as almost to surpass belief." He provides drawings of Orion and the Pleiades to illustrate the new stars he has observed within the constellations (47–48).

The Milky Way next claims his attention:

> The galaxy is, in fact, nothing but a congeries of innumerable stars grouped together in clusters. Upon whatever part of it the telescope is directed, a vast crowd of stars is immediately presented to view. Many of them are rather large and quite bright, while the number of smaller ones is quite beyond calculation.
>
> But it is not only in the Milky Way that whitish clouds are seen; several patches of similar aspect shine with faint light here and there throughout the aether, and if the telescope is turned upon any of these it confronts us with a tight mass of stars. And what is even more remarkable, the stars which have been called "nebulous" by every astronomer up to this time turn out to be groups of very small stars arranged in a wonderful manner. (49–50)

This and the earlier passages typify the quality of experiential presence Galileo maintains throughout the work. The emotion displayed in his diction convinces us that this is a natural response and not an artificial rhetorical ploy deliberately crafted to arouse belief. Such passages display a quality Glynn Norton finds characteristic of humanist writings. He calls it extemporaneity: flights of eloquence, made possible by training and talent, stimulated by the subject, and spontaneously written.[32]

The last topic Galileo discusses, his observations of the satellites of Jupiter, by contrast, conveys the experience in a carefully controlled prose. Since what he discloses is the most stunning discovery of all, he seems to want to emphasize the fact that he was completely unprepared for what he saw, that far from anticipating it, he simply stared with a patient, cool gaze: "On the seventh day of January in this present year 1610, at the first hour of night, when I was viewing the heavenly bodies with a telescope, Jupiter presented itself to me; and because I had prepared a very excellent instrument for my-

---

32. Professor Norton expressed these ideas in a Folger Shakespeare Library Seminar of 1988, "Eloquence Beside Itself: Extemporaneity and the Sublimation of Rhetoric in the Renaissance Text." Quintilian describes the training necessary for effective seizure of the moment in *Institutio Oratoria* 12.2.11–15. This ancient concept, *kairos*, is treated by James L. Kinneavy, "*Kairos*: A Neglected Concept in Classical Rhetoric," *Rhetoric and Praxis*, 79–105.

self, I perceived (as I had not before, on account of the weakness of my previous instrument) that beside the planet there were three starlets, small
indeed, but very bright" (51). He goes on to give a chronological account of
the viewings, which disclosed that these starlets were revolving around
Jupiter. The full import of this discovery emerges at the end of this description:

> Here we have a fine and elegant argument for quieting the doubts of those who,
> while accepting with tranquil mind the revolutions of the planets about the sun in
> the Copernican system, are mightily disturbed to have the moon alone revolve about
> the earth and accompany it in annual rotation about the sun. Some have believed that
> this structure of the universe should be rejected as impossible. But now we have not
> just one planet rotating about another while both run through a great orbit around
> the sun; our own eyes show us four stars which wander around Jupiter as does the
> moon around the earth, while all together trace out a grand revolution about the sun
> in the space of twelve years. (57)

Having once more registered his own views regarding the Copernican system, Galileo closes the little treatise with some speculations about the
changes in size of the satellites, which he had observed. He imputes their
enlargement to a shroud of vapor, which he thinks envelopes them, arguing
from similarities observed in the apparent enlargement of the moon and sun
at sunrise and sunset when they are seen through the vapor covering the
earth. He adds that the moon as well as the earth are probably surrounded
by such an envelope. More on this last subject, he promises, will be forthcoming in his *System*.

The *Sidereus nuncius* demonstrates clearly Galileo's mastery of rhetoric,
particularly in the dedicatory letter. Certainly, laudatory prefaces were common means of gaining, keeping, or increasing patronage, but Galileo's was
fabricated with far more artistry than most, and its effect was to bestow enduring fame on the prince.

Throughout the treatise proper, rhetoric appears in the expression of
feelings, embellishes conclusions, and presses for the Copernican thesis on
the strength of evidence to be detailed more fully in the promised book. But
Galileo's arguments concerning the phenomena are generally dialectical,
fully within the scholarly practice of the day. The ethos he presents is that of
a competent scholar, fully aware of the singularity of his revelations. Already, however, we can discern in this work a characteristic that will become
more pronounced in later writings: Galileo directs his appeals to like-
minded people with little concern for those who might find it difficult to
accept his discoveries and conclusions. This lack of sensitivity, or, as it will
become, contempt for his opponents, is apparent here not in what he says,
but in what he does not say. Unlike Copernicus, he does not court those he

knows will be skeptical, but rather he announces his discoveries as if no doubt could exist regarding them. Nevertheless, he was very much aware of the resistance his disclosures would meet, and for that reason he enlisted others to attest to his veracity and to his proficiency as an astronomer as soon as the book appeared.

Kepler's Response to the *Message*

Sidereus nuncius brought Galileo immediate fame. It reached a wide audience and aroused much controversy, for many refused to believe that the sights he described were actual phenomena, speculating that they were caused by flaws in the lenses, by spots magnified, or even that Galileo had fabricated them. The Paduan professor offered to display his telescope and invited people to look through it; many were eager, although some of his colleagues at Padua refused to do so.[33]

To gain support for his revelations Galileo sent a copy of the book to the Tuscan ambassador in Prague, Giuliano de Medici, asking him to request a written response to the volume from Johannes Kepler. Kepler was by then Mathematician to the Holy Roman Emperor, Rudolph II, at the capital of the Empire in Prague. The emperor himself was fascinated with astronomy and had even obtained a telescope and observed spots on the moon previous to Galileo's published account. He had obtained a copy of *Sidereus nuncius* before the one sent by Galileo arrived, and he had already asked for Kepler's reaction. On hearing from the ambassador concerning Galileo's request, Kepler agreed to send his opinion before the courier departed in less than a week's time. The letter was published within a month under the title of *Dissertatio cum Nuncio sidereo*, in Edward Rosen's translation, *Conversation with the Sidereal Messenger*. It soon became a best seller, spawning a pirated edition in Florence in the same year.[34]

Kepler's Dissertatio

That the *Conversation* is far more rhetorical in content than the *Sidereus nuncius* is not surprising since its purpose was rhetorical—a work intended to convince the public of the authenticity of Galileo's claims. Consequently, we will be attentive to the way in which Kepler weaves rhetoric into his evaluation of Galileo's observations and conclusions.

33. Galileo notes the refusals in a letter to Kepler, 19 August 1610, translated by Carola Baumgardt, *Johannes Kepler: Life and Letters* (New York: Philosophical Library, 1951), 86; He refers to philosophers at this "gimnasium" who refuse to look, *Opere* 11:423, and notes specifically Cesari Cremonini in a letter to Gualdo, 6 May 1611, *Opere* 11:99–101. Cremonini and Giulio Libri may have been the philosophers of 1610; see Drake, *Galileo at Work*, 165.

34. Rosen, *Kepler's Conversation*, xvii.

Kepler chose to keep the published response to Galileo's request in its original form, a letter, but he did include some additional material: a preface dedicating the work to Giuliano de Medici and a "Notice to the Reader."[35] The *Conversation* is marked throughout by an effervescent enthusiasm that Kepler evinces in so many of his writings. He obviously relishes the fact that Galileo has at last come out publicly in support of Copernicus, as he had urged him to twelve years before, and, more importantly, that he had also produced the kind of evidence needed to support the position.

Truth is the theme of the letter. Kepler identifies this as his and Galileo's sole concern. The preface announces the theme: "From this dedication learn my love for proclaiming the truth, and also the glory, which rests on truth alone, of the Medicean rule; in this latter activity the writings of Galileo take precedence" (4). In these words Kepler not only echoes Galileo's dedication of the *Sidereus nuncius* to Cosimo II, but he generously amplifies his homage to the Medici family. Kepler's sensitivity to the situation in which Galileo has placed himself through the revelations and opinions he expresses in the work is evident in the German's responses.

The emphasis on truth was intended to establish at the outset the purity of Galileo's motives. Kepler correctly perceived that the reality of the observations was to be the critical issue.

In the Notice to the Reader that follows the dedication, Kepler apologizes for the great haste apparent in the letter's composition, explaining that he had wanted to oblige those who had asked for his opinion of the book, and he thought the most efficient way of doing so was to publish the letter he had written to Galileo. This solution did not please many of his friends, he says, who raised a number of objections. These are especially interesting, for the responses he makes to them reveal his conception of his own ethos. Some found the letter too "unconventional"; others thought he ought to omit the introduction and greatly revise some expressions. To these criticisms Kepler replies that since each is entitled to his own tastes, he prefers to stick to his own style and to leaven his arguments with humor.

In addition, his critics also took him to task for having erroneously attributed antischolastic views to Johann Wackher, a friend who is mentioned in the work. He defends himself, saying that he likes to indulge in hyperbole, explaining that debaters often force others into defending absurd positions for the sake of practicing disputation.

To another objection that he had immoderately praised Galileo, he answers: "I do not think that Galileo, an Italian, has treated me, a German, so well that in return I must flatter him, with injury to the truth or to my deep-

---

35. Rosen's translation is used in this chapter; page references are noted in the text. A facsimile of the original is in *Opere* 3.1:97–126.

est convictions." For Kepler to assert that he has nothing to gain is a commonplace, but he may well have intended a litotes or understatement here as well, since he had reason to think himself ill-treated. Galileo had not answered his request for an opinion on the comets in an earlier letter and when Kepler asked for a response to his *Mysterium cosmographicum,* he received none.[36] Furthermore, in spite of Kepler's numerous works on topics of mutual interest, Galileo had not corresponded with him about them, nor is there evidence in the *Sidereus nuncius* that he considered the views expressed in them. The simple answer may be that Galileo had not read them and probably had no inclination to do so.

When Kepler mentions near the end of the Notice that he has taken advantage of the occasion to defend some of his own views, it may well be that he intended by this a mild revenge. Certainly he seems to enjoy the opportunity to correct Galileo and to recommend his writings to him.

Written in Latin, the letter to Galileo is organized in the traditional manner. The salutation is brief, but it matches the correspondents' prestige equally, by balancing the honorific attached to Galileo with the author's own official title:

To the Noble and Most Excellent Signor

GALILEO GALILEI,

PROFESSOR OF MATHEMATICS AT THE UNIVERSITY OF PADUA,

JOHANNES KEPLER,

HIS SACRED IMPERIAL MAJESTY'S MATHEMATICIAN,

sends his most cordial greetings.

The letter includes an exordium, which Kepler refers to by that term. As we have noted, it was this part that some of his friends had suggested he delete. The reason is not difficult to discover. It is "unconventional," too playful and personal for most tastes, given the mission Kepler has undertaken. Kepler begins with a description of the effect of the book's arrival, "For a long time I had stayed at home to rest, thinking of nothing but you." These thoughts of Galileo were prompted by the publication of his own book on Mars. Taking his cue from Mars, the God of War, Kepler likens the vacation he took after the strains of completing that work to recuperation following a military campaign. As he paused to reflect on the ordeal, he says he entertained the hope "that among others Galileo too, the most highly qualified of

36. Galileo wrote promptly to thank Kepler for his book, pleading insufficient time to respond in detail, 4 August 1597, *Opere* 10:67–68. Galileo apparently asked Kepler's messenger to send more copies, which request was met by Kepler with delight. He writes that he is sending two more books and explains that it is his custom to seek the candid opinion of scholars, a response that does not refrain from censure if that is merited; Kepler to Galileo, 13 October 1597, *Opere* 10:69–71; 69.5–9.

all, would discuss with me by mail the new kind of astronomy or celestial physics . . . and that he would resume our interrupted correspondence, which had begun twelve years before."

The excitement kindled by Galileo's discoveries permeates Kepler's account: "But behold, a surprise report about my Galileo is brought to Germany by the couriers around March 15th. Instead of reading a book by someone else, he has busied himself with a highly startling revelation," a revelation about the Medicean planets. The startling news of Galileo's discoveries was brought by his friend, Johann Wackher, who spoke to him from his carriage drawn up in front of Kepler's house. "He was so overcome with joy by the news, I with shame, both of us with laughter, that he scarcely managed to talk, and I to listen" (10). The feelings of shame, which Kepler reveals so openly, were evoked by fear that the discovery of the planets would throw into doubt his description of the composition of the universe in *Mysterium cosmographicum*. Wackher, he relates, "maintained that these new planets undoubtedly circulate around some of the fixed stars," echoing the views of Cardinal Cusanus and Giordano Bruno, which would tend to support the view that there is an infinity of worlds. This is probably one of the antischolastic opinions that Kepler's friends worried about.

When he was able to borrow the Emperor's copy of the book for a short time, Kepler says he was happy to see that Galileo addressed the work to " 'philosophers and astronomers'," and " 'lovers of true philosophy summoned to the commencement of great observations'," and only hoped that he might be included among them (11). Still caught up in the moment he writes: "I yearned to discuss with you the many undisclosed treasures of Jehovah the creator, which He reveals to us one after another. For who is permitted to remain silent at the news of such momentous developments? Who is not filled with a surging love of God, pouring itself copiously forth through tongue and pen?" (11–12).

Soon after his brief glimpse of the work, the letter from Galileo arrived, which, Kepler says happily, was "full of affection for me." He declares that he is moved by the "honor" accorded him in requesting his written opinion. He accepts, making clear that he does so, first, because he wants to, second, because his friends have urged him to, and, finally, because Galileo has requested it. Sustaining the military analogy, he ends the exordium with the hope that "it may bring you this advantage: against the obstinate critics of innovation, for whom anything unfamiliar is unbelievable, for whom anything outside the traditional boundaries of Aristotelian narrow-mindedness is wicked and abominable, you may advance reinforced by one partisan." The choice of figures was surely inspired by a premonition of the dangerous battles in store for supporters of the Copernican system.

The reference to Aristotelian narrow-mindedness is a commonplace in

Renaissance texts. Its very commonness does not mean that it is an empty complaint. On the contrary, the Aristotelians to whom Kepler refers were the conservative academicians who settled all questions by invoking the text of Aristotle. By the text they meant the Greek texts in which the real truth of Aristotle was thought to repose. Incessant conflicts arose over cryptic passages. The Latin and Arabic commentaries along with the scholastic tradition that had grown up around them had issued in many different schools whose various opinions served to keep the debate seething in Kepler's day. For Kepler and Galileo, as for other innovative natural philosophers, the textual scholars were the enemies who scorned anything in natural philosophy not already approved by "the philosopher."

Despite the debates over elements of Aristotle's natural philosophy, many of which were obviously out of date, his logical methodology still furnished the basic structure for inquiry, whether dialectical, demonstrative, or rhetorical. Kepler's defense against the carping of his friends serves to underscore his consciousness of the rhetorical elements in his writings. In addition, he is a very conscious dialectician, as his reference to techniques of argument in the "Notice" shows. An example of his exceptional skill in dialectic and rhetoric is A Defence of Tycho against Ursus (Apologia pro Tychone contra Ursum), recently analyzed by Nicholas Jardine.[37] The Defence is a classic piece of refutation, even more remarkable since it defends Tycho's explanation of celestial motion, which we know was not Kepler's view, in the face of Ursus's charges that astronomers cannot capture the reality of the matter.

Given Kepler's awareness of rhetorical techniques, it is not surprising that he has used the exordium to establish the ethos he wanted to convey, an ethos of spontaneous elation he hopes will capture the attention and good will of the reader. The rambling introduction, retained over the protestations of his friends, conveys the excitement of the moment and evokes a similar response in his readers. He must have hoped that the humor and simplicity of the account would disarm the skeptical. Evidently, he was also determined to have his own stature as an astronomer and mathematician acknowledged; this is revealed not only in the recitation of his title in the salutation, but even more by the later references to his own writings.

The purpose of the narration that follows is likewise apparent. Here, in a variety of ways Kepler focuses on Galileo's truthfulness and the validity of his observations. First he announces his own acceptance of what Galileo has reported: "Because he loves the truth, he does not hesitate to oppose even

37. Nicholas Jardine, The Birth of History and Philosophy of Science: Kepler's "A Defence of Tycho against Ursus" with Essays on its Provenance and Significance (Cambridge: Cambridge University Press, 1984). Although written about 1600, Kepler's work was not published until 1858. Jardine finds his defence of realism against scepticism a precursor of modern arguments for scientific realism.

the most familiar opinions and to bear the jeers of the crowd with equa-
nimity." Then he employs a series of rhetorical questions to underscore the
absurdity of doubting Galileo's disclosures. He asks why he should not believe "a most learned mathematician, whose very style attests to the soundness of his judgment?" Arguing from contraries expressed in an anaphoric
series, he forces the reader to accept his own conclusion—"Shall I with my
poor vision disparage him with his keen sight?" "Shall he with his equip-
ment of optical instruments be disparaged by me, who must use my naked
eyes? Shall I not have confidence . . . ?" (13). Drawing from the topoi of
contraries and the more and the less, Kepler asks whether Galileo would
mock his patron by naming fictions and claiming their reality. He wonders
why if Galileo were trying to deceive he would posit only four planets.

Turning from a seemingly endless store of enthymemic questions, Kepler
cites the corroborating testimony of none other than the "Most August Emperor," who in pursuing his interest in astronomy had also viewed spots on
the moon through a telescope and had wondered if the moon, like a mirror,
did not reflect earth's configurations. Kepler notes that these observations
are supported also by ancient authorities such as Pythagoras and Plutarch.
His and Maestlin's discussion on this subject, published six years earlier,
confirm Galileo's account, a point that he will take up in detail later.

Kepler concludes the narration by declaring that "mutually self-sup-
porting evidence" convinces him that there is no reason to question
Galileo's report about the moon, which predisposes him to accept the remainder of the book, including the satellites of Jupiter. Extrapolating from
his geometrical model of the universe, Kepler says that he wishes he could
help search for more satellites that he believes can be found: two for Mars,
six or eight for Saturn, and one possibly for Venus and for Mercury.

In the partitio that follows, the German astronomer says he will examine
every section of Galileo's work in parallel order so that nothing will frustrate
a reader "devoted to philosophy, either to deter him from having faith in
you, or to induce him to spurn the philosophy which has hitherto prevailed"
(14–15). Despite his plan for a systematic disposition of the subject matter,
Kepler's approach remains informal and conversational throughout.

Views on the Telescope

In considering the first part of Galileo's book, that concerned with the construction of the telescope, Kepler notes that the instrument is not actually a
new invention. Giovanni Battista Della Porta treated it many years ago, and
six years earlier he himself described the geometrical principles in his *Optical
Part of Astronomy*. Kepler says that he mentions these previous treatments
not to depreciate Galileo's technical discussion but only to point out the reliability of the instrument.

Nevertheless, some of Galileo's detractors were delighted with what to them was Kepler's revelation that Galileo was not the first to invent the telescope. Edward Rosen notes that Georg Fugger, agent of the Fugger family in Venice, wrote to thank Kepler for the information in the *Conversation*, saying that Galileo's "'mask has been torn away from him'." Fugger had previously written Kepler decrying Galileo for trying to preen himself with the "feathers" of others through his claims to have invented the telescope. The misunderstanding seems to have emerged from the description of the work that appears on its title page; Galileo is described as revealing what he saw with a "spyglass lately devised" by him, in van Helden's translation. The Latin *reperti* might mean either "invented" or "devised," van Helden notes. Rosen explains that Galileo never meant that he had invented the telescope and, furthermore, in the *Sidereus nuncius* he relates how he was inspired to make one on hearing of a Dutchman's instrument. So Kepler, far from seeking to discredit him as Fugger thought, meant to acclaim his improvements and contributions.[38]

With characteristic candor, Kepler acknowledges the reservations he had had about the ability of the instrument to penetrate what he had thought was a dense blue atmosphere around the earth. In addition, he had feared that since the moon was conceived to be composed of purer elements of the quintessence, it would be impossible to differentiate these from particles of the earth's atmosphere. Ironically, such philosophical suppositions had prevented Kepler, the mathematician and astronomer, from attempting observations with the telescope. But Galileo, who sought to be appointed philosopher and mathematician to the Grand Duke, preferred empirical investigation, seeking afterwards to fit his findings into a philosophical frame. As Kepler expresses it: "Putting aside all misgivings, you turned directly to visual experimentation. And indeed by your discoveries you caused the sun of truth to rise, you routed all the ghosts of perplexity together with their mother, the night, and by your achievement you showed what could be done" (18). This may have been one of those hyperbolic passages Kepler's friends deplored.

Kepler remarks that he would like to construct a telescope himself, but states that if he does attempt to do so he will try to increase the number of lenses. He then proceeds to give Galileo a detailed lesson in the optical effects which would result, referring him to specific pages in his *Optics* where the appropriate principles are discussed. At the end of his discussion of the telescope Kepler implores the Italian fellow-astronomer to send him one:

Would you like me to express my feelings? I want your instrument for the study of lunar eclipses, in the hope that it may furnish the most extraordinary aid in improv-

38. Rosen, *Kepler's Conversation*, xviii–xix.

ing, and where necessary in recasting, the whole of my "Hipparchus" or demonstration of the sizes and distances of the three bodies, sun, moon, and earth. For the variations in the solar and lunar diameters, and the portion of the moon that is eclipsed will be measured with precision only by the man who is equipped with your telescope and acquires skill in observing. (22)

Unfortunately, Galileo did not send him a telescope, and Kepler's impatience was only partially satisfied when four months after his letter to Galileo the Elector of Cologne lent him his.[39]

Kepler follows his request with the proposal of a mutual research program: "Therefore let Galileo take his stand by Kepler's side. Let the former observe the moon with his face turned skyward, while the latter studies the sun by looking down at a screen (lest the lens injure his eye). Let each employ his own device, and from this partnership may there some day arise an absolutely perfect theory of the distances" (22). In this way he thinks they could also make "accurate observations" of a comet's parallax. Then the question of the comet's position could be definitively resolved.

Kepler's invitation to Galileo to take up a place at his side is an interesting inversion of the order the reader might have expected. But Kepler must have wished that Galileo could join him in the research that he had already begun, which he mentioned previously—the Hipparchus project that Kepler ultimately left uncompleted.[40]

Galileo's observations of the moon are treated next. Kepler mentions here his own research on the moon's surface as recorded in *Optics*. In this regard, he suggests another joint research project in which he would compete with Galileo in examining small spots on the moon.

Kepler acknowledges that his own conception of the topography of the moon was wrong and credits Galileo with "brilliant and irrefutable logic" in arguing that there are mountains on the moon (26). He agrees with Galileo that the very dark spots are seas. One spot is particularly interesting to Kepler, and he fantasizes that it may have been the work of enormous lunar inhabitants, inferring their size from the moonscape which boasts higher mountains than earthly ones. He conjectures that the spot is actually an underground shelter the moon creatures have excavated to escape the sun's heat.

Returning to more supportable conjectures, he relates that both he and Maestlin think that the earth exhibits phases similar to the moon when viewed from the moon. In addition, he points out that twenty years earlier than Galileo Maestlin had proved that light is reflected from the earth to illuminate the moon when it is in its crescent phase. Again Kepler appears

39. Gingerich, "Kepler," 299.
40. Rosen, 96, n. 174.

to be motivated not by a desire to denigrate Galileo but to show that the Italian is not suggesting explanations that are unacceptable to respected astronomers.

On the other hand, Kepler's support for Galileo does not keep him from correcting the Italian astronomer when he thinks he is in error. In response to the latter's attribution of the red edges of the earth's shadow on the moon to a dawn-light phenomenon, he refers Galileo to his *Optics,* where he explains the phenomena as caused by a refraction of light by the atmosphere. He thinks the cause Galileo proposed would not restrict light to the edges but would spread it uniformly across the face. "A much more successful explanation" would be advisable, Kepler suggests, when Galileo writes his *System of the World.*

Stars and Planets

Coming next to the section of the *Sidereus nuncius* on other celestial bodies, Kepler takes Galileo to task for his discussion of the eye's perception of starlight: "I should like to ask you, Galileo, whether you are satisfied with the reasons for this effect, as presented by me in my discussion of the process of vision on page 217 and especially on page 221 of *Optics.* For if you find nothing amiss, you may hereafter discuss the matter correctly" (32). In his precise references to the *Optics* Kepler betrays more than a little pique at the fact that Galileo has ignored his careful study of optical phenomena. But he goes on to the next point with no further comment.

He congratulates him on his observation that the fixed stars twinkle and asks: "What other conclusion shall we draw from this difference, Galileo, than that the fixed stars generate their light from within, whereas the planets, being opaque, are illuminated from without; that is, to use Bruno's terms, the former are suns, the latter, moons or earths?" (34). Arguing from probabilities, he is quick to counter the view of Bruno that there are as many worlds as stars. Since Galileo has confirmed that there are more than 10,000 stars visible, this, says Kepler, simply reinforces the contention of his *Astronomia nova* that an infinity of worlds is impossible. He directs the reader to that work for the full argument.

Concerning the description of the Milky Way in the next section, Kepler declares that these observations have "conferred a blessing on astronomers and physicists" (36) by revealing the true character of the phenomena.

In considering the last section of Galileo's book, the new "planets," Kepler rejoices that Galileo has not discovered planets revolving around stars, for this would help to corroborate that "dreadful philosophy" of Bruno, whom his friend Wackher admires. In a curious and perhaps self-congratulatory passage, Kepler then notes how just it is that men are revered whose intellect anticipates sense evidence. He thereupon iterates a number of such feats, including Columbus's divination of the New World. And among mathematicians he notes Pythagoras, Plato, and Euclid, who thought that God must have created a world based on the five solids. "But they mistook the pattern." Copernicus, for his part, was "equipped with a mind that was not average, yet drew a picture of the universe virtually as it is seen by the eye" (38). Copernicus illumined only "the bare facts." Then Kepler adds: "Trailing far behind the ancients will be Kepler. From the visual outlook of the Copernican system he rises, as it were, from the facts to the causes, and to the same explanation as Plato from on high had set forth deductively so many centuries before. He shows that the Copernican system of the world exhibits the reason for the five Platonic solids" (38).

The value Kepler accords to his own quest for causes as the ground for demonstrations is very clear in this passage. It is this quest that permits him to be honored ever so slightly, and, at the same time, to imply that Galileo's contribution, like Copernicus's, is in the realm of sense evidence.

Kepler concludes his excursus with some further points about the human mind that Galileo will echo in the preface and text of the *Dialogue Concerning the Two Chief World Systems:*

It is not an act of folly or jealousy to set the ancients above the moderns; the very nature of the subject demands it. For the glory of the Creator of this world is greater than that of the student of the world, however ingenious. The former brought forth the structural design from within himself, whereas the latter, despite strenuous efforts, scarcely perceives the plan embodied in the structure. Surely those thinkers who intellectually grasp the causes of phenomena, before these are revealed to the senses, resemble the Creator more closely than the others, who speculate about the causes after the phenomena have been seen. (39)

In another splurge of fancy, Kepler suggests that there may be inhabitants on Jupiter. He suggests that some day brave men will venture there in "ships or sails adapted to the breezes of heaven." He invites Galileo to join him in preparing the way for such a voyage: "Let us establish the astronomy, Galileo, you of Jupiter, and me of the moon" (39).[41]

It is no accident that Kepler then turns to those who most threaten the discoveries and hypotheses of astronomers, the theologians. He declares that he wants to "tweak the ear of the higher philosophy" when he challenges theologians to consider why God has revealed these things to man at

41. Kepler did proceed with his astronomical study of the moon, even though Galileo's interest in Jupiter was not pursued with equal vigor. In his *Somnium,* Kepler continued the work of his 1593 student dissertation, Lunar Astronomy, revising and recasting it in the form of a dream. He added voluminous notes over the years from 1620–30, and the work was published posthumously in 1634. See Edward Rosen's Introduction and translation, *Kepler's Somnium: The Dream or Posthumous Work on Lunar Astronomy* (Madison and Milwaukee: University of Wisconsin Press, 1967).

this time. Perhaps he gradually leads man by the hand up the ladder of knowledge.

Finally, Kepler considers the conclusions that can be drawn from Galileo's discovery of the Medicean planets. Proceeding from the topos of final cause, he argues that the four little moons were created for Jupiter just as our moon was created for the earth. "From this line of reasoning we deduce with the highest degree of probability that Jupiter is inhabited." He adds that Tycho Brahe had come to that conclusion simply on the basis of the enormity of the planet.

In this context the German astronomer raises a problem that was to trouble those who pondered Galileo's discoveries. Should there be other earths, which one would be situated in the noblest position? And how can God be said to have created all of nature for man? Declining to consider that question further, he notes "we have not yet acquired all the relevant information" (43).

The last part of his *Conversation* outlines some of the arguments he thinks can be offered in favor of the earth's superior position. A number of these are based on the dialectical topos of what is proper and its rhetorical derivative, the argument from prepon, appropriateness. Kepler states that man lives on the globe "which by right belongs to the primary rational creature, the noblest of the (corporeal) creatures." For it is placed in the "bosom of the world around the heart of the universe, that is, the sun."

In further enthymemes he says that the sun is the center of the universe and that this is proper (*prepon*), since it reveals the power of God. The earth's placement properly serves to magnify the "opulence of God" and man's dependence on him. What better position for man to be in than to circle the sun and make his observations, for the telos of man is to contemplate God. He cannot do that as well on an earth at rest. Kepler crafts an apt analogy: "So surveyors, in measuring inaccessible objects, move from place to place for the purpose of obtaining from the distance between their positions an accurate base line for the triangulation."

He introduces yet another argument, this one suggested by Wackher and based on Kepler's conception of the geometrical structure of the universe. Geometry is "unique and eternal, and it shines in the mind of God." The fact that we share in it is one of the reasons man is said to be made in "the image of God." Our universe is composed of the most perfect geometrical figures, making more universes superfluous. Arguing from final cause, he asks, "For what is the use of an unlimited number of worlds, if every single one of them contains all of perfection within itself?"

Begging to be forgiven for his "diffuse and independent way" of discussing nature, Kepler ends his evaluation of the *Sidereus nuncius*.

Kepler's extensive response and Galileo's book have afforded an unusual opportunity to examine the way in which scientific theories gained credence among scientists. The arguments advanced ranged from demonstration to dialectics to rhetoric, depending upon the aims of the author. A scientific evaluation was requested of Kepler, and he gave one, but he knew that to lend credence to Galileo's account he must also convince the reader that the evidence offered in the *Sidereus nuncius* was credible. For this rhetoric was required. Kepler's rhetoric was neither thundering nor plaintive, but, as suited the man, humorous, inventive, and winsomely assertive.

CHAPTER FOUR

The Significance of the Sunspot Quarrel

The year following publication of *Sidereus nuncius* brought Galileo many tributes. Most importantly, the handsome compliment he had paid to Cosimo II by dedicating the work to the young heir and by naming the satellites of Jupiter after his family had the desired effect. The book, sent along with a telescope, gained for him the position of chief mathematician and philosopher at the court of the Medici in Florence he had so passionately sought.

In recognition of his marvelous discoveries, the Jesuit faculty of the Collegio Romano accorded him a day-long celebration, featuring an oration praising his book![1] The most illustrious of their members, Christopher Clavius, congratulated him, and he was also given an audience with Pope Paul V.

One of the most gratifying of these honors must have been Galileo's induction into the illustrious Accademia dei Lincei, the Academy of the Lynx-eyed—the beast possessed of phenomenal sight. Begun in Rome by Prince Federigo Cesi, its chief aim was to further the study of nature and mathematics. The Academy and its members were to play an important part in the publication of many of Galileo's later works. The objectives of the Academy expressed in its Constitution afford some insights concerning their interest in his work. The members sought are philosophers, yet the group pledges not to "neglect the ornaments of elegant literature and philology, which like graceful garments, adorn the whole body of science." They promise to "pass over in silence all political controversies and every kind of quarrels and wordy disputes, especially gratuitous ones which give occasion to deceit, unfriendliness and hatred."[2] The last passage ironically foreshadows the quality of the controversies in which they were to become engaged through Galileo's writings, in contradiction to their stated ideals.

The desiderata shed light on the quality of expression favored by Prince

---

1. Drake, *Discoveries,* 75.
2. Drake's translation in *Discoveries,* 77–78. Drake notes that the constitution is quoted with some modifications from a translation by John Elliot Drinkwater (-Bethune) in his *Life of Galileo,* 37.

Cesi and presumably other members of the nonacademic intelligentsia. An elegant style was desirable in the discourse of philosophers and mathematicians, but members should not flourish the Lincean seal in political or acrimonious disputes, the domain in which rhetoric generally held sway. The Prince had not foreseen that rhetoric could also invade disputes about natural philosophy. It did so, actually, with his encouragement, but perhaps without his recognition that this was the case.

Early Disputes at Florence

Fame meant also that Galileo's opinion on any controversial philosophical issue would gain immediate attention. Soon after Galileo's return from his triumph at Rome, he became involved in a dispute with Ludovico delle Colombe about the physics of floating bodies. Colombe, although not an academic, was an erudite and respected exponent of Aristotelian natural philosophy in Florence. Stillman Drake thinks that he was the inspiration for Simplicio, the conservative pedant of the *Dialogue on the Two Chief World Systems*.[3] The debate was exacerbated by a discussion during dinner at the Grand Duke Cosimo's palace. Two cardinals were present, one of whom was Maffeo Barberini, later to be Pope Urban VIII.

In arguing about why bodies float in water Barberini took Galileo's side, while Cardinal Ferdinand Gonzaga agreed with Galileo's opponents. At the Grand Duke's urging, Galileo published his arguments, and in so doing he vigorously refuted the other side. (Actually neither he nor his adversaries hit upon the principles of surface tension that govern the phenomena.)

In his *Discourse on Floating Bodies* (1612), Galileo disputed certain concepts of Aristotelian physics and claimed that he could demonstrate his position using mathematical reasoning and measurements. His opponents were adamant in their rejection of his proof because they believed that it was impossible to import mathematical reasoning into the realm of nature. As conservative Aristotelians, they held that mathematics could not be combined with physics; each had its own domain in which only principles proper to that science could be used to find true causes. Galileo, of course, maintained that a philosopher could also be a mathematician and could decide when quantitative data should be considered.[4]

A more serious quarrel erupted at the beginning of 1612 with a request for Galileo's views on a publication concerning sunspots by an anonymous scholar who wrote under the name of "Apelles." Again Galileo clashed with conservatives, whose ideas generally dominated contemporary opinion, and

3. Drake, *Discoveries*, 79.
4. Wallace, *Galileo and His Sources*, 284–88.

in this dispute he extrapolated from these discoveries, hazarding opinions on the motion of heavenly bodies.

The request for Galileo's opinion on the matter was one more instance of the extent of his renown. It came from Mark Welser, a banker and scholar living in Augsburg, who sent along a booklet of three short letters on sunspots he had published the previous year, *Tres epistolae de macularis solaribus.* In his letter to Galileo, Welser remarks that although the subject may not be new to him, he trusts that Galileo will be pleased to find that even in Germany an astronomer is emulating him.[5] This cheerful expectation, however, was not to be fulfilled. Welser's request prompted a series of letters by Galileo in which he contemptuously dismisses the German astronomer's efforts as inept at best. He uses the occasion to make known his own observations of the sunspots and to argue that these provide further evidence for the inadequacy of the old cosmology. His letters were subsequently published by the Lincean Academy in 1613 as the *Letters on Sunspots.*

Galileo soon learned that Welser's anonymous author was Christopher Scheiner, a German Jesuit who had studied at Ingolstadt and was teaching there. His reluctance to append his name to the work is explained by the Order's desire to have its members refrain from open participation in controversies. Apelles was an appropriate choice, since that legendary artist at the court of Philip II and Alexander the Great was said to have hidden behind his portraits to hear comments on his work. As the story goes, Apelles overheard a shoemaker making disparaging remarks about a shoe he had depicted, upon which the artist corrected the work after his departure. The next day the shoemaker returned to inspect the painting, but then he began to criticize the leg above the shoe. Upon hearing this Apelles stepped out and told him that he should stick to his trade.[6] Thus, the rest of the pseudonym, *latens post tabulam,* hidden behind the picture (or tablet), connotes a desire for anonymity but at the same time hints at the writer's interest in a response.

The allusion was not lost on Galileo. The author's desire to keep his identity hidden seems to have irritated him, possibly because he read into the pseudonym another admonition—"be careful for I hear your comments." In his third letter to Welser, Galileo says at one point that he has "no picture to conceal him from spectators" (129).[7] The content and style of Scheiner's letters show that the Jesuit wished primarily to announce his observations and describe the sunspots. He is very cautious in proposing explanations

5. Drake reprints the first letter from Welser in *Discoveries,* 89.
6. Pliny the Elder *Natural History* 35.84.
7. Drake's translation of Galileo's letters on the sunspots in *Discoveries* (89–144) is the source of these English quotations, while citations of the originals are from *Opere* 5:2–260.

about the phenomena, since these would have serious implications regarding the entire cosmic picture. The publication of the letters and Welser's request for Galileo's opinion caused him to become embroiled in a debate, which he did not seem to have anticipated or desired.

Ultimately Galileo was to quarrel with Scheiner over the priority of discovery of the sunspots, which may have contributed to the ill-will many in the Jesuit Order bore Galileo in the later years of his life. In a careful analysis of the evidence, Maria Luisa Righini-Bonelli concludes that it was Scheiner who made the first observation with the telescope.[8] But neither Scheiner nor Galileo were the first to have noticed the sunspots, as Drake notes, for Virgil and Charlemagne had done so centuries before, and Johann Fabricius in the summer of 1611 published an account of what had been reported over the centuries about the phenomena.[9]

Obviously even though Galileo had been asked in a letter to comment on the sunspots, he might have chosen any of several genres to respond: treatise, disputation, tract, or the recently invented essay form. The rhetorical possibilities offered by the letter genre in this case appeared to outweigh other choices.

The Place of Letters in Early Scientific Discourse

Galileo's letters to Welser generally fall within the literary genre of polite letters, as did Kepler's response to *Sidereus nuncius*. Galileo's response is more polished. He, too, intends his letters to be read by a greater audience, an expectation common among humanist authors.

In the early seventeenth century, letters on scientific subjects had not yet been fashioned into the subgenre of "scientific communications," such as are found in the *Philosophical Transactions* of the Royal Society in the last years of the century.[10] The first communications of that society retained traces of the parts of the formal letter (described in the second chapter in relation to Copernicus's dedicatory letter to the pope) and the informal tone of the "polite letter," so attractive to Renaissance authors. This form emerged after the recovery of Cicero's and Seneca's letters, where fewer

8. Maria Luisa Righini-Bonelli, "Le Posizioni Relative di Galileo e dello Scheiner nelle Scoperte delle Macchie Solari nelle Pubblicazioni Edite entro il 1612," *Physis* 12 (1970): 405–12.

9. Drake, *Discoveries,* 82–83.

10. Charles Bazerman describes the growth of the scientific report in *Shaping Written Knowledge: The Genre and Activity of the Experimental Article in Science* (Madison: University of Wisconsin Press, 1988). In addition, see my discussion of some early scientific letters, "Newton and the Jesuits in the *Philosophical Transactions,*" in G. V. Coyne et al., eds., *Newton and the New Direction in Science,* Proceedings of the Cracow Conference, May 25–28, 1987 (Vatican City: Vatican Observatory, 1988), 117–34.

parts—the body of the letter with only a brief salutation and closure—and a relaxed, conversational, yet consciously literary style predominated.[11]

Galileo himself commented on the style in which he wrote about philosophical and mathematical matters in a famous letter to Prince Leopold of Tuscany:

I am indeed unwilling to compress philosophical doctrines into the most narrow kind of space and to adopt that stiff, concise and graceless manner, that manner bare of any adornment which pure geometricians call their own, not uttering a single word that has not been given to them by strictly necessity. . . . I do not regard it as a fault to talk about many and diverse things, even in those treatises which have only a single topic . . . for I believe that what gives grandeur, nobility, and excellence to our deeds and inventions does not lie in what is necessary—though the absence of it would be a great mistake—but in what is not.[12]

The choice of genre was an important one from a rhetorical standpoint. The letter offers some peculiar advantages that make it a very expeditious means of airing opinion on controversial issues. Its form, however, has the potential of frustrating some readers. In appearing to address one person, a letter-writer ostensibly speaks to a particular audience and, for this reason, can assume a great deal about that audience: a similar mind set, mutual values, concerns, enemies, and, even, in some cases, a specialized vocabulary.[13] On

11. Ronald Witt describes the evolution of letter writing in general in "Medieval 'Ars Dictaminis' and the Beginnings of Humanism: A New Construction of the Problem," *Renaissance Quarterly* 35 (Spring 1982): 1–35. For the influence of *ars dictaminis* on Renaissance letters of this kind see Paul Oskar Kristeller, "The Scholar and his Public," 10–14, and "Humanism and Scholasticism in the Italian Renaissance," 85–105, in his *Renaissance Thought and its Sources*. The medieval art is discussed extensively by James J. Murphy, *Rhetoric in the Middle Ages* (Berkeley: University of California Press, 1974), ch. 5. Judith Rice Henderson describes the evolution from *ars dictaminis* to epistolography, "Erasmus on the Art of Letter-Writing," in James J. Murphy, ed., *Renaissance Eloquence* (Berkeley: University of California Press, 1983), 331–55.

12. The letter, dated 1640, is translated by Paul Feyerabend and appears in his *Against Method* (London: Redwood Burn, 1978), 69. Favaro includes the lengthy letter-treatise, 8:489–545. The quotation is from 491.18–29.

13. Chaim Perelman and L. Olbrechts-Tyteca have commented on characteristics of particular and universal audiences that are particularly illuminating for this discussion; see *The New Rhetoric* (Notre Dame: University of Notre Dame Press, 1971), 19–35. Seeing audience as the construction of a writer or speaker, the authors note that in addressing a particular audience "by the very fact of adapting to the views of his listeners, [the speaker] might rely on arguments that are foreign or even directly opposed to what is acceptable to persons other than those he is presently addressing" (31). The universal audience, which would include particular audiences, is "the whole of mankind, or at least of all normal, adult persons" (30). In inquiries about topics of natural philosophy, then, dialectical reasoning is assumed to be directed to a universal audience, for it should be understood by all reasonable people conversant with the terminology. As Perelman and Olbrechts-Tyteca put it, "The agreement of a universal audience is thus a matter, not of fact, but of right" (31).

issues where feelings run high, this privileged communication may be used to stir up emotions for and against ideas in a closed world. The very nature of a select audience could privilege a narrowed vision and limit consideration of the opposite point of view. This approach to an issue is markedly different from that demanded by a treatise or a disputation where both sides would have to be aired extensively.

In responding to Welser's request for his opinion about Scheiner's letters, Galileo was in a difficult position. He did not want to alienate Welser by expressing his opinion candidly, yet he obviously thought Apelles an incompetent astronomer. In addressing Welser as an intermediary he could carry on a continuous commentary on Apelles' arguments and on his methods. The distance provided by this indirect criticism finds him at times adopting a patronizing tone that would have been insulting had it been addressed to the author directly.

For Scheiner, the decision to communicate his discovery of the sunspots to Welser by means of a letter was originally pragmatic, not rhetorical. It was the most practical means of sending word of his discoveries, and retention of the original format in publication was the simplest and quickest solution. There were some rhetorical gains. Since Scheiner was little known and forced to publish anonymously, he might expect his work to command more attention through an address to a well-known public benefactor. But aside from the salutation and brief acknowledgements of his benefactor, Scheiner's letters are succinct, including for the most part only direct accounts of observations and conclusions about them. In this they have more in common with the correspondence emanating from literary and scientific academies in Europe and England later in the century.[14]

Since we are most concerned here with the use of rhetoric to advance scientific arguments, we will not analyze in detail the scientific content of the letters. Instead we shall consider the major theses and their refutations, noting the rhetorical nuances that were to have consequences outside the academic arguments. Galileo's letters receive more attention than Scheiner's in this discussion because they develop the implications of the issues at greater length and because his responses are much richer in rhetorical overtones.

The First Sally in the Sunspot Duel

Galileo's correspondence with Welser was published in 1613 under the title *Istoria e dimostrazioni intorno alle macchie solari e loro accidenti*. Stillman

---

14. For the activities of these early scientific societies in seventeenth-century Italy see Martha Ornstein, *The Role of Scientific Societies in the Seventeenth Century* (Chicago: University of Chicago Press, 1928), 73–82. See also Bazerman, *Shaping Written Knowledge*.

Drake has rendered the title as *History and Demonstrations Concerning Sunspots and their Phenomena,* but *accidenti* (i.e., accidents) carried a more specific meaning than phenomena to philosophers of the time. We noted in chapter 1 that accident, one of the predicables, finds frequent use as a topos in dialectical investigations. Galileo's title implies that he has found evidence of some ephemeral things that might or might not be associated with the sunspots. If these attributes come into being and disappear, then the old cosmology that held the heavens to be changeless would be called into doubt. As the first man in history to have discovered new things in the heavens by means of the telescope, Galileo surely felt himself to be the person most qualified to speak about such things. Welser gave him an opportunity to address the topic.

Welser had originally written to Galileo in Italian, and Galileo answered him in the same language. In a letter to Paolo Gualdo, dated 16 June 1612, a month after Galileo's first response to Welser, Galileo says that he chose to use his native tongue in replying to Welser so that "everyone" in Italy would be able to read his views. In order that knowledge of Italian should not limit the readership, however, he asked Gualdo and another acquaintance to translate his response into Latin quickly so that foreigners could also read it. He explained that he himself was too busy to undertake it.[15] Unfortunately the translations, if they were forthcoming, were not forwarded to Scheiner, who could not read Italian. Welser reported in his second letter to Galileo that Scheiner had great difficulty in finding Italian translators.

Apelles begins the first letter, dated 12 November 1611, with very little preamble, saying simply that he offers for Welser's interest observations of some new and incredible phenomena.[16] These surprising things he thinks will resolve some of the doubts and controversies of astronomers. After this brief exordium Apelles describes the sunspots he discovered seven or eight months before and the problems occasioned by his subsequent observations with the telescope. At first he was not sure whether his eyes or his instrument were deceiving him, or whether what he saw was actually in the earth's atmosphere or nearer the sun. In this period the reliability of the image revealed by the telescope was an important consideration, for many instru-

15. Stillman Drake discusses the correspondence, *Discoveries,* 84–85. The letter is included in *Opere* 11: 326.

16. William Shea, in *Galileo's Intellectual Revolution* (New York: Science History Publications, 1972), 49–51, provides an excellent summary analysis of Scheiner's *Tres Epistolae,* which I have followed in many instances. The facsimiles of these letters are in *Opere* 5:2–32, and the last letter to Welser by Scheiner on the subject, *De maculis solaribus et stellis circa Jovem errantibus accuratior disquisitio,* is reprinted in the same volume (37–70). Where *Opere* is referenced for a translation, the English is mine. The last letter of Apelles is also analyzed by Shea (52–53), as well as Galileo's responses to all of these (51–72).

ments were faulty and the unfamiliar process of peering through a lens proved to be difficult, as anyone who has ever looked into one and seen his own eyelashes or floating motes can verify.

After carefully analyzing the possibilities and noting the corroboration of eight other observers, Apelles concluded that the phenomena he saw were spots located in the vicinity of the sun. But he was loathe to leap beyond the received tradition concerning the purity and inalterability of the sun and infer that these were actually spots on its surface. Probable reasoning about the evidence led him instead to another solution:

I have always considered it inconvenient [*inconveniens*] to place spots darker than any ever seen on the moon (with the exception of one small spot) on the bright body of the sun. It is not plausible [*probabile*] to do so, for if they were on the sun their motion would imply that the sun rotates, and we should see the spots return in the same order and in the same position they had among themselves and with respect to the sun. So far, they have failed to reappear although other spots have followed the first ones across the solar disc. This is a clear argument that they are not on the sun. I do not think, therefore, that they are real spots, but rather bodies partly eclipsing the sun, namely stars located either between the sun and ourselves or revolving around the sun. (Shea, 49–50; *Opere* 5:26.20–29)

Although Scheiner begins with the "inconvenience" of placing spots on the sun (since such a phenomenon was precluded in Aristotelian cosmology), he does not dismiss this assumption out of hand, but rather seeks to see if that inconvenience is in fact supported by the observations he makes of the spots. The problem with his reasoning is, of course, that he assumes that the sun, being composed of the quintessence, has a hard surface on which spots would retain their patterns.

In the next very short letter of 19 December 1611, Scheiner describes his observations of Venus in its conjunction with the Sun, which he had hoped would furnish support for his conclusion that the sunspots were probably stars or planets. He notes that the conjunction had been predicted in an *Ephemerides* by the eminent mathematician Giovanni Antonio Magini. (The mention of that authority would not have impressed Galileo, for Magini had obtained a position at the University of Bologna Galileo had once sought, and, moreover, he had recently discounted Galileo's discovery of the Medicean satellites.) Since Magini had forecast a forty-hour period of conjunction, Scheiner thought he had ample time for his observations. He hoped the "window" would allow him to see Venus as a very large spot passing across the sun's surface. Unfortunately no such spot was visible, so Apelles' letter analyzes the various reasons for this failure. That his observations could have been in error or that Magini could have been wrong were

17. See Stillman Drake's discussion of the letter in *Galileo at Work*, 182.

inconceivable, Scheiner says. The one convincing explanation [*hoc uno evin-cetur*], is that Venus revolves around the sun and not the earth (28.21). Thus, it must have passed behind the sun, a conclusion that accords well with the Tychonian system.[17] Tycho Brahe had concluded that the planets did revolve around the sun but that the earth was not one of them. She remained unmoved while the whole panoply, sun and planets together, circled her. The solution this provided for the celestial observations contradicted the Ptolemaic system, but it did preserve the Church's position on the stability of the earth. If the old system must be thrown out, Brahe's was the only one open to Scheiner.

The third letter, written a week later on 26 December 1611, finds Apelles rejoicing, for he can report that the sun still reigns immaculate: "Amazing, how success can be brought about by courage. Remember those things which previously I timidly proposed; now I am not afraid to affirm them without fear, urged on by certain and well established reasonings [*certis et compertis rationibus nixus*], which I leave to your judgment: happily indeed, the body of the sun is completely free of injurious spots; I have been persuaded that an argument to this effect can be made" (*Opere* 5:28.37–40).

The argument which he heralds so rhetorically is a refined version of the earlier one: if spots actually are on the sun, they would be expected to rotate with it and return in the same conformation. Repeated observations do not disclose this, so he concludes that they cannot be there. But if not there, "Ubi ergo?" He considers various possibilities in turn: they are in the atmosphere, in the lunar sphere, offering "demonstrations" from parallax and observation that neither is the case. He dismisses the possibilities of their being in the spheres of Mercury or Venus. On the basis of the demonstrations he offered for the moon, he believes these planets are in orbit between the moon and the sun (*Opere* 5:29.9–29).

The possibility that the spots are clouds on the sun is likewise untenable, for who would put them there? Why would they be so large, and how could they move in such a uniform pattern? Nor can they be comets, Scheiner says. The final answer he proposes is that they are composed of a denser material philosophers say are stars. They appear to be solid and opaque. Lending support to this conclusion was a darkening at the edges of the spots he claimed to observe. This he said indicated that the stars have phases like Venus. He includes a diagram to support this possibility (*Opere* 5:30).

From the probable conclusions he has offered, Apelles draws a number of corollary hypotheses. Among these is the possibility that the satellites of Jupiter are actually stars and more numerous than the four described by Galileo. He also suggests that the odd oblong shape of Saturn is caused by companion stars.

In closing, he introduces a question he cannot resolve: whether these stars that appear as spots on the sun are wandering or fixed. He indicates that he is inclined to believe they are the former. Adding no embellishment, he ends the third letter with a simple, "Vale" (*Opere* 5:31).

Scheiner's letter permits us to see close at hand the reactions of a typical astronomer of the age who is confronted with the fearful prospect of a a worldview in jeopardy. He has made little use of rhetoric in arguing for his positions. The joy he feels in finding his cosmic picture still whole provides him with an opportunity for a little personal *rhetorices colores,* when the ethos of timidity is replaced by happy confidence. The desire to accommodate his findings to the received system led him to ignore other possibilities—that the spots might shift and rearrange themselves as Galileo will point out— but he does not propose authority as a step in the proof itself. Thus, strictly speaking, he recognizes that sense observation and probable argument are the appropriate path to astronomical conclusions. Yet his belief in the Aristotelian cosmos distorts his assessment of the evidence. The rhetorical color introduced to celebrate his conclusions he must have thought would stir a grateful echo in Welser and other readers. Some of these readers, moreover, because of his rhetoric might not consider opposing arguments that challenged their assumptions. Thus, in voicing this aside to Welser he does introduce emotional suasion, muddying slightly the dialectical and demonstrative character of his account. If he were challenged, he might claim that such an aside was permitted by the genre, the letter, and that he meant it simply as a personal reflection on what he had found.

The same problem is engendered by the "humanized" disputations on philosophical matters described in the final section of this book. Once the audience becomes an important consideration for an author, the aim of the dialectical process may become clouded. This is true of Galileo's responses to Sheiner's letters as well, although his rhetorical targets are quite different.

Galileo's Reaction to Scheiner's *Tres epistolae*

Five months elapsed before Galileo answered Welser's request to review Scheiner's letters. In his reply Galileo apologizes for the delay, stating that his own ill health has prevented him from making sustained observations. Throughout the three letters to Welser Galileo is always deferential, expressing his concern for the ailing man's health and responding extensively to his requests for information.

In the exordium, he invites Welser's sympathy by explaining that he did want to offer enlightening opinions on the sunspots, but that he had to do so with exceeding care, for any error would immediately be seized upon by his enemies: "As your excellency well knows, certain recent discoveries that depart from common and popular opinions have been noisily denied and impugned, obliging me to hide in silence every new idea of mine until I have more than proved it. Even the most trivial error is charged to me as a capital fault by the enemies of innovation" (90).

The price of fame had been high. As his work became more and more renowned, his critics became more vociferous. Galileo incurred stubborn resistance to his observations and the conclusions he drew from them. A few years later a similar complaint about his detractors is voiced in the exordium of his *Letter to Madame Christina, Grand Duchess of Tuscany,* wherein he considers the theological difficulties engendered by the Copernican thesis (see chapter 7).

Galileo's First Letter

In his first response to Scheiner's letters, Galileo opposes almost every conclusion Apelles offers regarding the sunspots. He initially concedes that Apelles is right in proposing that the sunspots are real objects and not illusions created by the medium of observation, but he contends that they move in a direction opposite to that Apelles attributes to them. Moreover, they are on or close to the surface of the sun, he declares, despite Apelles' objection to this opinion on the grounds that it violates the received view of the sun and the heavens. Galileo explains that he has also observed the phenomena for many months, and he thinks that the spots revolve with the sun. Their movement carries them around the sun from west to east in a path slanted south to north. This is a journey similar to that of Venus and Mercury and the other planets.

Turning in the second letter to the "inconvenience" of positing spots on a rotating sun and the joyful solution proposed by Apelles, Galileo's tone sharpens. Adopting a professorial stance, he reminds Apelles that conclusions about something's names and attributes—whether they are pure or lucid—should be assumed on the basis of their essence and not the other way around, "since things come first and names afterwards." The implication that received views must not furnish our guide to observation will be even more boldly stated later. Further, as we shall see, Galileo is really not concerned with discovering "essences" but with discerning secondary intentions, the properties and accidents of entities. He contests Apelles' opinion that the spots are much darker than those on the moon, saying that this is not so, that, in fact, they are brighter and only appear to be dark when one looks at them in relation to the sun itself.

Moreover, he informs Apelles that he has completely missed the point of the latest discoveries that he (Galileo) has made with the telescope. Apelles believed that one of the great benefits of his observations of the movement of the sunspots was that it enabled him to determine Venus's and Mercury's orbits, which he had concluded were below that of the sun's. Galileo retorts that since Venus undergoes phases like the moon, it must be lit by the sun. Thus Venus must circle the sun rather than the earth: "These things leave no room for doubt about the orbit of Venus. With absolute necessity we shall conclude, in agreement with the theories of the Pythagoreans and of Copernicus, that Venus revolves about the sun just as do all the other planets. Hence it is not necessary to wait for transits and occultations of Venus to make certain of so obvious a conclusion" (94). Observations regarding the path of Venus in front of (transit) or passage behind (occultation) the sun that Scheiner had thought so important are disparaged as corroborative but hardly worth the trouble. The necessary demonstration to which Galileo alludes provides proof that Venus orbits the sun.[18] The claim that demonstrations can be extended to the other planets, however, is not so plausible. In the discussion Galileo does not mention at all Tycho Brahe's solution, which indicates that he did not regard it seriously. To Welser he remarks, "No longer need we employ arguments that allow any answer, however feeble, from persons whose philosophy is badly upset by this new arrangement of the universe."

When he begins to comment on the third letter, Galileo tries to sweeten his critique by applauding the capabilities Apelles has shown in trying to accommodate "many novelties" in accord with "good and true philosophy." It may be that this masked author is too much impressed by what he has been taught to break away entirely from positing such things as "eccentrics, deferent, equants, epicycles and the like as if they were real, actual and distinct things" (96–97).

Galileo goes on to explain the difference between true philosophy and mathematical philosophy. The latter is concerned with "saving the appearances," while philosophical astronomy is that which investigates "the true constitution of the universe" (97). He adds: "For such a constitution exists; it is unique, true, real, and could not possibly be otherwise" (97). This is a powerful declaration that claims the "old" philosophy is untenable now that his observations have revealed the true structure of the universe.

His realist epistemology does not, however, compel him to concede that he still cannot offer the true and certain observations that fully prove the Copernican system superior to the Tychonian. To concede this would be a tactical error in his rhetorical strategy, and it is this omission that makes it effective rhetoric but less effective dialectic. Since Galileo was a master of both forms of argumentation, one must assume that he feared others might

18. Van Helden discusses the progress of Galileo's discovery that Venus has crescent phases like the moon, which shows that Venus is illuminated with borrowed light. The phases exhibited do not agree with the Ptolemaic theory but are in accord with both the Copernican and the Tychonian, *Sidereus nuncius*, 106–9.

not be persuaded by an a fortiori argument: that since he already had some incontrovertible evidence, all the more reason to believe that he would have the rest soon. He may have thought it better to pass over the missing fundament, not a glaring oversight in a letter, but a serious flaw in a disputation.

Returning to Apelles' discussion of the substance of the sunspots, he says that although he himself is not sure of their composition, he cannot agree with Apelles that the sunspots are stars that orbit the sun. For his part, Galileo prefers to posit a different possibility. He offers a dialectical argument based on similarities that suggests that in their movements and appearance the sunspots most resemble clouds. They change in shapes and density in much the same way, and they amass and disappear like clouds. He is careful to state that he does not mean that they are formed from water vapor, as are our clouds, but that their appearance closely resembles clouds. Perhaps they issue from fumes or vapors in the sun or are attracted to the sun from elsewhere.

Galileo then proceeds to offer some telling refutations of Apelles' hypothesis that the sunspots are stars, pointing out the differentiae, that stars do not appear to change shape, nor do they make irregular appearances, forming and disappearing. Their dark aspect is insufficient reason to classify them as stars. The sunspots, when lit on one side, may cast shadows as do all opaque bodies: dense clouds, Venus, and the moon, for example. But these, like sunspots, lack the "essential properties" of stars (100–101).

When Apelles bolstered his thesis by comparing the four Medicean stars to sunspots, saying that they display similar properties, he risked deeply offending Galileo, for he implied that the astronomer had not observed the satellites carefully. This imputation may have stimulated Galileo's reference at the beginning of his letter to the infinite pains he takes in recording his observations. In replying to Apelles' supposition, Galileo controls his anger and says only, "It grieves me to see Apelles enumerate the companions of Jupiter in this company" (101). But he quickly disposes of the charge with evidence of the periodicity of the satellites as they pass around Jupiter.

The arguments Galileo advances concerning the sunspots serve as a brilliant example of the efficacy of dialectical reasoning. Arguing from sense experience, similarities, and differences, he has offered a far more convincing hypothesis than Scheiner's. In concluding the discussion of Scheiner's position, he is deferential towards Welser but firm in stating his own views:

Apelles comes finally to the conclusion that the spots are planets rather than fixed stars, and that they lie precisely between the sun and Mercury or Venus, which are the only planets that ever appear between us and the sun. To this I say that I do not believe the spots to be planets, or fixed stars, or stars of any kind, nor that they move about the sun in circles separated and distant from it. If I may give my own opinion to a friend and patron, I shall say that the solar spots are produced and dissolve upon the surface of the sun and are contiguous to it, while the sun, rotating upon its axis in about one lunar month, carries them along, perhaps bringing back some of those that are of longer duration than a month, but so changed in shape and pattern that it is not easy for us to recognize them. This is as far as I am willing to hazard a guess at present, and I hope that Your Excellency will consider the matter closed by what I have suggested. (102)

Galileo, aware that he proposes a view that directly counters the natural philosophy of his day, appears convinced that intelligent men will see from the evidence and arguments that he is right. As William R. Shea points out, Scheiner, on the other hand, seems unwilling to recognize any alteration in the heavens beyond a rearrangement of parts because this would upset traditional philosophy. Since the Jesuit could find no evidence of parallax, the absence of which would mean the spots are located at a great distance, the only explanation he can entertains is that the spots he saw were stars, stars that were not "novelties" but only previously undetected.[19]

Galileo's deferential manner toward Welser, demonstrated in the uncharacteristic "If I may give my own opinion to a patron and friend" (102) in the passage above, shows a concern at this point not to risk losing possible supporters for his position by adopting an abrasive or supercilious posture.[20] In the remaining part of his conclusion, Galileo extends further compliments to Welser, apologizing for his loquaciousness. In a memorable metaphor he likens the task of examining alternative answers to tuning a reed in "this great discordant organ of our philosophy—an instrument on which I think I see many organists wearing themselves out trying vainly to get the whole thing into perfect harmony. Vainly, because they leave (or rather preserve) three or four of the principal reeds in discord, making it quite impossible for the others to respond in perfect tune" (103).

With a bow toward the high intelligence and love of truth shown by Apelles and the usual effusive compliment to a patron "kissing your hands with all reverence," Galileo concludes the first letter to Welser.

In his second letter Welser expresses deep gratitude for the famous Italian's expansive response to his request: "From the little which I can master of the subject it seems to me so well written and contains such good and well-founded arguments, set forth most modestly, that despite your having in the main contradicted the views of Apelles, he should consider himself much honored by it" (104). The ailing banker goes on to say that he regrets he is not in a city where Italian printers can be found so that he might entreat

19. Shea, *Galileo's Intellectual Revolution*, 51–54.
20. The Italian reads *"e se ad un amico e padrone dovessi dir in confidenza l'opinion mia"* (*Opere* 5:111.9–10). The expression would ordinarily be a commonplace, but Galileo, engaged in controversy, obviously intends more than a humble confidence.

Galileo to permit him to publish the letter. Indicating his understanding of Galileo's desire for caution in expressing his views, he says that he thinks publication "might safely be done since you proceed in so judicious and circumspect a manner that, even if something is discovered in these matters which we do not suspect at present, you could never be charged with precipitousness nor with having spoken positively about things that are doubtful" (105). Welser's desire to continue his support for such investigations was only increased by Galileo's letter. He adds: "It would be a public benefit for these little treatises concerning new discoveries to come out one by one, keeping things fresh in everyone's mind and inspiring others to apply their talents more to such things; for it is impossible that so great a framework should be sustained upon the shoulders of one man, however strong" (105). In closing, he promises to convey to Apelles the letter and diagrams Galileo has promised to send.

The public character of the sunspot issue indicated by Welser's remarks clearly highlights the reasons for rhetoric's increasing role in scientific debates. The airing of scientific subjects has, like political issues, become the responsibility of good citizen-scientists. He implies too that men of means have a duty to support such endeavors for the advancement of science as well as for the increase of public knowledge.

Further, Welser's admission that he did not entirely understand all that had been said but that he found Galileo's arguments persuasive, reveals something important about a large part of the anticipated audience for the letters. They have some understanding of the subject but are not experts. This is the kind of audience for whom rhetoric was invented. Obviously, when such a public has very strong opinions, authors will instinctively resort to rhetorical strategies to prepare the ground. Galileo's assertion of a modest responsible ethos was obviously helpful, as was his reputation, in inclining Welser to his side on the issue of the sunspots. Concerning the question of planetary orbits, he sought through the strength of his ethos simply to imply a topos of property and to extend his impressive proof for Venus's revolution around the sun to cover the other planets too, without having to cite the same kind of sense evidence. This kind of sleight of hand, permissible in a letter, was to be repeated in his *Letter to Christina* in 1615.

The Second Letter

Galileo continued his discussion of the *Tres epistolae* in a second letter to Welser, dated 14 August 1612, in which he included the promised diagrams. Here the astronomer takes an even stronger position on the location of the sunspots, basing his arguments on further observations and measurements. He notes that the spots move in concert in parallel paths and that their shapes denote their placement on a spherical body. This opinion is supported by the fact that the sunspots are narrow at the edges of the disk and fatter in the center, that they appear to move farther near the center and less at the edges, and, finally, that the distances between the spots grow as they near the center and narrow almost to the point of imperceptibility at the edges. These are the effects of foreshortening that would be expected if the spots were positioned on a rotating globe.

In an exemplary exhibition of dialectical argument, Galileo considers different possibilities: spots in the air between us and the sun, their placement on an orb around the sun, a motionless sun and spots moved by a medium, a motionless earth and a circumambient fluid. But he refutes each in turn. To the first possibility he replies that the spots would have to be very tiny since they are quite small in relation to the sun; yet they would have to be very dense to keep the sun from shining through them. He wonders why if they are in the air they are not observed elsewhere in the sky and how they could maintain such perfect order between them.

Besides the probable arguments he has offered, he says "necessary demonstrations" refute all objections that the spots must be contiguous with the sun.[21] The most important proof is the fact that the spots exhibit the same pattern viewed from various places on earth. The main line of argument is similar to that of the grand plan for the *Dialogue,* where Galileo shows that objections to the Copernican thesis are refutable and that objections to the Ptolemaic scheme are manifold.

The last hypothesis he considers is that the spots are caused by a fluid moving about a motionless sun. He says this is the most difficult to disprove, but he suggests that he can offer a stronger dialectical argument:

Yet to me it seems much more probable that the movement is of the solar globe than of its surroundings. I am led to believe this first because I think this circumambient substance to be very fluid and yielding—a proposition that appears quite novel in the ordinary philosophy, but which I am assured by seeing how easily the spots contained in it change their shapes, aggregate together, and divide up, which could not happen in a solid and consistent material. Now an orderly movement such as the universal motion of all the spots seems incapable of having its root and basis in a fluid substance, whose parts do not cohere, and which is therefore subject to commotions, disturbances, and other accidental movements. (112)

In an analysis related to his view that the spots are on or near the sun, he considers the possibility of an ambient surrounding a motionless sun, but refutes it with an earth-based physical explanation that adumbrates the principle of inertia. He contends that bodies not moved by violent motion or a downward force are indifferent and tend to remain in whatever state they

---

21. *Mia ci restano le dimostrazioni necessarie e che non ammettono risposta veruna* (*Opere* 5:128.16–17).

are placed, either moving or still, unless they are acted upon or impeded. He wonders how a globe could resist a moving ambient surrounding it and offers a hypothetical example, a thought experiment, to answer this: a ship once receiving an impetus on a calm sea would continue in motion around the earth without stopping; if it were stopped it would stay at rest. Arguing from analogy he says: "And the sun, a body of spherical shape suspended and balanced upon its own center, cannot fail to follow the motion of the ambient, having no intrinsic repugnance or extrinsic impediment to rotation" (114).

In discussing this image, which Galileo develops at length, Shea notes that Galileo "stops short" of the principle of inertia enunciated by Newton. In spite of his prescience Galileo still entertains circular motion for the planets, and was unable to relinquish the idea of natural motion.[22]

Galileo then provides instructions to be conveyed to Apelles on the best way of making accurate drawings of the sunspots. He explains that it is a method perfected by his former pupil, the Benedictine monk Benedetto Castelli.

The discussion that follows illuminates Galileo's perception of Aristotelian methodology and its efficacy for solving problems in natural philosophy. In defending his own divergence from tradition, Galileo cites the precedent of ancient authorities who differed from Aristotle on some of the issues, but he quickly adds that even Aristotle would have changed his mind had he the sense observations that are available now. He explains that when Aristotle argued for the inalterability of the heavens, he based his opinion on the fact that no changes had been observed for centuries. Galileo argues that he is in fact using the very method that Aristotle would have applauded: sense experience as the basis for conclusions about natural phenomena. He declares: "I contradict the doctrine of Aristotle much less than do those people who still want to keep the sky inalterable; for I am sure that he never took its inalterability to be as certain as the fact that all human reasoning must be placed second to direct experience [evidente esperienza]. Hence they will philosophize better who give assent to propositions that depend upon manifest observations, than they who persist in opinions repugnant to the senses and supported only by probable reasons" (118; 139.8–16). He claims that the method of Aristotle is his guide; thus the careful transcription of lecture notes on the Posterior Analytics (described in the first chapter) has not been forgotten. Of course it is politic for him to express such sentiments, but the truth of his assertion is borne out by our analysis. This dialectical strategy of turning Aristotle against the Aristotelians he was to repeat in his Letter to Christina and in the Dialogue Concerning the Two Chief World Systems. It

22. Shea discusses the arguments and the thought-experiment introduced by Galileo (64).

might be expected to mollify conservatives and remind them of the critical steps in their methodology.

Galileo supports his claim that new sense data are at hand by citing observations of comets and novae recorded by Tycho Brahe. These observations enabled Tycho to mount a geometrical demonstration proving that the comets are beyond the moon's sphere, and show that change does take place there. That Galileo cites Tycho's observations as evidence for his own position is curious, first, because, as Stillman Drake points out, this seems to be his only complimentary reference to Tycho and, second, because it indicates that he once had a very different opinion on the comets from that which he was to press in a later debate about the comets with another Jesuit, Orazio Grassi.[23]

Finally, Galileo returns to the subject of the drawings he sends along with his letter. Fortunately "divine Providence" has given us the means to arrive at conclusions about these spots, he says, for people in locations distant from each other can make drawings and compare them. He himself has received information from observers in Brussels and Rome, which is in exact accord with his own observations. "This argument alone should be enough to persuade anybody that such spots are a long way beyond the moon" (119). Before receiving Galileo's second letter and the diagrams, Welser wrote a brief note at the end of September, sending along a new work by Apelles on the sunspots. He plans to publish it, he explains, "chiefly for the observations, which I believe will be welcome to all lovers and investigators of truth." Pleading the torment of his illness, he provides no further evaluation of Scheiner's effort except to say, "For the rest, I hazard no decision one way or the other, as I have no zest to apply my mind to it properly" (120).

Shea stresses the Aristotelian character of Galileo's arguments in his evaluation of these passages. He notes that the astronomer did not say that all of his conclusions were demonstrated. Demonstration he claimed only for the "contiguity of the spots on the surface of the sun, but for the rotation of the solar body, he asserted only probability, and even less certainty for the hypothesis that vanishing spots returned after fifteen days" (58).

A week later Welser wrote again; this time to thank Galileo for his additional reflections on the matter, which, although he had not time to read carefully, he termed "manna from heaven." His arguments, he says, "proceed with great plausibility and probability [*molto verisimilitudine e probabilità*]" (121; 184.10–11). Feeling that the effects of his illness have marred his own ability to grapple with the issues, Welser's frustration leads him to voice the Augustinian view that whether Galileo's arguments "arrive precisely at the truth" we cannot be sure until God allows us "to look down

---

23. Drake, *Discoveries*, 119, n. 11. The debate with Grassi is treated in part 3 of this study.

from on high upon that which we now contemplate from this vale of misery." Welser had been recently elected to the Lincean Academy, showing that his intellectual acumen was widely respected, but he suffered greatly from gout, an illness that was to lead him to take his own life two years later.

Welser's reluctance to take sides in the earlier letter and the epistemological skepticism of the next may have spurred Galileo to greater effort in soliciting his favor, which is quite noticeable in his last letter to the banker on the sunspots.

Apelles' Second Attempt

Apelles' second work on the sunspots, *De maculis solaribus et stellis circa Jovem errantibus accuratior disquisitio,* is composed of three more letters Welser published in the summer of 1612. By this time Scheiner had received Galileo's initial response to his *Tres epistolae,* but he had not yet been apprised of the contents of the second letter. Perhaps taking his cue from Galileo's more graceful epistolary style, Scheiner begins his letter of 12 January 1612 with an expanded exordium. There he develops the most effusive compliment to Welser in his correspondence, explaining in ponderous neo-classical Latin how his patron's great fame and authority demands more careful viniculture to produce a wine suitable for him.

In the body of the letter Scheiner then proceeds to advance elaborate geometrical proofs for his contention that Venus does revolve around the sun. He bases his mathematical argument on the tables of that planet's movements set forth in Magini's *Ephemerides.* He also adds arguments from authority, citing well-known ancient and modern philosophers and mathematicians who have agreed that Venus could eclipse the sun. Besides Plato and Ptolemy he mentions his fellow Jesuit Christopher Clavius. Far down in his compilation of proofs for the thesis, he proposes observations of the phases of Venus made by Galileo and other mathematicians in Rome. The "others in Rome" appears to have been added to underscore the point that Galileo is not the only one making such observations. The mathematicians referred to were probably other Jesuits at the Collegio Romano (*Opere* 5:46.32–34).

He includes in this letter an account also of observations of the movement of particular sunspots and the length of time of their appearance on the sun. Through a geometrical demonstration he attempts to show how solid bodies can produce the effects of sunspots.

In the second letter, a brief note of 14 April, Apelles reveals that he has discovered a fifth satellite moving around Jupiter, and he includes as testimony to the fact six drawings of observations made from 20 March through 8 April 1612. In a crowning rhetorical touch, he names the fifth satellite after Welser and his family. (*Opere* 5:56.10–11). Perhaps the purported discovery and the compliment were responsible for the new note of skepticism in Welser's remarks to Galileo.

The arguments of the first two letters are marshalled to support Scheiner's view that the spots must be stars. Venus, a wandering star, might appear as a spot on the sun when it crosses in front of it, and the satellites of Jupiter from his observations appear to be erratic in their paths, resembling planetary movement.

In the last letter, dated 25 July 1612, Scheiner attempts to answer the objections readers have made to his assertions about the sunspots. The resistance to belief in their existence has been so strong that Scheiner decides to fall back on the time-honored custom of appealing to doxographical evidence. He begins with the names of distinguished scholars and prelates whom he says have shown interest in the phenomena: Cardinal Borromeo, the Archbishop of Milan, Andreas Chioccus, a Veronese medical doctor, Antonius Maginus, Johannes Kepler, Johannes Praetorius, and Johannes Ziegler of the Society of Jesus. These men, he says, disagree with me in some opinions, but they do agree that the phenomena exist and that they are not being deceived by the lenses or their eyes (*Opere* 5:62.20–30).

Next, Scheiner carefully airs the opinions of contemporaries as to the nature and location of the sunspots. The conclusions he draws are generally in accord with those previously outlined in the first of his *Tres epistolae*. One great difference is apparent, however. He no longer claims that the sunspots are unchanging bodies. Mentioning Galileo's observations as recorded in the Italian's his first letter to Welser, Apelles says that there is much evidence to support the view that the spots are quite irregular and seem to change their shapes.

Approaching the subject from a different angle, Scheiner describes his observations of an eclipse of the sun by the moon. As the moon began to cover the sun, he could see the sunspots. They were swarthy in color, while the moon was even darker beside them. As the eclipse progressed, he says, "we saw what we searched for" in that the moon appeared to be transparent, "pellucid as a crystal or some other transparent glass, but uneven in this quality." It appeared to be whiter and denser in some portions than others. When the moon completely covered the sun, the sun "shone through with a much weaker whiteness" (*Opere* 5:67.25–28). This observation is of course quite different from that reported in Galileo's *Sidereus nuncius,* where Galileo describes mountainous terrain resembling the earth's. (In his annotations on this passage Galileo wonders why there are such dark shadows if they are not mountains (*Opere* 5:68.25–30)).

Scheiner concludes that the sunspots are "no less dense or opaque than the moon" (*Opere* 5:68.2). This, he thinks, would seem to indicate that both are composed of the same materials. In a corollary speculation, Scheiner also departs from Galileo's view that the moon shines with reflected light from the earth. The Jesuit thinks there is no need for this explanation because it is apparent that the moon shines from the reflected light of the sun. Although he refrains from mentioning Galileo in these instances of disagreement, he is at pains to note that his own line of reasoning agrees with that of optics and philosophy.

While he is not sure about the precise location of the sunspots or their nature, Scheiner is sure that the old view concerning the nature of the heavens will have to be revised, "especially in the regions of the sun and Jupiter." We should heed the advice of Clavius, who, in light of the new discoveries, thought we should consider some other cosmic system to explain them (Shea 53; *Opere* 5:69.4–8).

Scheiner ends the letter with an awkward, almost bizarre, complimentary close: "What we disclosed in the first picture we still maintain: [that there is] some substance which is either generated and ended, or eternal; about this we have inquired with sagacity and industry, at least as much as man is able. And you meanwhile, eminent man, enjoy those things which have thus far been sufficiently displayed. Farewell to God, to you, to you Apelles, your house, and all the literary community" (*Opere* 5:69.33–70.4).

Scheiner wants to be noted in history as having determined the fact of the sunspots. Beyond that he claims simply to have provided what he thought to be new observations and speculations about them. He seems now to want to serve notice that he has given up the enterprise, resigned to the fact that full knowledge of creation must remain uncertain in this life. Yet his final words add an unexpected dash of bravado as he bids farewell to an audience that includes God, the literary world, and the legendary Apelles.

The second set of Scheiner's letters shows that although he argues dialectically about the sunspot's properties from sense evidence, he still maintains some conclusions that differ from those of his more famous reviewer. It may be that the sense evidence available to them differed. The better quality of Galileo's telescope may have permitted more refined observations.[24] But the basic difference between them lay in Scheiner's acceptance of the dichotomy between the celestial and terrestrial regions and of the quintessence as the constituent element of the celestial orbs. For Galileo the telescope had de-

24. Shea concludes that in the observations of the eclipse Scheiner is led by his own expectations to make an erroneous conclusion about the light of the sun. From what Scheiner was able to observe with his instrument, however, the conclusions do not seem forced. Galileo's observations of the moon in which he saw evidence of mountains were made over time and during different phases of the moon. His telescope were probably much better than Scheiner's. And, as Shea points out, Galileo was to be led by a predilection to misinterpret the comets (52–53).

stroyed the old philosophy, and it provided clear evidence that Copernicus
was right.

Galileo's Final Response to Apelles

The close attention Galileo gives to Welser's letters in his final discussion of
the subject, dated 1 December 1612, displays a sensitivity to both the mood
and opinions of his admirer. After acknowledging receipt of Apelles' work,
Galileo thanks Welser for the favorable reception accorded the arguments
expressed in his most recent letter. But he says that "since even the most
clear-minded of men may at first glance" praise a work influenced by "some
special affection for the author" he will await a second, more impartial, reading. He echoes the tone of Welser's words regarding the fallibility of the human mind in this world, amplifying the benefit he thinks a second reading
will supply: "That, when it comes, will serve me until the knowledge for
which we now search almost like blind men in the impure and material sun
shall come to us from the true, pure, and immaculate Sun, together with all
other truths in Him, as Your Excellency very prudently says." Galileo then
turns the theme to his own advantage with a brilliant disquisition on knowledge. Not one to concede that human fallibility deprives one of any hope of
attaining knowledge in this world, Galileo immediately shifts the focus to
what one may hope to learn.

We can learn about the properties of things, things on earth and in the
heavens. Paradoxically, we have more precise knowledge about the heavens
than about the earth. For example, the shape of the moon was known before
that of the earth, and the periodic movements of planets are better known
than are ocean currents. While we do not yet agree about whether the earth
moves or not, we do know a lot about the movements of the stars (124).

This is a particularly adroit argumentative strategy, validating his and the
attempts of others to seek certain knowledge about the heavens, given an
Augustinian skepticism about the capability of human minds. Galileo carries the point further: "Hence I should infer that although it may be vain to
seek to determine the true substance of the sunspots, still it does not follow
that we cannot know some properties of them, such as their location, motion, shape, size, opacity, mutability, generation, and dissolution" (124).
Thus, philosophers concerned with the heavens may not be able to treat of
substances, a major concern of metaphysicians and theologians, but they
still have much to occupy them.

Continuing to resonate to the mood of Welser's letter, he points out that
the ultimate goal of astronomy is love of the "divine Artificer," which will
"keep us steadfast in the hope that we shall learn every other truth in Him,
the source of all light and verity." Galileo has risen to the challenge pre-

sented by Welser's doubts and Scheiner's pessimism. Since their views would be shared by many he senses that they must be overcome. Obtaining recognition of the fact that astronomy has the power to uncover some truths about the sunspots is part of a larger campaign to establish the truth of the Copernican thesis. In this passage he simply applies the principles uttered in the first letter concerning the difference between mathematical and true philosophy.

He concludes this contemplative passage with a gracious acknowledgement of his debt to Welser for whatever new truths may have been discovered at the Augsburgian's behest. With a characteristic twist, he adds, "Let this [Welser's desire for his opinion] also be my excuse if I fail to get to the heart of so novel and difficult an enterprise." The gentle teasing nourishes the intimacy Galileo maintains throughout the correspondence.

Rhetorical Considerations

Galileo's conception of his approach to the issue of the sunspots and the larger audience he projected for the letters is revealed in the next point he makes. Welser mentioned in his last letter that he had heard with delight that their mutual friend Prince Cesi was planning to publish Galileo's letters on the sunspots. Galileo explains that he had indeed sent copies of his letters to Cesi but not necessarily with a view to having them published. He has been reluctant to do so, he says, for fear that in writing his opinions to Welser who is a sympathetic reader he may have failed to circumvent objections of severer critics. He had also hoped to work out some further problems regarding the sunspots before considering publication. Nevertheless, he has agreed to let Prince Cesi decide, crediting his judgment about their readiness for dissemination.

Galileo's statement concerning the argument's limitations is quite intriguing, given our concern with the rhetorical elements in his work. He seems to have recognized the fact that in directing his discourse to a particular audience he has failed to develop the issue dialectically. This temptation is endemic to the letter genre.

The passage becomes doubly interesting in view of a letter to Paolo Gualdo six months earlier in which Galileo mentions that he prefers to write in Italian in order to reach a larger audience. Two possible explanations for the seeming contradiction in his remarks to Welser might be entertained. Perhaps he had second thoughts about the desirability of publication after he got caught up in his correspondence with Welser, and that prompted his remarks to the banker, or perhaps he expected to have his letters published all along, since Scheiner's were, but he wanted to enter this caveat in the record so as to disarm his opponents.

If one accepts the first explanation—that he changed his mind about the letters' suitability for publication as he warmed to the subject—respect for Galileo's foresight and his rhetorical and dialectical acumen is diminished, but it does place his character in a better light than the other alternative. Nonetheless, Galileo did choose to have the letters published, and the friendly conspiratorial stance we have seen in the previous letters remains characteristic of his argumentative style. He seems always to be speaking in a confidential way to some brilliant friend or friends who are sure to admire the perspicuity of his observations and conclusions. Unfortunately these "conversations" often take place in a charged atmosphere, which his some-times acrimonious comments exacerbate. The letters were said to have greatly offended Scheiner, and they appear to be partly responsible for the growing enmity of the Jesuit community. While Galileo does not attempt to meet the objections he anticipated from his opponents, he does make some overtures to Scheiner at the beginning of the letter.

After the preliminary digressions, Galileo notes with regret that Apelles had not seen his second letter before he began the latest work. He explains that part of the reason for the delay in sending his last letter was because his Venetian friend Giovan Francesco Sagredo wished to have a copy made. Galileo makes a belated apology as well for his continued use of the vernacu-lar in view of the fact that Apelles cannot read it. He explains that he does so mainly because he has such a great appreciation for his native language and also because the Academicians of Florence prefer it in discussions of this sort. Moreover, he says, he and his friends take great delight in receiving Welser's replies in that language. Another reason Galileo did not shift to Latin, although he does not mention it, is that he was not as skilled in Latin, as his request to Gualdo and his friends to translate for him indicates. While the Academicians in Florence may have preferred Italian, nevertheless Latin was still the universal academic language. The choice of Italian had the effect of accentuating the particularity of Galileo's address to Welser and to the select audience beyond him.

Following these polite gestures, Galileo again digresses to discuss the re-ception of his treatise on bodies that float in water, a copy of which he had sent to Welser earlier and to which Welser had given a favorable response. The excursus is interesting for what it reveals of Galileo's conception of his opponents. He mentions how pleased he was to hear Welser's account of his reaction to that work, in which the banker relates that at first he found the tract perplexing but was won over by the conclusions which were "clearly demonstrated" (126). He mentions that others like Welser "who have the reputation of good judgment and sound reasoning" have had similar re-sponses. Those who do not agree with his conclusions Galileo characterizes as "stern defenders of every minute point of the Peripatetics" (126). They are slavish followers of the text of Aristotle who "never wish to raise their eyes from those pages—as if this great book of the universe had been writ-ten to be read by nobody but Aristotle, and his eyes had been destined to see for all posterity" (127). This estimate of his opponents is to figure promi-nently in all of the defenses of Copernicanism that Galileo writes.

His patience having expired, he tells Welser that he has decided not to reply to the two published criticisms of the work because it would have no effect on his opponents. In a similar vein, he was to tell Father Piero Dini in May 1615 that he knew that many proofs could be used to refute his oppo-nents' rejection of the evidence for Copernicanism, but that he thought these men too obtuse to follow his reasoning.[25] That excuse appears to be behind his truncated treatment of the proofs in the *Letter to Christina,* a point to be discussed in the next part of this study.

With a sympathetic comment on Welser's illness and a reference to his own infirmities, he returns to the main business of the letter. He says that he finds Apelles' opinions contrary to his own, but that he feels a reply must be made since he, Galileo, has "no picture to conceal me from spectators."

Scheiner's Inadequacies

The first part of Scheiner's letter, as we have seen, contains elaborate proofs of his conclusion that Venus revolves around the sun. Galileo objects to the German's geometrical proof, saying that it would not "entirely satisfy a scru-pulous mathematician," and he finds the entire argument unnecessary in view of Copernicus's excellent treatment of Venus in *De revolutionibus.* He also asserts that Scheiner is very wrong in his estimations of the size of Venus and, thus, would be wrong in his assumption that it should cast a large spot.

This criticism, and the one after it in which he questions the adequacy of the authorities that Scheiner cites in this regard, serve to undermine the Jesuit's credibility as an astronomer.[26] Through contrary evidence and au-thorities, Galileo carefully refutes each of the proofs Sheiner makes, even citing Clavius against him. The decisive evidence for Venus's orbit around the sun, he states, is that provided by his own observations of the phases exhibited by Venus. But instead of according that evidence the privileged place it deserved, Apelles has relegated it to a proof of lesser importance.

In summing up the Venus arguments, Galileo comments that Apelles seems to be less sure of his views and may eventually come around to agree-ing with his own judgments fully. But the change will occur, he says, not because of the arguments he (Galileo), has made, since Apelles cannot read

25. *Opere* 12:184.20–28.
26. Van Helden thinks that Galileo is unfair in his criticism of Scheiner's authorities here. After having excused them as authorities for not having a telescope, Galileo says they should have observed Venus with their naked eyes in the daytime and not at night and have known that their daylight observations would be more accurate. See *Measuring the Universe,* 70–71.

them, but because the masked author will find on reflection that there are really no other conclusions possible. He adds that in a number of minor instances Apelles has already changed his mind for the better.

The patronizing tone is more pervasive in this letter than in the first, where Galileo presented a similar evaluation of Scheiner's arguments. Earlier, Galileo appears to have been concerned to establish his solidarity with Welser and so was more careful not to disparage Welser's protege. Now that Apelles' patron has praised Galileo's acumen, and has strongly implied that he has greater sympathy for his view than for Apelles', Galileo becomes freer in his refutation.

The tenor of his critique and his basic disagreement with Scheiner's methodology is disclosed in this next point:

I readily agree with Apelles in believing that the spots are not immersed within the sun's substance, but not on the strength of his arguments. First he assumes something which would undoubtedly be denied by anyone who wished to take the contrary side, as no one would be so simple as to maintain that the spots are within the solar substance, admit their changes of shape, and still assert the sun to be solid and inalterable. Any adversary would resolutely reject this last assumption as well as the proof Apelles adduces for it, which is that such is the prevailing opinion (according to him) among philosophers and mathematicians. And there would be good reason to reject this, for in the sciences the authority of thousands of opinions is not worth as much as one tiny spark of reason in an individual man. Besides, the modern observations deprive all former writers of any authority, since if they had seen what we see, they would have judged as we judge. As a matter of fact those authorities who did not believe the sun could be yielding and changeable were still farther from believing that it is sprinkled with dark spots. And now that its supposed immaculacy must yield to observation, it is vain to run to such men asking for support of the opinion that the sun is hard and unchangeable. As to the mathematicians I do not know that any of them have ever discussed the hardness and immutability of the sun, or even that mathematical science is adequate for proving such properties. (134–35)

This is a devastating critique. Not only does it point out the logical inadequacies of Scheiner's argumentation, it also asserts that he is ignorant of the authorities he proposes, and it chides his lack of appreciation for the value of contemporary telescopic observations.

For Welser, Galileo may have appeared to be quite modest in admitting that his arguments would probably not have an immediate effect on Apelles, but that in time the German mathematician would eventually arrive at the same solutions. To Scheiner and his confreres this may have seemed the height of presumption. The comment does convey Galileo's supreme confidence in his own grasp of the truth.

Being described as "running to such men to ask for support" casts Scheiner as a desperate neophyte. Since he was a mathematician and astron-

omer, he had grounds for disagreeing with Galileo's limitation of their speculations. Although he may have eventually come to the same conclusions, as
Galileo predicted, still as a Jesuit trained in the most respected educational
system of the Church, this kind of criticism must have earned its author little
good will. The proofs which Scheiner offers are mathematical and supported by geometrical diagrams. The authorities he notes in his discussion
of Venus are Giovanni Antonius Magini, Christopher Clavius, Tycho
Brahe, and Johannes Kepler, all mathematicians and reputable scholars of
the day.

In a passage that follows closely on the above, Galileo attempts to mitigate the withering remarks by interjecting a soothing aside, "I really wish I
knew some way of denying this without offending Apelles, whom I wish
always to respect" (135–36). His real intent, however, is expressed to Prince
Cesi in a letter of 4 November 1612: "I intend to show in what a silly way
the matter has been dealt with by the J[esuit], and to reprimand him as he
deserves. But to do this without offending the Signor Welser is no small
undertaking."[27]

The benign comment to Welser was made in the context of Galileo's refutation of Scheiner's view that the sunspots are located at some distance from
the sun. In proof of his contention Scheiner had asserted that the sunspots
remain visible for unequal periods, which would indicate they are not governed by any supposed rotation of the sun. Galileo denies this, saying that
after repeated observations he thinks the only conclusion he can make with
any certainty is that the spots stay on the sun's surface for about fourteen
days. Whatever we may wish, Nature does not shape herself to our desires.
Admonishing Apelles he adds, "We must take care that no passion—either
toward others or ourselves—bend us away from our aim of pure truth"
(136).

In the final part of the letter Galileo is concerned to show that Apelles is
wrong in thinking the sunspots are stars. Again he charges Apelles with altering reality to suit his beliefs. In this case, the German astronomer gives
the stars the same attributes as the sunspots, saying that they too suddenly
appear, change shape, and disappear. To bolster this view he notes the appearance of the satellites of Jupiter. They come and go and one has great
difficulty in determining their periods. Galileo replies in magisterial enthymemes: "Well, I should not like to have Apelles think that I am so vain
and light a man as to have offered to the world some spots and shadows as
stars, nor that I would have dedicated to so great a prince as the Most Serene
Grand Duke and to his regal house things that are so transitory" (138).

Galileo then states that the Medicean stars are real and that they have reg-

27. Shea quotes the letter (53).

ular periods and do not appear and disappear except when passing each other or Jupiter. He suggests that Apelles look again and study them, for there are only four, not five. The fifth was probably a star that Apelles failed to distinguish. The errors Apelles makes Galileo attributes to imperfections in his telescope and his method of computing their distances. (What Galileo does not know is that his own telescope lacked the power to show that there are, in fact, more.)[28]

Returning to a theme of his first letter, Galileo brilliantly dispatches Apelles' conception of the sunspots. Names given to things do not much matter, says Galileo, unless we attribute properties to things by the names we give them. We must be particularly careful in this matter because spots and stars mean different things:

Stars are never seen except luminous; spots are always dark; the first are either motionless or most regular in motion; the others have but a single common motion though they are affected by myriads of irregularities; the stars are arranged at varying distances from the sun; sunspots are all contiguous to it or imperceptibly removed from its surface; we see the former only if far to one side of the sun, the latter only in line with the sun; the former are most probably made of dense and very opaque matter, the latter being rarefied in the manner of clouds and smoke. (139–40)

The clear distinctions Galileo draws here from his own observations show how close he came to what modern astronomers hold. But the very differences in the observations between Scheiner and Galileo simply point up the imperfect nature of the instruments, and these discrepancies give further reason for the disbelief many had in the phenomena the instruments purported to disclose. Galileo inveighs against the Peripatetics in his own country who suggest that these things are illusions. With amusing but caustic wit he notes:

It is about time for us to jest right back at these men and say that they likewise have become invisible and inaudible. They go about defending the inalterability of the sky, a view which perhaps Aristotle himself would abandon in our age. Their view of sunspots resembles that of Apelles, save that where he puts a single star for each spot, these fellows make the spots a congeries of many minute stars which gather together in greater or smaller numbers to form spots of irregular and varying shapes. (140–41)

Then Galileo shows how the implications of such an explanation contradict the cosmology espoused. To admit that the spots are congeries of stars would be to introduce erratic motion into the heavens.

The astronomers to whom he referred seem to be those who had recently taken part in a debate at the Collegio Romano, news of which had come to

---

28. Drake notes the nineteenth-century discoveries; see *Discoveries*, 138, n. 19.

Galileo from Prince Ceci. The Jesuits had supported the view of their confrere Scheiner against a Dominican who argued for Galileo's explanation of the sunspots. The irregular shape of the sunspots the Jesuits thought could be explained by a cluster of stars.[29] The Fathers thus give evidence of the beginning of their estrangement from the man they had honored just two years before. Then they had welcomed him and verified his sightings of the moons of Jupiter. But the acrimony in Galileo's response aroused the opposition of some within their ranks even before the General of the Order commanded its members to fall into line and present a common front behind Aquinas in theology and Aristotle in philosophy.[30]

The refusal of the Peripatetics to admit change in the heavens Galileo could not understand. "Alteration" is not "annihilation," he argues. One does not call the process by which an egg hatches a chicken "corruption." The problem is that "these men are forced into their strange fancies by attempting to measure the whole universe by means of their tiny scale" (42). The ridicule is cleverly designed to deflate whatever prestige these academics may have had in Welser's eyes. He continues with a general dismissal of the Peripatetics: "People like this, it seems to me, give us reason to suspect that they have not so much plumbed the profundity of the Peripatetic arguments as they have conserved the imperious authority of Aristotle. It would be enough for them, and would save them a great deal of trouble, if they were to avoid these really dangerous arguments; for it is easier to consult indexes and look up texts than to investigate conclusions and form new and conclusive proofs" (142). In what is to become a familiar theme regarding the power of the human mind in later writings he adds: "Besides, it seems to me that we abase our own status too much and do this not without some offense to Nature (and I might add to divine Providence), when we attempt to learn from Aristotle that which he neither knew nor could find out, rather than consult our own senses and reason. For she, in order to aid our understanding of her great works, has given us two thousand more years of observations, and sight twenty times as acute as that which she gave Aristotle" (142–43).

The argument is indeed persuasive. But his strategy in lumping all those who disagreed with him into one category labeled Peripatetics is fallacious.

29. Drake, *Discoveries*, 141; *Opere* 11: 395.

30. The General of the Society of Jesus, Claudio Acquaviva, first wrote a letter to the Society stressing the need for uniformity in the teaching of theology and philosophy on 24 May 1611. The implementation of his request seems not to have been very effective, as Richard Blackwell points out, for Acquaviva wrote a second letter 14 December 1613 spelling out concrete methods of insuring this; see Blackwell, *Galileo, Bellarmine, and the Bible* (Notre Dame: University of Notre Dame Press, 1991), 139–42. His first letter seems not to have affected Scheiner's thinking, since he notes Clavius's call for a reexamination of cosmology.

As he well knew, not all of his opponents were slavish followers of Aristotle. Neither Scheiner nor the mathematicians and astronomers at the Collegio were justly placed among the philologists who comb the text of Aristotle for answers to problems. Although they sought to preserve some of the principles of the old cosmology, the evidence was not patently clear and thus did not justify the charge that they willfully ignored it.

Pointing up the great difficult in these early observations, Galileo ends the letter with an admission that he too may have been wrong in what he reported in observing Saturn. When he first looked at the planet he thought he saw two other stars on either side of it, but in turning his telescope again upon it two years later he found that the other stars had disappeared. He asks whether Saturn devoured his children or whether his lenses deceived him. In a bit of wry humor he suggests that this admission offers encouragement to those who earlier found his discoveries in the heavens incredible.

Galileo ends by asking Welser to offer his excuses to Apelles if he has been too vehement in his objections. Demonstrating the acceptance he enjoys in a warm circle of patrician admirers, he closes by extending to Welser the greetings of Filippo Salviati, a member of the Crusca academy, at whose villa Galileo notes he is staying while continuing his celestial observations.

The letter provides ample evidence of Galileo's skill in debate. He frames his arguments carefully and clearly, demolishing the opinions of his opponent with more reasonable, although novel, premises. The freedom of his intellect, the daring that permitted him to grasp the import of what he saw, enabled him to reason more acutely than Scheiner apparently had. The letters also demonstrate Galileo's rhetorical prowess and at the same time the weakness of his strength. He knew how to shatter an adversary's image through disparaging analogies, how to destroy an argument by characterizing its assumptions as hopelessly antiquarian. The ingenious comparisons, the lively progression of each clause and sentence, the apt phrase, all contrive to build a marvel of rhetorical refutation. Yet his artistry would have its best effect on those who already admired the author; it would sting and madden some opponents, and at the very least frighten those who found it difficult to give up a world view that had served the faith so well.

PART TWO

THE HERMENEUTICAL CRISIS

CHAPTER FIVE

Interpreting Scripture

In this section we turn to the theological difficulties engendered by the Copernican thesis and the arguments developed by its proponents as they tried to accommodate the new system to the prevailing interpretations of Scripture. The two disciplines of philosophy and theology were in this issue inextricably entangled. Much of the energy of the opponents to Copernicanism in the philosophical debates was stimulated by the implications of the system for religious teaching.

Not only was the literal meaning of certain biblical texts in jeopardy, but these texts reflected a time-honored cosmology. Both heaven and earth were turned about by this new system, and God and the angels seemed to be further and further removed from its daily operation. The spheres were now in doubt, and so the prime mover was no longer needed to set the first one in motion. The celestial orbs were thought to be guided by some natural causes other than the ministration of angels, and they no longer turned around earth; rather earth joined them to travel around the sun. Earth and man were literally no longer at the center of the cosmic plan. A reinterpretation of the Scriptures meant more than a mere revaluation of words.

Since the Copernican issue now intruded upon the domain of theology, our concern in this part of the book will be to discover whether rhetoric entered into the theological arguments concerning heliocentrism in much the same manner as it had in the philosophical ones. Theologians had traditionally made extensive use of dialectical argument, but they had excluded rhetoric from the science of theology, preferring to use it only in its practical application, homiletics.

Zuñiga on the Earth's Motion

The first hermeneutical effort of note that attempted to reconcile the Copernican thesis with traditional interpretations of the Scriptures was that of the Spanish priest, Diego de Zuñiga—or Didacus a Stunica, as the name is written in Latin. Zuñiga, a professor of theology and philosophy at Salamanca, was active in the reform of scriptural exegesis and wrote a lengthy *Commentary on Job*, which was published in 1584. It contained one novel interpretation the Church could not condone: his explication of the sixth verse of the ninth chapter, "Who shaketh the Earth out of her place, and the Pillars thereof Tremble." The passage in the commentary was later singled out for correction by the Church in 1616. Our analysis of Zuñiga, as of all but the last of the other exegetical pieces in this chapter, is based primarily on the English translation made by Thomas Salusbury in the mid-1650s, which he published in his *Mathematical Collections and Translations* (1661).[1] Since Salusbury was far closer to the authors in time and to contemporary knowledge of mathematics and astronomy, his work seems appropriate to our purpose. His role in the publication of Galileo's writings in England is described in chapter 10.

In addition to being a professor of philosophy and theology, Zuñiga was a member of the Hermits of St. Augustine. He taught at the University of Osuna as well as at Salamanca and published an earlier work on free will. Having studied at Salamanca under Fray Luis de Leon, who thought that both Hebrew and Greek should be used in solving difficult passages of Scripture, he brought to the study of Scripture a more liberal approach than many of his day.[2] When he wrote his *Commentary on Job* he was also much impressed by the arguments offered in *De revolutionibus*, for he did not fear to state that Copernicus demonstrated planetary movement by means of his thesis.

Unlike most of the commentators to follow, Zuñiga takes a literal view of the ambiguous passage from Job, saying that the movement of the earth is best explained by the "Opinion of the Pythagoreans." Showing great familiarity with the astronomical literature on the subject, he names other authorities who have held the earth to move: Philolaus, Heraclides Ponticus, Plato, and Hypocrates among the ancients, and Copernicus among contemporaries. He states unequivocally that this opinion is the only one to explain adequately the movements of the planets, and points out that Ptolemy himself admitted that he could not accurately predict the motions of the equinoxes or the beginning of the years. He adds that the Alphonsines and Thebith Ben Core (Thabit ibn Qurra) attempted to do so, yet positions among them differed.

Turning to possible objections based on other scriptural texts that seem

1. The page references in the writings of Zuñiga, Kepler, and Foscarini are to the Salusbury translations in volume 1 of *Mathematical Collections and Translations* (1661). A recent translation of the excerpt from Zuñiga and Foscarini's *Letter* appears in Blackwell, in Appendices 2 and 6 respectively.

2. Robert Westman supplies details of Zuñiga's life and education in "The Copernicans and the Churches," 86–87, 92–93. F. Jordan Gallego discusses Zuñiga's efforts and their relation to the Council of Trent in "La Metafisca de Diego de Zuñiga (1536–1597) y la Reforma Tridentina de los Estudios Ecclesiasticos," *Estudio Agustiniano* 9, no. 1 (1974): 3–60.

to show the immobility of the earth, Zuñiga applies the principle of accommodation—that the Holy Spirit accommodated its expression of truths to the language and concepts of the common man—a view Kepler and Galileo were also to adopt. This principle, so widely used by those who favored reinterpretation, became a "special topos" of the dialectics of hermeneutics.

Zuñiga says that Copernicus "most plainly explained and demonstrated" the equinoxes from the movements of the earth, and this opinion is "not in the least contradicted by what Solomon saith in Ecclesiastes: But the Earth abideth for ever" (469). He reasons, as Kepler was to do, that the Holy Spirit here emphasizes the fact that the earth has endured through the ages "without any sensible alteration." These and other passages, he argues, should have no effect against Copernicus's position because they simply speak in common parlance. Even Copernicus and his followers often speak of the revolution of the sun, he remarks.

In the concluding part of the explication Zuñiga mounts a powerful argument: "No place can be produced out of Holy Scripture which so clearly speaks the Earths Immobility, as this doth its Mobility" (469–70). Arguing from the absence of an opposite teaching, he claims the text provides scriptural support for heliocentrism.

His bold position was hard to sustain. More than a decade later, no doubt under considerable pressure, Zuñiga revised his view, saying in his *Philosophia* of 1597 that the position of Copernicus is difficult to accept and that the earth's motion seemed to him to be "absurd."[3]

The passage was vigorously confuted by the Spanish Jesuit Juan de Piñeda, who asserted that Zuñiga was patently wrong in his assessment of the inadequacy of Ptolemaic astronomy. Piñeda used strong rhetorical language to attack the Copernican opinion, terming it "foolish, frivolous, reckless and dangerous to the faith."[4] In his discussion he appeals to the authority of the most cited conservative mathematician, Christopher Clavius. Clavius had indeed accepted the Ptolemaic system earlier, but as we have seen he also expressed a need for a reexamination of it in view of the new evidence provided by the novas of 1572, 1600, and 1604, and the telescope of Galileo.[5] Many apologists for Copernicus and Brahe cite the authority of Clavius in their discussions.

What Zuñiga thought to be a convincing reference to heliocentrism in Job is controverted by Piñeda, who maintains that the movement referred to

3. The text of the later work is quoted by Westman in "The Copernicans and the Churches," 109, n. 54.

4. Blackwell's translation, 26.

5. For Clavius see Lynn Thorndike, *The Sphere of Sacrobosco and its Commentators* (Chicago: University of Chicago Press, 1949), 42. The changes in his views are noted by Westman (95).

was simply a shaking of the earth in its place, not its removal from it.[6] The most important opinion of Zuñiga's view was pronounced long after his death in 1598. In its 1616 decree ordering emendation of Copernicus's *De revolutionibus,* the Holy Office of the Congregation of the Faith also demanded that the offending passage be expunged from the *Commentary.*

Kepler on the Scriptures and Copernicus

During the controversy over Zuñiga's interpretation of Job, Kepler was at work on his own solution to scriptural barriers to Copernicanism. The difficulties were dealt with in the original introduction to the *Mysterium cosmographicum* of 1596, but were omitted from the published version. They did not find their way into print until the publication of *Astronomia nova* in 1609. Kepler does not discuss hermeneutics in general in the excerpt, but he does introduce one strong principle, a variation on the accommodation topos. In the opening passage he points out that people often speak in an unqualified way from the standpoint of their visual experience, not expecting their words to be taken in the literal sense:

It must be confessed, that there are very many who are devoted to Holinesse, that dissent from the Judgment of *Copernicus,* fearing to give the Lye to the Holy Ghost speaking in the Scriptures, if they should say, that the Earth moveth, and the Sun stands still. But let such consider, that since we judge of very many, and those the most principal things by the Sense of Seeing, it is impossible that we should alienate our Speech from this Sense of our Eyes. Therefore many things daily occur, of which we speak according to the Sense of Sight, when as we certainly know that the things themselves are otherwise. An Example whereof we have in that Verse of Virgil;

> *Provehimur portu, Terraque urbesque recedunt.* [When we leave port,
> the land recedes.]

So when we come forth of the narrow straight of some Valley, we say that a large Field discovereth it self. So Christ to *Peter, Duc in altum;* (Launch forth into the Deep, or on high,) as if the Sea were higher than its Shores; For so it seemeth to the Eye, but the Opticks shew the cause of this fallacy. Yet Christ useth the most received Speech, although it proceed from this delusion of the Eyes. (461)

Kepler has chosen a disarming group of examples: first from Virgil, next from common speech, and then from Christ himself. Lest this seem audacious, he notes that optical principles can explain what might appear to be a fallacious statement. In the next paragraph he mentions similar passages in his own work and Copernicus's that speak about celestial phenomena in terms of common understanding. It is a widespread practice of scholars, he says, and then asks rhetorically: "What wonder is it then, if the Scripture

6. Piñeda and his excoriation of Zuñiga are noted by Westman, 94–95.

speaks according to mans apprehension, at such time when the Truth of things doth dissent from the Conception that all men, whether Learned or Unlearned have of them?" (462).

He next points out the use of figurative speech in texts to which various interpretations can be given, such as the "Poetical Allusion" of Psalm 19, in which the sun is spoken of as coming forth from his "Tabernacle of the Horizon, as a Bridegroom out of his Chamber." Whether Christ's journeys or the Gospel's development is referred to is not clear. The important thing is that the Psalmist knew that the sun does not go out of the horizon in that manner, but that this was what his eye disclosed to him. Mentioning other similar passages, he comes to the one most often cited in these defenses of the earth's movement, the text in which Joshua orders the sun to stand still. Kepler points out that Joshua had simply prayed that the mountains not hide the sun from him, but he expressed it in words that suited his senses. Even if someone had informed him that the sun could not really do so, he would probably have said that he did not care how it was done, he only wished to have the day prolonged: "But God easily understood by Joshua's words what he asked for, and by arresting the Earths Motion, made the Sun in his apprehension seem to stand still" (463). For their further edification, Kepler refers his readers to the tenth chapter of his *Optics,* where he explains that the sun appears to move because to us it seems so small and the earth so large. He argues that only reason informed by instruction in such matters can convince us that it is the earth that moves. In another example, Kepler asks if Scripture really means us to believe that man should not attempt to investigate heaven and earth when it asks man "Whether he can finde out the height of Heaven above, or depth of the Earth beneath." In answering his rhetorical question he declares: "no man that is in his right mind will by these words circumscribe and bound the diligence of Astronomers, whether in demonstrating the most contemptible Minuity of the Earth, in comparison of Heaven, or in searching out Astronomical Distances; Since those words speak not of the Rational, but real Dimention; which to a Humane Body whilst confin'd to the Earth, and breathing in the open Air is altogether impossible" (464).

Kepler moves then to passages that assert the immobility of the earth. To counter these he argues that Scripture offers moral, not physical instruction. Thus, in Ecclesiastes Solomon certainly did not mean to dispute with astronomers about the earth's stability but to remind man of his "Mutability; when as the earth, Mankindes habitation, doth alwaies remain the same" (464). Similarly in Psalm 104, where God is said "To have laid the Foundations of the Earth, that it should not be removed for ever," the Psalmist is not concerned with physical causes. Kepler suggests the Psalmist's deeper intent: "He goeth not about to teach men what they do not know, but putteth them in mind of what they neglect, to wit, the Greatnesse and Power of God in creating so huge a Mass so firm and steadfast. If an Astronomer should teach that the Earth is placed among the Planets, he overthroweth not what the Psalmist here saith, nor doth he contradict Common Experience; for it is true notwithstanding that the Earth, the Structure of God its Architect doth not decay (as our Buildings are wont to do) by age" (465–66). In a clever extrapolation on this point, Kepler says that it is obvious that the Psalmist was not acting as an astronomer or he would have mentioned the creation of the five planets, of whose motion "nothing is more admirable, nothing more excellent, nothing that can more evidently set forth the Wisdome of the creator amongst the Learned."

In closing this dialectical argument, Kepler uses an appealing figure: "And I do also beseech my Reader, not forgetting the Divine Goodnesse conferred on Mankind; the consideration of which the Psalmist doth chiefly urge, that when he returneth from the Temple, and enters into the School of Astronomy, he would with me praise and admire the Wisdome and Greatnesse of the Creator, which I discover to him by a more narrow explication of the Worlds Form, the Disquisition of Causes, and Detection of the Errours of Sight" (466–67). Kepler then turns from dialectic to the rhetoric of blame to excoriate those who stubbornly continue to oppose the new astronomy on theological bases:

But he who is so stupid as not to comprehend the Science of *Astronomy*, or so weak and scrupulous as to think it an offence of Piety to adhere to *Copernicus,* him I advise, that leaving the Study of *Astronomy*, and censuring the opinions of Philosophers at pleasure, he betake himself to his own concerns and that desisting from further pursuit of these intricate Studies, he keep at home and manure his own Ground; and with his own Eyes wherewith alone he seeth, being elevated towards this to be admired Heaven, let him pour forth his whole heart in thanks and praises to God the Creator; and assure himself that he shall therin perform as much Worship to God, as the *Astronomer,* on whom God hath bestowed this Gift; that though he seeth more clearly with the Eye of his Understanding; yet whatever he hath attained to, he is both able and willing to extoll his God above it. (467)

Not only has Kepler dispatched those who oppose him on pious grounds, showing them that they do not understand the grounds of their own arguments, he has claimed for astronomers a superior intellectual gift.

Kepler does not restrict himself to the scriptural arguments leveled against Copernicus but takes on the detractors' use of the Fathers of the Church, pointing out that theology proceeds by the "weight of Authority" and philosophy by the "weight of Wisdom" (467). In a brilliant rhetorical stroke he destroys the value of the Saints as critics of astronomy without denying their sanctity:

Therefore Sacred was *Lactantius,* who denied the Earths rotundity; Sacred was *Augustine,* who granted the Earth to be round, but denyed the *Antipodes;* Sacred is the Liturgy of our Moderns, who admit the smallnesse of the Earth, but deny its Motion: But to me more sacred than all these is Truth, who with respect to the Doctors of the Church, do demonstrate from Philosophy that the Earth is both round, circumhabited by Antipodes, of a most contemptible smalnesse, and in a word, that it is ranked amongst the Planets. (467)

The antithetical and climactic order of the passage demolishes the philosophical opinions of these revered figures as it elevates the highest authority, Truth.

Foscarini's *Lettera*

The obvious rhetorical character of the last part of Kepler's discussion is in sharp contrast to the approach of Paolo Foscarini, whose *Lettera . . . Sopra l'Opinione de Pittagorici e del Copernico* (1615) is the most elaborate and the most important of early attempts to reconcile Scriptures with heliocentrism.[7] Foscarini, born about 1580, became a Carmelite of some distinction, rising to serve as provincial twice. He taught philosophy at Messina and was highly respected for his careful discussion of the philosophical and theological problems in the *Lettera,* which he had addressed to Sebastiano Fantoni, the General of his Order. Foscarini was invited to Rome to defend Copernicanism before the Church arrived at its opinion in 1616.

Like Galileo's *Letter to Christina,* Foscarini's *Lettera* is more a tract than an epistle. It is thirty-three pages long. Writing in Italian in a pleasant conversational style, he explains that the work originated at the "command" of Signore Vincenzo Carraffa, a Knight of St. John of Jerusalem, who asked him to undertake an apologia for Copernicanism.

Turning to the task at hand, Foscarini first seeks to establish the credibility of the Copernican system. He begins by conceding its apparent absurdity:

The Foundations on which this Opinion may be grounded, least [lest], whilst otherwise it is favored with much probability, it be found in reality to be extreamly repug-

7. The long title of Foscarini's letter is *Lettera del R.P.M. Paolo Antonio Foscarini Carmelitano Sopra l'Opinione de' Pittagorici e del Copernico, della Mobilita della Terra e Stabilita del Sole, e del Nuovo Pittagorico Sistema del Mondo* (Naples 1615). Richard Blackwell notes that Salusbury's translation was based on a Latin edition of the letter, published in Protestant Europe by David Lotaeus, who excised a few comments about papal authority (88, n. 3). I have compared the Salusbury version with the original Italian and the recent English translation by Blackwell, who finds the Salusbury prose "archaic." I have preferred to retain the seventeenth-century language, which seems to me to be closer to the Ciceronian style of the day. The Italian text is in the earlier edition of Galileo's *Opere,* ed. d'Alberti (Firenza 1842–56), 5:466–94.

nant (as at first sight it seems) not onely to Physical Reasons, and Common Principles received on all hands (which cannot do so much harm) but also (which would be of far worse consequence) to many Authorities of sacred Scripture: Upon which many at their first looking into it, explode it as the most fond Paradox and Monstrous Capriccio that ever was heard of. (473–74)

The physical and philosophical challenges the new system presents are not nearly as daunting as is the theological one. The "fond Paradox" is mild enough, a perplexing puzzle of the sort beloved of sophistic debate, but the second is more disturbing. "Monstrous" added to "Capriccio" converts a caprice or whim into a malicious trick.

Foscarini laments the fact that antiquated opinion is so entrenched that it resists any change and prefers to judge physical truths by authority. He thinks that such opinions ought to be evaluated by "better Reasons lately found out, or from Sense it self" (474).

A rhetorical question makes explicit the "better reasons": "But . . . have not the several Experiments of Moderns, in many things, stopped the mouth of Venerable Antiquity, and proved many of their greatest and weightiest Opinions, to be vain and false?" (474). He reminds his readers of the erroneous notions formerly held about the Antipodes and the uninhabitability of the Torrid Zone, as likewise "many Dreams" of Aristotle and other philosophers. But in defense of the Ancients, Foscarini offers what is to become a commonplace: if the observations and arguments of our moderns had been known to these ancients, they would have changed their opinions and accepted their "manifest Truth" (475). Their errors should then teach us not to have such faith in the opinions of authorities that we take them for divine truth.

The heart of the matter is divine authority, he acknowledges, which is not to be questioned. The difficulty is in knowing how to determine its prescriptions in passages that are not "so plain." In conceding the difficulty the new opinions present for sense experience and for the faith, he crafts a metaphor, which he repeats later: "Wherefore since this Opinion of Pythagoras and Copernicus hath entered upon the Stage of the World in so strange a Dress and at first appearance (besides the rest) doth seem to oppose sundry Authorities of Sacred Scripture, it hath (this being granted) been justly rejected of all men as meer absurdity" (475). The image is similar to the more famous one used by Galileo in the preface of his *Two Chief World Systems*, when he says that he has decided "to appear openly upon the Theatre of the World as a Witness of the naked Truth." Perhaps Foscarini's text inspired him, or perhaps he and Foscarini were acquainted with Juan Luis Vives' *Fabula de homine* (1518). There Vives depicts man as a creation of Jupiter, intended to play many roles in the theater of the world for the amusement of the gods.

Continuing to lay the ground work for his claim regarding the new thesis, Foscarini shows himself to have considerable knowledge of astronomy, including the recent discoveries announced in the *Sidereus nuncius*. He explains the drawbacks of the old system: its multiplication of devices such as eccentrics and epicycles and the fact that it does not seek to describe reality but simply to save the phenomena. He states firmly that recent observations with the telescope show that some of the old ideas about the heavens are probably untrue. Mountains on the moon, additional bodies around Venus, Saturn, and Jupiter, the multitude of stars that make up the Milky Way, all show the inadequacies of the earlier conceptions. He also reviews some of the dialectical arguments in support of the Copernican view that we have seen in Galileo's work of 1610 and the *Sunspot Letters:* the probability that Mercury and Venus rotate around the sun rather than the earth, which encourages the inference that the earth, rather than being immobile, also moves around the sun.

To make his contention more palatable, Foscarini, like Copernicus and Galileo, offers a list of authorities whose opinions favor heliocentrism. Among the moderns he names "Father Clavius, a most learned Jesuite," who, though rejecting the Pythagorean opinion, urges men to search for some other.

In light of these views, Foscarini says he decided to search out the truth for his own satisfaction. He approaches the task unafraid because he knows that the new opinion could not conflict with Scripture; for one truth does not contradict another—one of the foundational principles of Galileo's *Letter to Christina*. Foscarini thus concludes his lengthy narration.

Scriptural Exegesis

Seemingly unaware of Kepler's brief effort, Foscarini says that since no one else has yet attempted such a task he will do so, knowing it would certainly be welcomed by Galileo Galilei, Johannes Kepler, and the illustrious Lyncean Academy. The Carmelite provides a comprehensive analysis of the problem texts and classifies them into six groups: passages that: (1) assume the earth to stand still; (2) say that the sun moves and rotates around the earth; (3) declare that heaven is above and the earth below; (4) place hell in the center of the earth; (5) speak of the relationship of heaven and earth as being in opposition; (6) say that at the Judgment the sun will stand still in the east and the moon in the west, for it would seem that if the earth does move it should have been referred to as standing still during the Day of Judgment (478–81). The sixth classification, unlike the rest, comprises texts that are found in the Fathers and the writings of theologians.

To refute the arguments against Copernicus that rest upon these six classifications, Foscarini poses six maxims or principles. Submitting them to the "Judgment of Holy Church," he vows to retract any of his contentions if reason and experiments should show something else to be more probable.

First, all attributions to God or his creatures in the Scriptures that seem unfitting are explained in one or more of four ways: as metaphoric language, as an accommodation to our understanding, as common parlance, and in similitude to mankind.

Foscarini illustrates the first kind of attribution with examples from Scripture. Quoting Aristotle in support of the principle, he says that the Philosopher seems to have hinted at this when he said, "Some things are more intelligible to us; others by nature, or *secundum se.*"

He then analyzes the account of creation in Genesis, showing the extent to which the Scripture seeks to be in accord with the vulgar understanding and the appearance of things. Light is said to have been made first, yet the text speaks of the morning and the evening making up the first day. Obviously the Scripture is speaking of the light in relation to us, otherwise it would simply have referred to the movement of light around the heavens. To speak of days and nights is to accommodate to our understanding. He mentions in passing the passage from Virgil when Aeneas mentions the earth receding as the boat pulled away from shore.

He applies this principle to the text that was to challenge all the exegetes, including Galileo: the command of Joshua, "Sun stand thou still." He conjectures the sun's light stood still on the earth as the earth stopped moving. The passage from Isaiah regarding the sun's having moved ten degrees backward may also be explained in the same manner. In this context Foscarini makes the observation common to most of the interpreters that the purpose of the Scripture is to teach us what is useful for salvation—God's law—not whether there are eccentrics and epicycles. In perhaps an unpolitic corollary to this point, he states that the Church and the pope make no mistakes in matters of faith but can err in practical affairs and philosophy.[8]

The rest of the principles move from the realm of theology to philosophy to elucidate Scripture in light of basic philosophical principles. The applications of the arguments are dialectical in force. The second, a metaphysical principle, states that all spiritual and material, eternal and corruptible, movable and immovable things are governed by God's "perpetual, unchangeable, and inviolable Law" that constitutes their essence. God's handiwork and his law is eternal, Scripture tells us. On this basis the heavens are said to be changeless and eternal. It follows then that the heavens themselves ought to be said to be immovable and unchanging and the earth amenable to

8. Blackwell, *Galileo, Bellarmine, and the Bible,* 235–36. This passage was one of those missing from the Latin translation on which Salusbury's version was based.

movement and change. This principle, he says, could remove all the opposition based on the passages in the first group where the earth is said to stand still.

The next three principles apply common Peripatetic teachings in natural philosophy to the texts. The first of these states that when a part of something is moved and not the whole, it cannot be said to move absolutely but *per accidens*. The stability referred to is that of the whole. Thus the earth considered as a whole is not changeable, since it is not generated or corrupted except in its parts. This, he says, must be the sense of the text in Ecclesiastes 1.4, "One Generation passeth away, and another Generation cometh, but the Earth abideth for ever." The next maxim assumes the doctrine of natural place. It affirms the teaching that no corporeal thing whether movable or immovable, having been given a natural place, can be removed from it except by violent motion, after which it tends to return to its place. Nothing can be removed from its natural place *"secundum Totum"* without "most great and dreadful mischiefs" (492). So the earth could be said to move around the sun as within its own natural place in a location between Mars and Venus. Thus Earth could still be said to be stable and immovable in her course.

Another principle follows from the previous one: that "things created by God may have parts that can be *ab invicem,* or by turns, separated from themselves" and these disjointed parts may not be treated collectively. "In this sense . . . the Earth is said to be Immoveable, and Immutable: yea even the Sea, Aire, Heaven, and any other thing" can be said to be unchanging. Parts may be separated per accidens and then if the impediment is removed they will return to their place spontaneously. The earlier maxim treated the parts in relation to place, while this one treats of part in relation to the whole. Foscarini departs from received philosophy when he expands the discussion to speak of the new view of gravity, saying that it is nothing but the "certain power and appetite of the Parts to rejoyn with their Whole, and there to rest as in their proper place" (493). The same sort of power is thought to reside in the celestial bodies also. This is what causes their spherical shape. Levity is also explained here by Foscarini as the "Extrusion and Exclusion of a more tenuose and thin Body from the Commerce of one more Solid and dense, that is heterogeneal to it, by vertue of Heat." Just as heavy bodies are "compressive" in nature so light ones are "extensive," for heat dilates and rarefies things to which it is applied (493).

Continuing to develop the implications of this principle, Foscarini mentions that gravity and levity extend to the heavenly bodies, which are not composed of the fifth essence but are of the same matter as the elements. Nor is heaven impenetrable and dense, for comets are seen to be generated in it. He then digresses to take note of Galileo's recent treatise on the sun-spots, in which he says the Florentine has "most excellently and accurately spoken" and has observed that these spots probably have their cause in the exhalation or attraction of vapor.

The reference implies that Foscarini believes Galileo's name will be well received by his readers. Knowing that some of the points he develops here contradict the cosmology of Aristotle, Foscarini may have felt he needed the weight Galileo's sense evidence could give these probable conclusions. He first addresses the conflict with Scripture, pointing to the principle he had enunciated earlier regarding the analogical nature of many biblical texts. He notes that when Scripture speaks of solidity it should be understood to refer to the fact that there are no vacuums in nature, whether on earth or in heaven. Although the heavens are the rarest in their composition, they probably have the same proportion to air as air does to water.

Passing then to Aristotle's *De caelo*, he contends that Aristotle is quite wrong in his writings concerning the kinds of motion appropriate to heaven and earth. For simple bodies he says there is one simple motion, and this is of two kinds: right (or rectilinear) and circular. Earthly motion is "right," being either at right angles upward from the middle or downward; the first is true of light bodies, and the second of heavy. Circular motion, motion around the middle, is limited to heavenly bodies and is not governed by gravity or levity.

A New Philosophy

Basing his views on what Galileo has taught in the *Letters on Sunspots* and the *Sidereus nuncius*, Foscarini then gives a summary of what he terms the "new opinion" in philosophy. Although conceding that there is but one simple motion for simple bodies, the new philosophy holds that this motion is circular and occurs on earth as well as in the heavens. By such motion bodies are kept in their natural place, move around their own axes in a place, and reside in it. In this sense bodies can be said to be immovable. Right or straight motion can only be attributed to things dislodged from their natural place, for instance, a part separated from the whole. In this case the part is not in conformity with the nature and form of the universe. Thus it is destitute of the perfection that belongs to the proper nature of the universe. Again, in moving back to its natural place the object becomes once more immovable. In this sense, then, straight motion is not uniform or simple, because of the uncertainty of the gravity or levity of the body. Things that fall begin slowly and then move faster; such motion is not uniform. Things that rise also move at different rates of speed and they also do not do so evenly. In straight motion, then, there is an occult quality of circular movement that lurks in a part, a tendency to conform to the whole. Only circular motion is "Simple, Uniform and Aequable and of the same tenor [or rate] for that it is never destitute of its interne Cause: whereas on the contrary,

Right Motion, (which pertains to things both Heavy and Light) hath a Cause that is imperfect and deficient, yea that ariseth from Defect it self, and that tendeth to, and seeketh after nothing else but the end and termination of it self" (494–95).

In this way of thinking, circular motion is "proper to the Whole and Right Motion to the Parts." These distinctions, however, are only formal, not real. Thus the new opinions contest both Aristotelian philosophy and received cosmography. Foscarini ends this reflection about the implications of the fifth maxim by an equivocal statement of his personal view: "For as to the truth or falsehood of these foregoing Positions (although I conceive them very probable) I am resolved to determine nothing at present, neither shall I make any farther enquiry into them" (496).

The sixth and final principle states that things are only defined relative to all things or a great number of things, not just in relation to a small number. So the earth cannot be said to occupy the lowest place; it cannot be placed absolutely high or low except in relation to some part of the universe. It should be denominated in relation to all of the heavenly bodies. He explains that this maxim can be of great use to us in clearing away the difficulties that prevent our understanding the theological truths revealed in the Scriptures. For example, in texts that speak of heaven in opposition to earth, the reference to heaven may be understood to mean the empyrean heaven. Other passages wherein ascent to heaven from earth is mentioned may be understood to mean ascension from the part of heaven near the center to that at the circumference. And when St. Paul spoke of being taken up into the third heaven, we may interpret that as being the empyrean heaven, "the Seat of the Blessed." The second would be the realm of the fixed stars, which has commonly been called the eighth sphere or the firmament. The first heaven we may assume to be "that immense Space of Erratick and Moveable Bodies illuminated by the Sun, in which are comprehended the Planets, as also the Earth moveable, and the Sun immoveable" (497). This view of the three heavens based on Scripture was shared by Bellarmine himself, although Bellarmine would transpose the sun and the earth.[9]

Taking his cue from Copernicus, Foscarini goes on to extol the sun as the center of the universe, endowing his metaphor with even more rhetorical power than did the Polish astronomer. He ascribes quasi divine power to the sun: "Who like a King upon his August Tribunal, sits with venerable Majestie immoveable and constant in Centre of all the Sphaeres, and, with his Divine Beames, doth bountifully exhilerate all Coelestial Bodies that stand in need of his vital Light, for which they cravingly wander about him; and doth liberally and on every side comfort and illustrate the Theatre of the

---

9. See Blackwell's description of Bellarmine's cosmology, 41.

whole World, and all its parts, even the very least, like an immortal and perpetual Lamp of high and unspeakable value" (497). Similarly, God is the center of spiritual beings as Christ of mixed spiritual and corporeal beings. The truth contained in this sixth principle regarding the relativity of things, he thinks, can answer all the arguments that might be advanced against the Copernican thesis on the basis of the third, fourth, and fifth categories of scriptural texts mentioned earlier—those passages that speak of the relationship of the heavens to the earth.

Foscarini then goes on to reinterpret other texts. He discusses the position of earth relative to that of sun, Mercury, and Venus, and its position in relation to the universe as a whole, showing that in one respect the sun may be said to be above and not below earth, yet in another the earth may be said to be beneath the sun in respect to the latter's place in the center. In this way may be saved the passage from Ecclesiastes that refers to the deeds of man "Which are done, or which are under the Sun."

In a similar fashion he disposes of the vexing problem of where hell may be located if the earth moves around the sun and passes through the heavens. Its position is relative also. Some argue that in the new cosmic system hell would move above the sun into heaven, and thus would be in heaven. But in either system hell is in the center of the earth, a fitting place for the punishment of the damned. In the common understanding the heavens surround hell also, so that heaven and paradise would be both beneath and above it. Summing up the import of what he has offered to free the Copernican system from scriptural objections, Foscarini notes that he thinks these principles have made the system quite probable. Since faith need not require us to believe Copernicus wrong if we can interpret Scripture in the ways that have been proposed, then indeed his new system is the most probable solution to the problems that have plagued our common understanding of the heavens.

Foscarini is careful in the language he uses to describe the two rival systems, which he puts at the end of the major arguments of his epistle. He calls both "Opinions," emphasizing the fact that they are the probable conclusions of dialectical arguments and mathematical demonstrations, which therefore would not have the status of certainty. Again in a somewhat equivocal fashion he says that the Copernican opinion is "so probable, that its [sic] possible it may exceed even the Ptolemaick in probability." Its principal merit is that it allows deductions of a more ordered system. Once more he makes clear that the Scriptures need not hinder the view if they can be "opportunely and appositely . . . reconciled with it," as he has tried to do (499–500).

In the remainder of the letter Foscarini takes the offensive. He offers new interpretations of some ambiguous scriptural passages that he thinks may have been intended to give man a hint of the cosmic system described by Copernicus. Thus, Foscarini moves at the end from a hermeneutics that would remove impediments from the Scriptures to a hermeneutics that would provide support for the system. Galileo adopts a similar rhetorical strategy in his *Letter to Christina,* using different texts.

The Scriptures Foscarini selects contain mysterious images that have puzzled exegetes for millennia. The first of these is in Exodus 25.31. Speaking of the candlestick that was to be placed in the temple of God, the Carmelite explains that the Holy Spirit may have meant "to shaddow forth unto us the Systeme of the Universe, and more especially of the Planets": "Thou shalt make a Candle stick of pure Gold, . . . of beaten work shall it be made: his Shaft, and his Branches, his Bowls, his Knops, and his Flowers shall be of the same. . . . Six Branches shall come out of the sides of it: three Branches out of the one side, and three Branches out of the other side" (500). The six branches may refer to the six planets, whose periods of revolution around the sun are described beginning with Saturn: "the slowest and most remote of all finisheth his course about the Sun thorrow all the twelve Signes of the Zodiack in thirty Years." Jupiter, Mars, Earth, Venus, and Mercury are named next in that order with their periods accurately noted.

Foscarini finds the verses that follow more difficult to understand, but he invents various ingenious analogies in an attempt to tie them to the new cosmology and thus validate it. In a very curious section he explores the mysteries he believes are contained in the fruits mentioned in the Scriptures—apples, figs, and pomegranates. The fig and the pomegranate seem to him to yield particularly to cosmic exegesis—the fig because it has many kernels, each with a center, and because it is solid and hard in general, but near the circumference is composed of more "tenuouse substance." In this, he says, it resembles the earth, which at the center is hard and stony yet nearer its circumference its parts are "more rare and tenuouse," and it comprises in addition the even rarer parts of water and air. He continues in the same vein with further hypotheses, explaining that he does so because the Scripture ordains that this fruit be embroidered on the vestments of priests. Consequently, its parts must be especially significant.

Just as some of the obscure sayings of the prophets will be understood only when their prophecies have been fulfilled, Foscarini thinks that we may expect that these enigmatic figures will finally be made clear to us "once the true Systeme of the Universe is found out."

In the letter's conclusion, Foscarini directly addresses his Superior, saying that he hopes his exposition will be acceptable to him in light of the "love and diligence wherewith you persue Virtue and Learning." He concludes with a summary of the other works he is completing and sends along a tract, *Concerning Natural Cosmological Divination,* which, he explains, treats of

"Natural Prognosticks, and Presages of the Changes of Weather, and other things which fall within the compasse of Nature." Praying that God will grant to his Reverence "all Happinesse," he ends his discourse. The date is 6 January 1615, and the place the Convent of the Carmelites at Naples. The work carries the imprimatur: Imprimatur, P. Ant. Ghibert, Vic. Gen. / Ioannes Longus Can. & Cur. Archiep. / Neap. Theol. Vidit.

The bland, mundane conclusion and the narration with which he began give no indication that Foscarini expected anything but appreciation for his exegetical efforts. The hermeneutical methodology he recommended was not new. His attempt to reconcile the problematic passages, however, meant that he had to defend the new cosmological picture the Copernican system entailed. To do so he utilized new principles of natural philosophy which he employed as suppositions in his exegesis. This was a daring move. But even more than that, the whole of his attempt rests on the assumption that Copernicus offered the most probable explanation of celestial movement. Yet he seems, unlike Galileo and Kepler, and even Copernicus, to believe that whoever reads his work will accept it benignly, and see it simply as a dutiful attempt to use the best tools of both philosophy and theology to solve the cosmological problem.

Richard Blackwell has uncovered a theologian's negative judgment of the *Lettera,* apparently made shortly after its publication, in which the critic declared that the work supported the "rash" opinion of Copernicus.[10] Foscarini wrote a *Defense,* appealing to a threefold definition of rash in the writings of the respected Dominican theologian, Melchior Cano, also Bishop of the Canary Islands. He found that none of the senses of rash applied: (a) things that occur not by intent but by chance, such as things hastily and heatedly spoken; (b) insolent and impertinent statements; or (c) propositions that contradict the definitive doctrine of the Church.[11]

Foscarini's *Defense* is far more aggressive a work than the letter, and it contains some very interesting passages, which in content and tone are remarkably close to the thought and expression in Galileo's *Letter to Christina.* Although Blackwell does not suggest it, Foscarini's reply might provide an answer to the question of where Galileo garnered the erudite references to the Church Fathers and Scripture he applies in the *Letter.* We know that his former pupil Castelli had promised to send some references to him, but it is also possible that Foscarini may have sent him his *Defense,* or one of Galileo's friends in Rome may have done so since Foscarini was in Rome at the time. The timing would support such a possibility. The *Lettera* appeared in March 1615. The denunciation must have been written shortly afterwards, for

10. Blackwell, 98, 99, and Appendix 7 A and B.
11. Blackwell, 255.

Foscarini sent both his *Defense* and the *Lettera* to Bellarmine at the end of March or early April. The cardinal's response to Foscarini is dated 12 April 1615. Galileo was at work on the *Letter to Christina* during the summer of 1615.

Whatever the case concerning the influence of Foscarini's *Defense* on Galileo, it had no positive effect on the opinion of the Church. The *Lettera* was condemned the following year by the Congregation of the Index in its decree of 5 March 1616. Foscarini died three months later—his spirit broken?

Bellarmine's Response to Foscarini

Evidently in naive hope for a positive reading, Foscarini sent a copy of the published letter, along with a copy of his *Defense,* to Cardinal Bellarmine, asking him for his reaction to his work. "To the pure all things are pure," might well be said of Foscarini. The innocent goodwill of the Carmelite seems to have influenced Bellarmine, for he wrote a courteous, complimentary response, dated 12 April 1615.[12]

Writing also in Italian, Bellarmine thanked Foscarini for the letter and the other piece he had sent and complimented him on the ingenuity and erudition of both.

To answer the friar's request, the cardinal offers a tripart opinion. In a pleasant vein, but with an admonitory undertone, he says that first of all he was glad that both he and Galileo "are proceeding prudently by limiting yourselves to speaking suppositionally and not absolutely [*ex suppositione e non assolutamente*], as I have always believed that Copernicus spoke" (67). He mentions Galileo no doubt because of Foscarini's citation of the astronomer's observations in the proofs for his arguments.

Bellarmine elaborates on the point, stating that he sees no danger in saying that the assumption of heliocentrism "saves all of the appearances better than by postulating eccentrics and epicycles." This is all right for mathematicians. But to go further and "to want to affirm" [*volere affermare*] this "in reality" [*realemente*] is "a very dangerous thing" [*molto pericolosa*] and sure to upset "all scholastic philosophers and theologians." (In light of the events that are to follow this is clearly the most portentous statement in all of the Galilean literature.) Such a position Bellarmine says will in effect "harm the Holy Faith by rendering Holy Scripture false." He thinks the reverend father's principles of explication of the Scriptures are demonstrated well, but

12. The text of the letter is preserved by Favaro in *Opere* 12:171–72; Stillman Drake's translation is in *Discoveries,* 162–64; Finocchiaro translates it in *Galileo Affair,* 67–69, and Blackwell includes it in Appendix 8. I cite Finocchiaro's version in what follows.

that he has not tried to apply them to particular texts and imagines that he would have "great difficulties" should he attempt to explain all of the passages (67). In this regard, Bellarmine shows a surprisingly tolerant attitude, seeming to turn a blind eye to the few passages Foscarini has explicated and his claim that the principles he has set forth will take care of all the problematic texts. Perhaps Foscarini's dialectical approach to the validity of the Copernican system and his own enigmatic position has mollified the cardinal.

In his second point Bellarmine notes the position of the Council of Trent concerning interpretations that are commonly agreed upon by the Fathers. Where there is a consensus one must accept that opinion. Bellarmine observes that such is the case regarding interpretations of Genesis, Psalms, Ecclesiastes, and Joshua. All agree, including modern commentators, that texts referring to the sun's movement and the earth's immobility should be taken literally. While it might be contested that these subjects are not matters of faith, the point is that it is not the subjects but the word of the Holy Fathers that is a matter of faith. Revealing clearly the reason for the reluctance of so many ecclesiastics to entertain new interpretations, he asks Foscarini to consider "with your sense of prudence, whether the Church can tolerate giving Scripture a meaning contrary to the Holy Fathers and to all the Greek and Latin commentators" (68).

The final point is one that goes to the heart of the issue we are treating in this study, the certainty of the heliocentric position:

I say that if there were a true demonstration that the sun is at the center of the world and the earth in the third heaven, and that the sun does not circle the earth but the earth circles the sun, then one would have to proceed with great care in explaining the Scriptures that appear contrary, and say rather that we do not understand them than that what is demonstrated is false. But I will not believe that there is such a demonstration until it is shown me. Nor is it the same to demonstrate that by supposing the sun to be at the center and the earth in heaven one can save the appearances, and to demonstrate that in truth the sun is at the center and the earth in heaven; for I believe the first demonstration may be available, but I have very great doubts about the second, and in case of doubt one must not abandon the Holy Scripture as interpreted by the Holy Fathers. (68)

After this unequivocal statement of the grounds on which new interpretations can be made, Bellarmine cites the authority of Solomon for the prevailing view of the stability of the earth. Solomon said that "the sun also rises and returns to the place from which he rose, etc." [Eccles. 1.5]. The great king spoke not from the unenlightened view of the common man but with the inspiration of the Holy Spirit and from a knowledge of all the sciences. Bellarmine says that if Foscarini thinks that Solomon may have spoken only of the appearance of things, he should take into consideration the phenomena Foscarini himself had mentioned. When we proceed in a ship away from the shore, it appears that the shore is receding but we know this not to be so; but in the case of the movement of the sun and the moon, no sage is needed to correct the error because one "clearly experiences that the earth stands still and that the eye is not in error when it judges that the sun moves, as it also is not in error when it judges that the moon and the stars move" (69). Thus, Bellarmine places his confidence in the authority of the Church in these matters and in the evidence of his senses. He admits that the Copernican system better serves to explain the appearances, but since no unequivocal proof of it exists, he believes the Church must conserve the traditional interpretations.

Although Bellarmine may sound blindly obstinate to us, the fact is that many in his day and even in later years thought that the Copernican thesis was absurd. As Dorothy Stimson and Giorgio de Santillana point out, Francis Bacon had dismissed the view and Robert Boyle thought it merely a brilliant supposition. More significantly, one of the greatest astronomers, Tycho Brahe, was unconvinced. His major concern was the same as Bellarmine's: Where is the physical proof?[13]

Concerning the question posed at the beginning of this chapter as to whether rhetoric intrudes into the theological arguments, we would have to conclude that it rarely does so. It emerges primarily in regard to the intransigence of opponents, such as in the few passages where Kepler rails against the obtuseness and willful contentiousness of the detractors of Copernicus and where he points out the errors of the ancients. Foscarini's rhetorical appeal is his conciliatory tone, his air of sweet reason. No bitter passages mar the tone, and he appears oblivious to the possibility of strong opposition. When he writes of those who might be pleased by his work, he cites Kepler freely along with Galileo, not expecting opposition because of the German's Protestant connections, but approbation for his having sought to serve the interests of these famous astronomers. The remainder of the arguments of these exegetes are dialectical.

13. Stimson also notes that Richard Burton and Sir Thomas Browne rejected it (71, 88–89). Santillana mentions these sceptics in the preface to his edition of Salusbury's translation of Galileo's *Dialogue on the Great World Systems* (Chicago: University of Chicago Press, 1953), xv.

CHAPTER SIX

Dominicans on the Side of Galileo

By 1615 the Roman Congregation of the Holy Office, the Inquisition, realized that it had to make a decision about the status of Copernicus's *De revolutionibus*. Letters of complaint about the book and about Galileo's espousal
of the thesis had reached the Inquisition. Before the Congregation made its
decision, it consulted knowledgeable scholars on both sides of the issue, but
rumors were rampant that the book would soon be banned and the opinion
declared heretical.

The role of the Holy Office in monitoring opinion on the Copernican
issue is important because of its influence upon the Congregation of the Index, whose authority it was to place on the Index books deemed to contain
false or heretical teachings. The decrees of the latter were made under norms
developed at the behest of the fathers who met at the Council of Trent from
1545 to 1563. The Council had likewise provided that no book on religious
matters could be printed without the approval of Church officials, the famous Imprimatur. This same Council had also proclaimed the Vulgate to be
the only approved version of the Scriptures.

In this chapter and the next our concerns move beyond interpretations of
individual texts to the larger questions of whether Copernicus's book threatened the security of the faith by maintaining that the system of the universe
is different from what the Scriptures describe. Foscarini, among others, was
consulted on the issue by officials of the Church in Rome. As we shall see,
Galileo also came to Rome, but missed Foscarini, who had already left when
he arrived.[1]

Campanella also appears to have been consulted for his views on the matter. His opinions are the first to be examined in this chapter. The second are
those of Giordano Bruno, who precedes in time the exegetical writings previously discussed. His tragic execution in 1600 for heresy was to loom over
those who took part in debate over the Copernican thesis, but since his read-

---

1. Stillman Drake (*Galileo at Work*, 448) relates that Foscarini and Galileo had hoped to
meet shortly after their visits to Rome. Their plans were cut short either by the edict or by
Foscarini's death three months later.

ing of Copernicus is so different from, and unrelated to, the others being examined we have accorded him a place out of time in our examination.

Campanella's *Apologia pro Galilaeo*

Tommaso Campanella, like Foscarini, was one of the few members of religious orders who publicly came to Galileo's defense when the temperature of the debate began to rise after the publication of *Sidereus nuncius*. The fact that the friar had been imprisoned twice for his heretical views and for his political radicalism before he began to write the *Apologia*, which espouses the cause of intellectual freedom, endows the work with particular significance. That he was a member of the Dominican Order, noted for its philosophical and theological contributions to the Church in its early period and for its reputed enmity to Galileo in the seventeenth century, makes his writings even more salient for our concerns.

A native of Calabria, Campanella entered the Dominican Order in 1583 at fifteen, and early in his student days evinced great love for the Order's brightest intellectual light, Thomas Aquinas. He was later to shift his allegiance to the naturalistic philosophy of Bernardino Telesio. Although he still demonstrated a lingering respect for Aquinas in the *Apologia*, he thought the greatest barrier to the progress of knowledge was the stubbornness of the Peripatetic worshippers of Aristotle, the philosopher whose thought St. Thomas had reconciled with Christian theology.

Apologia pro Galilaeo, mathematico fiorentino was the title under which Campanella published his essay in 1622, but it was not the original title. Campanella had called it *Apologeticus pro Galilaeo*, indicating that it was a disputation in *utramque partem*.[2] The difference is important. The first would be a justification of Galileo's view, a kind of speech designed to convince others of its truth, whereas the second would indicate an affirmative position in a debate airing both sides of the issue, putting it in the disputation genre. The published title has the effect of defining the work as a strong defense of Galileo, when in actuality it is a disputation whose qualified conclusions support only his right to be heard. Thus the mere change of a word in the title gives to the work a rhetorical "spin" not originally intended.

Campanella's *Apologia* (we shall refer to it by its published title) is not a well-known work; most people who mention it know it by hearsay. The only

2. Bernardino M. Bonansea discusses the two titles in "Campanella's Defense of Galileo," in William A. Wallace, ed., *Reinterpreting Galileo* (Washington: The Catholic University of America Press, 1986), 207. Luigi Firpo's *Apologia di Galileo* (Turin: Unione Tipografica Editrice Torinese, 1968) is an important work on the topic and on Campanella himself.

English translation is that by Grant McColley, which first appeared in 1932 and received some very unfavorable reviews.[3] More recently Bernardino M. Bonansea has provided a helpful summary and appreciation of the work that supplements his biography of the friar.[4]

Apart from these two studies little attention has been given to Campanella's *Apologia* on this side of the Atlantic. The reason for its relative obscurity is not difficult to surmise: it is often tedious and its disputational format does nothing to enliven one's interest. Nevertheless, for a variety of reasons it merits further consideration. From the standpoint of form it illustrates the classic disputation, specifically its use of dialectical argument to probe both sides of the question while being partisan to one. It allows us to see whether, in this example at least, rhetoric entered into a dialectical exercise. In terms of content the work allows us to monitor prevalent opinions prior to the decision by the Church to censor *De revolutionibus,* or at least we can learn what one very knowlegeable man thought were the major views. It gives us another perspective to complement Galileo's, for both men argued against the prohibition of further investigation of the subject. It also permits a thorough examination of the arguments ranged for and against Copernicus's thesis and the new philosophy heralded by Galileo—topics that were then uppermost in the minds of readers.

Like the writings of Galileo we have examined, the book was intended for an educated public outside academic circles. It is addressed to a cardinal but presumes an audience beyond ecclesiastical circles. And like the letter of Galileo to the Grand Duchess it is actually directed to theologians. But while Campanella chooses to instruct theologians in the considerations that should be taken into account in judging the issue, Galileo generally browbeats them for not following the proper principles of hermeneutics. Neither, of course, had any effect on the outcome, but the complexity of the task and the multiplicity of the tools they employed are impressive.

Those few who have studied Campanella's work have argued about the date of its composition as well as its import. The date is a question that has

---

3. Grant McColley, "The Defense of Galileo of Thomas Campanella," *Smith College Studies in History* 22 (April–July 1937): 1–93. See Firpo's review of the translation and its reception in "Cinquant'anni di studi sul Campanella (1901–1950)," *Rinascimento* 6, no. 483 (1955): 300. Bonansea (208–9, n. 7) finds the translation inadequate but McColley's notes and introduction worthwhile. The original was published in Latin. A modern reprint of the Latin text, edited by Salvatore Femiano, is the definitive version (Milan: Marzorati, 1971); it is reproduced in Firpo's *Apologia,* 135–92.

4. Bonansea, "Campanella's Defense of Galileo." Bonansea's biography is *Tomasso Campanella: Renaissance Pioneer of Modern Thought* (Washington, D.C.: The Catholic University of America Press, 1969).

implications not only for Campanella's character but for the Church's handling of the issue. Luigi Firpo thinks that Campanella wrote the *Apologia* after the decree of 5 March 1616 banning the advocacy of Copernicanism.[5] This interpretation would endow Campanella with almost foolhearty courage given his previous difficulties with the Church, but Firpo makes the possibility seem somewhat credible in view of the Dominican's quixotic nature. Such a reading, however, would also make Campanella guilty of gross deception in the manner in which he presents the work.

On the other hand, Bonansea, citing evidence given by Salvatore Femiano in his edition of the text, presents a convincing rebuttal to Firpo, arguing that Campanella wrote the disputation before the condemnation.[6] Femiano's evidence, besides the wording of the *Apologia* itself, includes various letters attesting to the fact that the work was composed, as Campanella said it was, at the request of Cardinal Caetani, who was gathering opinions to aid the Church in making its decision. Looking at the letter of transmittal from Campanella to the cardinal published with the *Apologia*, one is persuaded that Femiano's contention is correct.

Bonansea translates the letter as follows:

> To the Most Illustrious and Reverend
> Lord Cardinal Boniface Caetani,
> Most Honorable Patron of the Italian Muses
> Friar Thomas Campanella
> Wishes Health and Peace
>
> I herewith send you, Most Reverend Lord, a dissertation wrought by your order, where I discuss the motion of the earth and the stability of the heavenly sphere, as well as the principle of the Copernican system, in relation to Sacred Scripture. You can judge for yourself what is rightly said and what should be defended or rejected, since you have been empowered to do so by order of the Holy Senate. I submit my own opinion not only to the Holy Church, but also to anyone better informed than myself, and especially to you, patron of the Italian muses. These latter will never perish as long as you are alive. May you, therefore, live forever. Amen. (208)

From the prima facie evidence the letter was written during the time the Church was weighing its decision and, thus, more than a year before 29 June 1617, the day of Caetani's death. This would also have been the period during which Galileo offered his views in Rome.[7]

5. Bonansea, "Campanella's Defense," citing Firpo's *Apologia*, 4, 19, 193.
6. Bonansea, "Campanella's Defense," 208–13.
7. Galileo says in the preface to the *Dialogue Concerning the Two Chief World Systems* that he was received by the "most eminent prelates at Rome," presumably when he went there in December 1615 and January 1616 to protect himself from the attacks on him there rumored to be growing in intensity.

It is a matter of record that a Papal Commission was established to look into the matter, so that the claim that Caetani was active in a canvas of opinion prior to 5 March 1616 is justified. Nevertheless, Firpo argues that Caetani would never have consulted someone as controversial as Campanella.[8] His contention that Campanella wrote the *Apologia* in defiance of the decree out of affection for Galileo would mean that Campanella intended from the outset of its composition to deceive, and chose the wording to convey the impression that he was simply preparing the piece for the cardinal's consideration. In a number of places he speaks of the pending decision, once, in fact, presenting the whole disputation as prevenient.

To Firpo the letter is a ruse added by Campanella to avert ill will, since news of the contents of the *Apologia* had reached high places. Yet if this were the case it is surprising that Roman prelates who knew of Caetani's mission did not denounce Campanella. But in fact a letter of Virginio Cesarini, a well-respected scholar, states that the *Apologia* was written while deliberations were taking place in the Holy Office before the decree against Copernicanism. Campanella himself, in a letter to Galileo of 3 November 1616, mentions that he had asked Cardinal Caetani to convey a copy of the *Apologia* to him, and wonders how he liked it. That evidence alone would seem to be enough to demolish Firpo's claims. Moreover, in a letter to Pope Urban VIII in 1628, Campanella discussed his views on Copernicanism saying that he wrote the *Apologia* in response to Caetani's request, but that he had accepted the judgment of the Church when the condemnation was announced.[9]

An examination of the rhetorical structure of the *Apologia* and its contents also induces agreement that the work was composed before the decree. The choice of the disputation as the proper genre to convey the arguments underscores the tentative character of the work. Its tone, coupled with the proofs, further demonstrate its aim: to show that careful consideration of both sides of the issue leads to the conclusion that Galileo should be at liberty to explore further the Copernican question. As Femiano concludes, the tract would have little purpose after the decision. Thus, the answer to the dating question is supplied by the rhetorical purpose of the tract.

The printer's note prefacing the volume does not vitiate the earlier date of composition. The reason for printing it in 1622 in Frankfurt was, of course, to feed the great interest in Copernicanism, which the ban against it had increased.

---

8. Bonansea, "Campanella's Defense," 209.
9. Ibid., 211–12, citing Femiano's discussion, 21–24.

The Basic Argument

Campanella sets forth the perimeters of his discussion clearly in the Pro-emium of the book. As he frames it, the question to be examined is: "Whether the way of reasoning that Galileo practises is reconcilable with the Scripture or not?" (42).[10] The focus of the disputation is on Galileo as the principal exponent of the Copernican thesis, following upon his discoveries with the telescope.

Writing in Latin, Campanella respects the usual order of the disputation. He first treats the arguments contra and then those pro. Following this he lays down the principles on which a decision should be made, treats opposing arguments, and finally offers his conclusions regarding the affirmative arguments. The arguments for the two sides are presented formally in a parallel format and treated in barebone fashion for the most part. The principles and the conclusions are discussed at greater length.

Since Campanella was in prison during the time he wrote the *Apologia*, he had limited resources at his command. He mentions having read Galileo's *Sidereus nuncius* and Kepler's writings, and he exhibits familiarity with other astronomical literature of the period, but in the main he must have relied on his prodigious memory for the citations from authorities that lavishly adorn the work. The disputation would be remarkable enough for the extent of its coverage of the pertinent literature even if the friar had had the well- stocked quires of a Dominican library at his disposal.

Arguments against Galileo

Following the plan he has announced, Campanella's first chapter includes eleven arguments raised against Galileo. In some of these he vividly captures the tone of the prevailing criticisms. Since these will be discussed fully in Campanella's refutation of them, we simply summarize them here. The first two charges concern the "novelties" introduced by Galileo that conflict with scholastic doctrines and the teachings of the Holy Fathers. Six others treat texts of Scripture that Galileo's work "clearly contradicts"—those that speak of the earth's stability or the movement of the sun and the stars (44). Most of these have been covered in the writings discussed in the previous chapter.

Three more arguments deplore the contradictions to scholastic philosophy posed by Galileo's discoveries and the inferences he drew from them. By

10. Page numbers after quotations are references to the Latin text of Femiano's edition of Campanella's *Apologia*, unless otherwise noted. Femiano's edition includes both a Latin and an Italian version. The English translations are generally from Bonsanea's "Campanella's Defense"; occasionally, I have translated other portions, correcting McColley's version when it seems in need of emendation.

asserting that there are mountains and water on the moon and the planets, Galileo extends the four-element theory to the stars and so suggests that there is a plurality of worlds. This view jeopardizes the teaching about heaven and has caused some to say that the doctrine of Christ's universal atonement would be cast into doubt. The final argument takes up scriptural texts that advise man not to investigate things that are "higher" and find out more than is "proper" for him to know. Galileo with his telescope has ignored this advice and "fabricates the whole structure of the universe to his desires" (48).

Arguments for Galileo

In the next chapter, Campanella frames the same number of arguments in support of Galileo. Eight of the eleven proofs are based on authority. The first rests on the testimony of those who permitted the publication of *De revolutionibus,* thinking it did not threaten Catholic faith. Campanella cites the approval of Pope Paul III and that of the cardinal [Nicholas Schönberg] who thought so highly of the work that (as the preface to *De revolutionibus* states) he offered to underwrite its publication himself. A roster of contemporary astronomers and philosophers who have approved of Copernicus's book follows: Erasmus Reinhold, Johannes Stadius, Michael Maestlin, Christopher Rothman. He mentions in addition the favorable opinions on the hypothesis of Cardinal Cusanus, Giordano Bruno, Johannes Kepler, William Gilbert, and Giovanni Magini. Campanella also names Christopher Clavius as one who, after upholding the Ptolemaic hypothesis for years, finally said that it was not sufficient to explain the phenomena. In a sly dig at the Jesuit Scheiner, Campanellea mentions that Apelles has followed Clavius' advice to look for another explanation.

Turning to the ancients in his sixth and most important argument for Galileo, Campanella seeks to legitimate heliocentrism. He recites the commonplace of its origin in the philosophy of Pythagoras. But he enlists an even more formidable proponent when he alleges that Moses also held this view. Seeking then to make Pythagoras more acceptable, he claims that the philosopher was actually a Jew who taught in Greece and then Italy, noting particularly that he visited Crotona and Calabria—Campanella's birthplace. To further strengthen the position of Pythagoras as a worthy sage, he resorts to a rhetorical "poisoning of the well" by citing the testimony of Pico della Mirandola and Saint Ambrose to the effect that Aristotle ridiculed Pythagoras in the same way as he had the Book of Moses, "for he could not grasp their lofty recondite reasoning and mystery with his logic" (52). After mentioning the aspersions cast upon Moses by the Greek Aristotle, Campanella adds, "this testimony of our great men [Pico and St. Ambrose] vindicates Galileo from the insults of the Greeks" (52).

Now the friar raises a related question regarding the proper teaching in Christian schools in view of this. By playing up Aristotle's denigration of Moses, he has magnified the Stagyrite's "heathen" state, his inability to understand Jewish mysticism. The implication is that he would have understood Christian mysteries even less. At the same time a whisper of insinuation can be heard in this argument, to the effect that Aristotle's present-day disciples might similarly be lacking in that kind of understanding.

In posing the question in such a way so as to require a decision between the Greek Aristotle and pagan philosophy as opposed to Christian philosophers who follow the new opinions, Campanella has pinioned the opposition with a false dilemma. The strategy is dialectical, but the reasoning is fallacious in that Campanella seeks to blacken all Aristotelians with the same brush. The friar goes on to amplify the point with other proofs from authority and a clever rhetorical appeal to national pride that was to be similarly used by Galileo in the preface to his *Dialogue*. Among the supporters of Pythagoras Campanella mentions Ovid and the disciple of Pythagoras, Numa Pompilis, who was respected as the wisest Roman Emperor. Pliny records truthfully, asserts the friar, that the Roman Senate erected a statue dedicated to Pythagoras following the Delphic oracle's command to so honor the wisest of the Greeks. Campanella summarizes the dilemma, surely with tongue in cheek: "Those who attack the doctrine and science of Galileo criticize Moses, Rome, and Italy. They place Aristotle above Pythagoras when long-buried truth is finally shining forth" (52–54). Presumably, Pythagoras is numbered among the Italians because he fled his native Greece for the Italian peninsula, where he lived until his death.

In this the longest of the arguments he advances, Campanella laces his dialectical proofs with rhetoric. The appeal to national pride is irrelevant to the argument, but it stimulates *pathos,* as does the complaint about the pagan Aristotle. In the latter case the friar plays on fear of heretical pagan doctrines; these recently had prompted criticisms of certain Averroist views, such as those taught by Cremonini, who was then under surveillance by the Inquisition.

In another proof supportive of Galileo, Campanella states that no theologians have condemned the Copernican thesis until the present day, and he adds that there are many who still would not do so. In this Campanella erred, probably unwittingly, for, as mentioned above in chapter 2, an important theologian and member of his own Order, Giovanni Tolosani, had actually found *De revolutionibus* heretical. The evaluation seems not to have been widely known.[11]

---

11. See Eugenio Garin's account in "Alle origini della polemica anticopernicana," *Colloquia Copernicana,* vol. 2. Studia Copernicana 6 (Cracow: Ossolineum, 1975): 31–42. Garin

Tolosani's opinion is particularly germane to our discussion of the argumentation advanced and so requires a brief digression. His critique is directed at the conflicts engendered by the system described in *De revolutionibus* with both Aristotelian natural philosophy and the traditional understanding of Scripture. Tolosani castigates Copernicus for violating the rules of dialectical reasoning and for mixing the disciplinary realms of physics and astronomy. He requests that men who understand these things be asked for their opinion: "Certainly they will find that his arguments have no force and can very easily be resolved. For it is stupid to contradict an opinion accepted by everyone over a very long time for the strongest reasons, unless the impugner uses more powerful and incontrovertible demonstrations and completely dissolves the opposed reasons. But he does not do this in the least."[12]

Whether Tolosani's work was very influential is doubtful, but the same criticism regarding the absence of demonstrations was prominent in Bellarmine's letter to Foscarini. The problem of the mixed sciences, the question of whether mathematical astronomy is a valid science, although not specifically mentioned by Campanella or Bellarmine, was a recurrent issue for the Peripatetics.

In his final arguments for the affirmative, Campanella devotes his attention to the objections based on Scripture. He notes that in calling the heavens the "firmament" the Scriptures meant something that does not move. If the heavens are immmobile, then the earth must move and the sun remain still. The spots on the sun, the comets, and the stars must all be composed of the same matter, for otherwise the Mosaic text is difficult to understand. St. Justin and other Church Fathers agreed with Moses that the vault of heaven is stationary whereas the pagans held that it is spherical and moves.

Principles for Judgment of the Issue

In the third and lengthiest chapter, Campanella presents three hypotheses or principles that he says must be established before the question can be resolved. The principles are worth a close look because Campanella sees them as supporting the criteria by which truth must be evaluated: authority, nature, and consensus. Each of the principles are buttressed by proofs, largely arguments offered by authorities.

---

includes a transcription of Tolosani's work in this article. A full discussion of Tolosani will be found in Salvatore I. Camporeale, "Giovanmaria dei Tolosani O.P.: 1530–1546, Umanesimo, Riforma e Teologia controversista," *Memorie Domenicane* 17 (1986): 145–252.

12. Westman translates the passage that I have quoted from Garin in "The Copernicans and the Churches," 36.

The first "hypothesis," as the original Latin terms it, is that the judges of a religious issue must love God above all, but love of God alone is not sufficient. It must be supplemented by knowledge of Scripture and of the sciences, especially knowledge of the subject about which a decision has to be made. In the proof that follows, Campanella cites, as did Kepler, the errors of Lactantius and Augustine regarding the Antipodes and St. Thomas's opinion on the impossibility of man inhabiting the equatorial regions. St. Thomas held this, says Campanella, because of his allegiance to Aristotle even though Albert the Great and Avicenna had furnished convincing evidence to the contrary.

The second "hypothesis" is quite complicated, for in it Campanella specifies six kinds of knowledge expected of a just judge, supporting each of these with various combinations of arguments from the Church Fathers, quotations from Scripture, and references to Galileo's discoveries.

To begin with, the proper judge must be a speculative theologian. As such he is expected to have a philosophical understanding of "both celestial and terrestrial things in order to dispute against sectarians" (60). The obligation of theologians is greater than that of artisans and scientists, who are expected to know "lower causes." Theologians must know *all the sciences,* including that of God, so that they can know all his works, for they must be able to defend divine truth with good arguments when man's knowledge seems to contradict it. Campanella bolsters this hypothesis with a reference to Aquinas's *Contra impugnantes religionem,* in which he argued against those who ridiculed the friars for studying secular science and eloquence. Even though theology needs no proofs from human science, such knowledge of nature and sensible things aids our understanding of the supernatural.

This principle is significant because it plays an important role in Galileo's *Letter to Madame Christina of Lorraine, Grand Duchess of Tuscany.* There Galileo also demands that theologians possess extensive knowledge and excoriates those conservative theologians who criticize Copernicus but do not demonstrate that he was wrong. Throughout the proof for the principle Campanella cites a mixture of secular and religious authorities who praise the search for knowledge, and in particular, knowledge of astronomy. He ends by attempting to stir fear in his readers when he suggests that by forbidding observation of the heavens we might miss the signs that foretell the end of the world.

A second necessity for the judge is his recognition that at present knowledge of astronomy has not been developed sufficiently. No philosopher or theologian has given enough attention to the subject and come up with an entirely satisfactory explanation of the nature, order, or motion of the heavens. This is not surprising, since Scripture relates that it is impossible to arrive at such an explanation, and the fact that so many theories are advanced attests to that truth. "For this reason they are mad who claim that Aristotle constructed the truth about the heavens, and nothing more need be investigated" (68).

Campanella's sympathies are with Galileo and the new philosophy; thus, he is at pains to expose the weaknesses in Aristotle's system. He begins by playing off the De caelo against the Metaphysics as follows. In the first work Aristotle posits eight spheres and a primum mobile responsible for turning the spheres by violent motion. In the second, he has the spheres moved instead by intelligences. Since some of the spheres move in an eastward direction and some west or north, the explanation requires these angelic intelligences to move in opposition to God and to one another. Thus, says Campanella, Aristotle makes war in heaven, placing the angels not only in opposition to each other but to God. The great philosopher gives no reason for these movements, nor does he explain the eccentrics and the equinoxes, nor why the heavens must be composed of the fifth essence. Campanella cites Aquinas and Simplicius as sources for his analysis, but much of the criticism is his own. His imaginative description of a war in heaven is, of course, not in Aristotle but in Campanella's reconstruction. The friar ends with a resounding deduction: "Therefore, it is proper to say that his astronomy is completely false because it does not admit those things that are proved by the senses and instruments that are very reliable" (70).

Returning to the pagan character of Aristotle's philosophy, Campanella reviews St. Thomas's critique of Aristotle's position on the eternal motion of the heavens, where the Angelic Doctor says that to state absolutely and not hypothetically that the heavens have an eternal motion is to deny the existence of God (71).

After delineating the imperfections of Aristotle, Campanella describes the progress made by Copernicus, who detected the difficulties in the Ptolemaic system, finding in time a better solution in the Pythagorean view. Galileo afterwards discovered new planets and another center of rotation. Thus, Campanella says, those who continue to extol Aristotle's teachings on the heavens are "insane and ignorant" (72).

This is dangerous ground for Campanella. He must have known that the Jesuits particularly had been charged with saving the teachings of Aristotle. The friar's unrelenting attack on all things Aristotelian may have been fueled partially by the Dominican Order's bitter rivalry with the Jesuits and a resentment of the power they had gained in Rome. But one also recognizes in Campanella a free spirit who would detest any proscription of thought and investigation.

In an appendix to this proof Campanella considers what may be learned from the evidence he has aired. It is here that he argues forcefully for intellectual freedom. He maintains that even if absolute truth about celestial matters might be impossible, as Scripture declares (Job 38.33, 37; Eccles. 3.11), still the search is not vain but a natural and laudable effort. The Fathers teach that the world is God's handiwork, so we should study it. In this way our souls are strengthened. He enlarges upon the theme, saying that those who claim we should be content with the answers of Aristotle are those who would keep man from discovering more about God.

In his third requirement for what a judge should know, Campanella attempts to refute a position popular with some conservative theologians: that Scripture implies that man should not pry into nature's secrets. Campanella provides a lengthy proof, first proving the opposite through the most important authorities. Neither Moses nor our Lord proscribed our delving into physics and astronomy. The testimony of St. Bernard adds weight. He once remarked that St. Peter and St. Paul "did not teach the art of fishing or of tentmaking and like matters; they taught me not how to read Plato or to be involved with the subtleties of Aristotle, but simply how to live" (76).

From testimony Campanella turns to deductive proof. It is clear, he says, that God has given us a rational mind and five senses as a method of investigating his world. Theologians agree that these are not clouded by the Fall. Solomon observed that "God left the world to the disputations of man," and "he himself investigated all things, not merely through the book of Moses, but by inspecting the world of nature" (76).

In a final point he invokes the accommodation topos: Scripture accommodates its teachings to the capacity of common men, for which he cites the authority of St. John Chrysostom and St. Thomas.

The fourth requisite for judges of Galileo's case is their recognition of the principle that whatever "forbids Christians the study of philosophy and of knowledge forbids them to be Christians; only the Christian law commends all the sciences to them, for it does not fear falsity" (60). In explaining and supporting this bold principle, Campanella invokes the wisdom of Leo X at the Lateran Council and others who have declared that one truth cannot be contradicted by another. Thus the truth of Scripture cannot be overturned by the truth of nature. He notes how superior this outlook is to that of the Mohammedans who have forbidden science because many Muslim philosophers, such as Averroes, Avicenna, and Alfarabi, have attacked their faith. The Greeks also forbade searching into the nature of the gods. The Christian, however, need not fear knowledge. In fact the sciences are sought in support of the faith. So St. Thomas taught in his *Summa contra gentiles* and *Contra impugnantes Dei cultum et religionem*. The Lateran Council, the Second Council of Nicea, and the Articles of Paris attest to this as well.

Campanella founds this principle upon the scriptural declaration that

Christ is "the power and wisdom of God" (80). Bonansea notes that Campanella believed deeply in the Neoplatonic-Augustinian tenet that "the Word of God is the supreme reason of all things" and that "a Christian is wise and rational only to the extent that he participates in the wisdom of Christid."[13]

Playing again upon the theme that a pagan Aristotle is an unreliable source of knowledge, Campanella finds support in the teaching of St. Thomas. The Angelic Doctor thought that Christian scientists are much more to be trusted than pagan ones because the Christian is perfected by grace. He adds the testimony of St. Jerome, who borrowed a figure from the Old Testament in commenting on the appropriation of pagan philosophy: "If you should love an alien woman cut her hair and cleanse her nails" (84).

But would it perhaps be better to adopt a new philosophy altogether, Campanella wonders. In responding to the question, he finds apt analogies in the Old Testament, where the rejection of Hagar for her haughtiness and the casting aside of alien wives is described. But Galileo has urged us to study the book of nature to renew the sciences. In accepting this new philosophy, says Campanella, we do not have to cut the hair or cleanse the nails of the foreigner (86). The conclusion of the enthymeme is obvious.

The fifth requirement rests on the previous one. It asks that judges realize that although those opposed to gaining knowledge through reason and experiment act as if they are proceeding from the doctrine of the Christian faith, they cannot find such methods forbidden in the Scriptures. In the proof Campanella inveighs against taking a single or private interpretation of the Scripture and insisting upon its application to a particular case. He cites St. Augustine, who states in *De Genesi ad litteram* that the Scriptures are to be interpreted in many ways lest they be subject to the derision of unbelievers. St. Thomas also notes that passages of the Bible can be given many interpretations just as long as they do not conflict with other biblical texts. The authority of Augustine and Aquinas is again used by Campanella to bring out the correlative point that the teachings of philosophy must be clearly distinguished from doctrines of the faith based on Scriptures. Campanella warns prophetically that theologians should not be quick to condemn Galileo's views, for if he "triumphs our theologians will bring upon our Roman faith no small mockery in the eyes of the heretics, for all his doctrine and the telescope have found avid acceptance in Germany, France, England, Poland, Denmark, and Sweden." On the other hand, if the opinion of Galileo is false it will not disturb theological teachings (92).

He returns to the point that not all that is false is necessarily harmful to the faith, noting that the mistakes of the saints in natural philosophy did not

13. Bonansea, "Campanella's Defense," 223.

cause them to be numbered with the heretics: "Besides, if he [Galileo] is wrong his view will not perdure. Therefore, I think this mode of philosophizing ought not to be prohibited, for then it will be highly esteemed by the heretics and we will become objects of derision" (92). Galileo was also to voice his fear of this possibility in the preface to the *Dialogue* when he defends his decision to take up the Copernican issue after the prohibition against its teachings.

Campanella then reminds Cardinal Bellarmine himself of the manner in which the Ultramontanes reacted to the decrees of the Council of Trent, warning him that he can expect an even stronger reaction if the Church should come out against physics and astronomy.

The last knowledge requirement for judges stresses the necessity of an acceptance of sense knowledge. Campanella emphasizes in the process Galileo's superiority over the Church Fathers in matters of natural philosophy. Although erroneous opinions derived from pagan works have been adopted by Christian theologians, in Galileo's case "no errors can be detected for he proceeds from sense observations in the book of the world, and not from opinion" (94). In other words, dialectical arguments based on opinion may err but those based on sense experience can be trusted.

Proceeding at last to the third "hypothesis" or prerequisite for reaching the proper conclusion in the disputation, Campanella repeats much that has been said but adds further requirements. Judges must have extensive knowledge of the range of exegesis on critical passages and also bring to bear sufficient knowledge of philosophy and mathematics. Aristotle, or any other philosopher, must not be taken as the sole authority.

Lastly, judges should not be led by allegiances or emotions to prejudiced conclusions. Campanella suggests that Galileo's detractors are jealous of him. They do not wish to be made students again when they are now termed masters. He quotes Horace: "They regard nothing as right, except what pleases them; / They consider it shameful to yield to younger men; / And to acknowledge when old their youthful learning is in ruins" (94). In suggesting that Galileo's opponents are blinded by emotion, Campanella, too, seeks to arouse emotion—indignation—in his readers. Again rhetoric comes to the aid of dialectic.

Campanella's Counterarguments

Having laid the foundation in the previous extensive chapter, Campanella turns in the fourth chapter to reply to the arguments against Galileo. To introduce what follows he refers his readers to an *opusculum*, written in 1609, in which he considers whether Christians should be permitted to develop a new philosophy that could replace the old philosophy of the Gentiles, who had not the benefit of revelation.[14] The *Sidereus nuncius,* coming a year later, must have seemed the answer to the friar's fondest hopes, since it provided evidence that the old philosophy must indeed be cast out.

Replies to Arguments against Galileo

The subject of the first argument, Campanella explains, has already been treated, that is, the dangers to the reigning philosophy posed by the novelties introduced by Galileo. Emboldened perhaps by his remembrance of the hopes he had so recently described in the *opusculum,* he now claims that it is heresy to demand that philosophy be founded on Aristotle. Continuing to hammer on the wedge he has inserted between a pagan Aristotle and a Christian Church, he says that it is obvious that in preferring Aristotle one condemns Augustine, Ambrose, and other Church Fathers because they opposed Aristotle's metaphysics and also many of his physical teachings, favoring instead Plato and the Stoics. If we do not condemn Aristotle, we will end by accepting his heresies. He names these: the doctrine of eternal motion, the immortality of a single soul for all men, the indifference of God towards all his creatures, God's contest with the angels in rotating the spheres, the myth of hell, the necessity of God's actions, the abrogation of Providence by chance. The two alternatives are the old philosophy of Aristotle, a pagan whose teachings carry the taint of heresy, and the new philosophy of Galileo, a philosopher loyal to the Church, who accepts its teachings and bases his observations about nature on sense experience.

The second argument made against Galileo, that he overturns the teaching of the Church Fathers and the scholastics, Campanella emphatically denies, asking that the same treatment be given to Galileo's evidence as was given to that of Columbus and Magellan when they contradicted the evaluations of the Fathers' opinions regarding the earth.

Campanella refutes the argument by making three counterassertions: (a) some theologians have embraced doctrines more repugnant to Scripture and the Fathers than Galileo's; (b) many of the Church Fathers and scholastics hold opinions in accord with Galileo's; and (c) Scripture is actually more in favor of Galileo's opinion than it is of his adversaries (102).

As evidence in support of the first assertion, Campanella points out that some theologians accept certain opinions of Simplicius and Alexander, commentators on Aristotle, even though these contradict Scriptures. He mentions also people who confuted errors of the Church Fathers who were

14. Bonansea (ibid., 226 and n. 80) corrects a misreading of McColley at this juncture; the latter mistakes Campanella's reference to "in questione praecedenti" as a reference to a preceding question rather than as a reference to an opusculum. Although Campanella does not say so, the friar was probably referring to his opusculum entitled *De gentilismo non retinendo,* published with another work, *Atheismus triumphatus* (1636).

not damned for doing so. In fact their opinions are now accepted and defended on the basis of sense experience. Such provide strong arguments for
Galileo. In support of his second allegation Campanella offers the testimony
of Aquinas that an opinion favoring the earth's mobility does not threaten
Church doctrine. He notes too that many authorities among the saints differ in their views on the location of the earth and of hell, but they all agree
regarding the scriptural teaching on the immobility of the heavens. Furthermore, they do not hold that the earth must stand immobile in the center of
the world. For this reason Campanella finds it difficult to understand why
some theologians say that Galileo's opinion is different from the received
one. These people ignore contrary opinions, offer no support from divine
revelation, and deny sense experience.

Man's theories cannot determine the order of the universe, which is cre-
ated and ordered by God, says Aquinas. To think otherwise is inimical to
Scripture. Thus, whether the earth moves or not cannot be a matter of the
Christian faith.

In a clever move Campanella provides examples of passages in which the
Church Fathers cite texts of Scripture that speak of the immobility of the
heavens and thus imply the mobility of the earth. He concludes that Scripture does not clash with Scripture. In sum, St. Thomas, Peter Lombard, and
the scholastics are more in agreement with Galileo than with those who oppose him.

To support the third assertion, Campanella interpolates a lengthy exegetical proof. He refutes a succession of five arguments in which texts from
Scripture are used to oppose Galileo's contention about the earth's motion
or the central position of the sun.

Campanella generally argues from the intention of the Scriptures, not
their literal sense. But in the first text from Psalm 92, where the psalmist says
that God established the earth not to be moved, he cites philosophical and
metaphysical arguments from St. Thomas, using them in a somewhat sophistical way to prove that St. Thomas allowed for a mobile earth. Aquinas
had argued that if hell is in the center of the earth, then the earth must be
hot and in motion above the center, since fiery things by nature are mobile. But here he was arguing about the position of hell and maintaining that
since the earth is cold, hell cannot be in its center.

Regarding texts declaring that the earth stands firm forever, noted by the
earlier exegetes, Campanella explains that there are many interpretations,
but he contends that it is much better to avoid ridicule by the German
astronomers who have ascertained that the earth moves and the sun is still.

Coming to the miracles of Joshua and Hezekiah raised in the fifth and
sixth arguments against Galileo, Campanella asserts that these are indeed
miracles, but they need not be taken literally. Rather the wording should be

explained by the desire to accommodate common belief. Thus the earth stood still, not the sun, and the sun seemed to go backwards. One need not say that the senses erred; rather it is our interpretation of the appearances.

The eighth charge against Galileo, his claim that water exists on the moon and the planets and thus forcing abandonment of Aristotle's doctrine of the incorruptibility of the heavens, Campanella says, is of little importance. After all, the account of creation in Genesis renders Aristotle's teaching difficult to maintain, and Galileo's contention does not contradict the Scriptures.

Campanella's resolution of these problems is ingenious. He recounts the Florentine's explanation of the discoveries, and comes to the conclusion that the Scriptures agree much more with Galileo than with Aristotle. Galileo, says the friar, "ought to be praised for he has at last by sense experience vindicated the Scripture from ridicule and distortion" (132).

In refuting the ninth argument the friar continues the same strategy, comparing Galileo favorably with the theologians. Campanella forcefully denies the charge that Galileo teaches a plurality of worlds; in fact, he says that all the systems exist under one order in an immense heaven. It is the theologians who must be charged with saying that there is more than one world; he recites the speculations of St. Basil and Clement regarding the existence of the elementary, celestial, and spiritual worlds. Galileo, on the other hand, describes what was disclosed by his "marvelous instrument." There follows an ecstatic recital of these wonders of God's handiwork.

Nevertheless, says Campanella, the Church holds that theological opinion cannot deny the possibility of the plurality of worlds. St. Thomas denied that there could be other worlds, but this view was corrected in the Condemnations of 1277. To claim the impossibility is to limit the power of God. Campanella goes on to defend the intention of St. Thomas in this regard. Using other passages from Aquinas he shows that he did not restrict the power of God. Finally, he points out, that Scripture does not mention the subject of the plurality of worlds does not mean that we should assume that it denies the possibility. It is fallacious logic to argue from the topos of negative authority, he says. Campanella closes with a convincing observation: "Moses is silent about all this, because having given the law to our world, he was not writing a physics of the whole system, indeed, not even of ours, except as it was needed for the law" (138).

The existence of creatures like ourselves in other worlds is not entailed in Galileo's observations about the movements of the orbs, as Campanella rightly concludes. Galileo actually denies that men could exist elsewhere in his *Letters on Sunspots*. But he does not rule out the possibility of other beings. He would not go so far, however, as to entertain the playful hypothesis of Kepler in his *Conversation with the Heavenly Messenger*.

The penultimate argument accuses Galileo of perpetrating scandalous opinions prohibited by the Gospels, which, of course, Campanella denies, saying that, on the contrary, Galileo simply "pursues the truth, which God commands and praises" (138). And, as Saint Gregory observed in his commentary on Ezekiel, "if truth be the cause of scandal, then it would be more useful to permit the birth of scandal than to relinquish truth" (138). Again he asserts that Galileo's conclusions do not conflict with the Scriptures or the Fathers, but with Aristotle.

The final argument ranged against Galileo's position is that Scripture counsels us not to seek knowledge beyond what is necessary. Campanella says that he has already answered this adequately in the first assertion of the second prerequisite for judgment of the case, where he shows that acquisition of knowledge is praised by the Fathers. The point is buttressed with a reference to verses in Ovid where the poet extols those who seek to know things thought to be beyond our ken. This he contrasts with the attitude of Cato, who thought mortals should be satisfied with mortal things. Citations to the Psalms of David that praise the beauty of God's handiwork in the heavens round out the proof.

The argument is interesting because Campanella thinks it convincing to incorporate the testimony of a pagan poet to support his refutation of a charge based on Scripture, but at the same time he wishes to arouse public opinion against heathen philosophy. The reason must lie in prevalent taste. Classical poetry is held in high esteem while regard for Peripatetic philosophy has begun to crumble, and Campanella wants to hasten its destruction.

Replies to Arguments for Galileo

After the multitude of proofs Campanella has amassed in favor of Galileo, one might expect him to use them in the last chapter of his *Apologia* to argue for him. Such is not the case, nor is it the proper method of disputation. He must end by considering objections that might be made by opponents to the arguments he has advanced in favor of Galileo.

In the introduction to this chapter Campanella recounts the confusion he has experienced in arriving at his own opinions on the matter. He first thought that the heavens and all the stars were composed of fire, as Telesio, Augustine, Basil, and other Church Fathers had believed. The observations of Tycho and Galileo revealing new stars and comets in the sphere above the moon and the discovery of sunspots all combined to cause him to suspend his judgment. He also doubted that the whole firmament could move around the earth at the rate of a thousand miles in one moment. The movements of the Medicean planets around Jupiter soon showed him that a single center for the universe cannot be posited. Certain similarities in the outward appearance of the stars, the planets, and other objects, however, make suspect some opinions about the sun expressed by Galileo and others. He concludes, anticlimactically, that he thinks the arguments for Galileo can be confuted only with difficulty. Having illustrated the tentative state of knowledge of the heavens by his own example, Campanella counters the first five arguments and the seventh with an assertion that the opinions of Copernicus and Galileo are only probable, not certain, a position which, he says, all theology recognizes. Arguing for sufferance for the inconclusiveness of these arguments, he points to the fact that Pope Paul III permitted the printing of *De revolutionibus,* but that did not mean he gave it his approval. Even the teachings of sacred theology do not gain the approval of the pope, he says, but they are recommended as useful and worthy readings. Moreover, if a thesis is allowed by the pope and the doctors of theology, Campanella maintains, this means only that the teaching is probably not contrary to Scriptures, not that it is necessarily so.

In response to the eighth, ninth, and tenth arguments that Galileo's views do not overturn Scriptures, he says that theologians evade definite answers about the heavens, but certainly the opinions are at odds with Aristotle. In examining them he has found that the Scriptures support Galileo's doctrine no less than that of other philosophers. But ultimately he prefers the stance of theologians: "there let the physicist see, but the Church be the judge, whether Galileo should be permitted to write and dispute on these matters" (144). Campanella does not reply to the eleventh argument presumably because it is fairly well covered in previous considerations.

To the sixth argument, however, he gives further attention, since it must be accepted lest it threaten his primary contention concerning the Christian character of the new philosophy. In it he has alleged the existence of support for heliocentricism among acceptable ancients in opposition to the gentiles or Greeks who favored geocentrism. Campanella mentions again that the principal exponent of heliocentrism, Pythagoras, was said by St. Ambrose to have been a Jew, which would remove from him the odium of being a "gentile." Campanella argues for acceptance of Pythagoras's opinion from the authority of the saint's testimony and from probability, saying that the philosopher's views most resembled those of the Hebrews. Moreover, he concludes that, since Aristotle derided Moses and Pythagoras, we should try to show that they were right through our unerring instruments and argumentations.

Campanella asks rhetorically, "Why do we murmur much as the Jews once did against Moses, who defended them against the injustice of the Egyptians?" (148–49). Ancient rabbis, he says, supported this same philosophy, as Dionysius the Carthusian tells us, and, he adds, Mohammed also speaks of seas, air, and mountains in the heavens. Campanella cites the

Koran to show that Mohammed accepted many things from the rabbis, such as the plurality of worlds and systems above our heaven. Muslim thought here supports his establishment of another valid philosophical tradition that rivals and surpasses the Greek. Campanella's premier argument was surely inspired by Ficino's Hermetic philosophy, which proposed a *prisca theologia* with roots through Pythagoras stretching back to Trismegistus and the Egyptians, who preceded the Greeks in time and thus in thought.[15] This inclination to Hermeticism with its Platonic overtones found in the two Dominicans, Campanella and Bruno, is not surprising. A strain of mystical Neoplatonism and Augustinian thought is evident throughout their history. We need only cite the cases of the Dominicans Meister Eckhardt and Johannes Tauler, who came under grave suspicion for their own Neoplatonism in the late Middle Ages.

In a last plea for liberty of thought Campanella refers once again to the teaching of St. Thomas and St. Augustine that "we should not risk the danger of our Scriptures being laughed at or, even more, of the suspicion that like the pagans we believe things contrary to the Scriptures" (148). He adds parenthetically that Bellarmine has said that the heretics do not dare refute our Roman theology at present.

The implication is clear: rash action to curtail philosophical advances would eventually cause attacks on theology. Given the dangers lurking in suppression, he asks that the Church "not prohibit the studies of Galileo, nor suppress his writings, especially on this account that our enemies would seize upon it avidly and celebrate it" (148).

In closing, he commends his work to "the censure and better judgment of Holy Mother the Roman Church," and wishes good health to Cardinal Caetani, patron of Italian *virtù*. The quality of *virtù* emphasized here seems particularly chosen to evoke remembrance of the character of Galileo. Campanella's *Apology*, then, does not contend that Galileo has demonstrated his position, nor does he ask that the astronomer be permitted to continue to express his views. Rather the Dominican leaves the resolution to the Church. One reason the friar does not press for endorsement of the heliocentric view is that he never fully accepted it himself. But he thought Galileo should be permitted to investigate it without being accused of heresy.

Bonansea points out that the decree against Copernicanism made shortly thereafter was never given the status of infallible dogma. It was a decree of the Congregation of the Index that received papal approval "only in *forma communi,* as distinct from an official endorsement of the pope speaking *ex*

---

15. Yates examines this view in her *Giordano Bruno and the Hermetic Tradition* (Chicago: University of Chicago Press, 1964), 12–17, henceforth referred to as *Bruno*.

cathedra." The latter would have required adoption of the position as a matter of faith. As it was, the decision was "fallible" but still had "binding force on all Catholics."[16]

The disputation has allowed Campanella to gather all the forces of probable reasoning he could muster to show that his solution is the most convincing one. We have noted the occasional resort to rhetoric and sophistical reasoning, but for the most part he carries out his intention in the traditional manner. One might conjecture that were the disputation presented as a determination (i.e., an examination for a degree), the masters would surely have passed him, but they might have chided him for breaking the rules.

Giordano Bruno for Copernicus

To pass from an examination of Tommaso Campanella's disputation to an analysis of Giordano Bruno's dialogue is like stepping from the Middle Ages to the Renaissance. Yet the more modern work of Bruno precedes his confrere's by almost thirty years. The differences are those not only of format but of content and style. Where Campanella provides a particularly apt illustration of dialectical reasoning, Bruno gives us an example of Renaissance prose in one of the period's favorite genres, the dialogue.

In Campanella's *Apologia* one hardly recognizes the author of the lively utopian fantasy, *La Città del Sole: Dialogo Poetico*. Campanella could also write in a literary vein, even though he might not equal Bruno in eloquence. But medieval genres continued to coexist within the Renaissance; *La Città* was written seven years before the *Apologia*. Like Bruno's dialogue, Campanella's also shows a fascination with Hermeticism; his city is planned to harness the powers of the stars to enhance the life and work of its citizens.

For Campanella, though, the Copernican issue and the occasion demanded a dialectical treatment to argue forcefully that intellectual liberty would bring the greater advantage to the science of astronomy and to the faith. In contrast, Bruno uses the occasion of an actual disputation on the Copernican thesis to develop a defense that is more rhetorical than it is dialectical. Bruno's case for the Copernican position, which he published in 1586 with the title *Cena de le ceneri,* or the *Ash Wednesday Supper,* was not a work that won many advocates for Copernicus, nor did it contain many arguments of continuing value in the Copernican debate. It did attempt to reconcile Copernicanism with Scripture, but because of Bruno's unorthodox views and his later execution for heresy, the attempt did not benefit

16. Bononsea, "Campanella's Defense," 236.

the cause. Frances Yates remarks that if Copernicus had seen the work "he might well have bought up and destroyed all copies."[17]

The chief merit of the work is that it probably served as a prototype for Galileo's *Dialogue*. It anticipates in some of the particulars of organization, attitudes, and characters of the more famous book.

Bruno's story and especially his place in the Hermetic tradition has been told so well by Frances Yates that it is unnecessary to rehearse it extensively here.[18] Our interest is in the visit he made to England in 1583 at the age of 34 and the debate he had with Oxford dons on the Copernican question. That experience forms the basis for the dialogue. The account of his visit, which he published soon after, so insulted the English that Bruno was moved to write a sequel wherein he attempted to apologize with an encomium for Oxford as it was before the Reformation.[19] Stanley Jaki in the introduction to his translation of the *Cena* finds the scientific content of the Oxford dialogue wanting. He declares that "both Copernicianism and science were badly shortchanged in the *Cena*."[20]

Campanella himself mentions the Nolan—the appellation comes from Bruno's birthplace, Nola—in his *Apologia*, despite Bruno's fate. He does so no doubt because Bruno was an early exponent of Copernicus. His name occurs twice, but only in the company of more orthodox authorities. His execution in 1600 (ironically, on the morning following Ash Wednesday) was too fresh in people's minds; his reputation would not help Campanella's argument nor would the Hermetic dynamic he there describes. Campanella astutely leaves out Hermetic arguments.

In discussing the style of the *Cena* Jaki finds Bruno careless with syntax and punctuation, but notes that his mastery of Italian and his creativity formed many eloquent passages.[21] Bruno's knowledge of astronomy was not small, but he used it to launch his own notion of the cosmos, an organic infinite unity composed of animated clusters of entities that fill space. These entities he thought to be governed by a cyclical law of return which they acted out among the eternal stars and planets. Influenced by works at-

17. Yates, *Bruno*, 297.

18. See Yates, "Giordano Bruno's Conflict with Oxford," *Journal of the Warburg Institute* 2 (1938–39): 227–42; and Dorothea Waley Singer, *Giordano Bruno: His Life and Thought, with Annotated Translation of his Work "On the Infinite Universe and Worlds"* (New York: Henry Schuman, 1950).

19. Yates, *Bruno*, 168.

20. Giordano Bruno, *The Ash Wednesday Supper*, translated with introduction and notes by Stanley Jaki (The Hague: Mouton, 1975), 14. Jaki provides a biographical account of Bruno's life and an evaluation of the work from the standpoint of modern science in his introduction. References to this translation are given by page numbers in the text.

21. Ibid., 24.

tributed to Hermes Trismegistus, Bruno's world was filled with magic and cabalistic symbolism.

The Preliminary Dialogues

The dialogues, five in number, begin with a poem that sets the tone of the work. The first three stanzas capture the spirit:

> Should my cynical teeth have pierced you through,
> Blame only yourself, you vicious canine;
> In vain you show me your stick and swagger,
> If you guard not against despising me.
> Since you have confronted me with injustice,
> I shall stretch and pull your skin all over;
> And should my body too fall to the ground,
> Your shame will be recorded in hard diamond.
> Go not naked to the beehive for honey,
> Bite not, if you know not the bread from stone,
> Walk not barefoot while disseminating thorns. (42)

The poem continues in like vein, adding more aphorisms and threats in the remaining two stanzas. The dedicatory epistle addressed to his French patron, Seigneur de Mauvissiere, the Ambassador to England, adds further sarcastic comments, except in regard to his own contribution, where he uses oxymorons to describe a "repast so grand and small, so magisterial and schoolish, so sacrilegious and religious" (43).

The First Dialogue

The characters are introduced in the first of the five dialogues: Smith, an educated layman; Theophil (God-loving), Bruno's surrogate; Prudenzio, a pedant; Frulla, Prudenzio's protege. Two other figures are described but do not participate, two Oxford dons, Doctors Nundinio and Torquato. They are depicted as beer-loving, pretentious pedants whose debate with the Nolan on the Copernican system is the focus of the work.

Prudenzio asks his three companions to sit down so that they can carry on their "tetralogue" devoted to the success of the Nolan's debate, which in the fictionalized account takes place in London. In answer to Frulla's query as to why this will be a "tetralogue," Prudenzio replies condescendingly that a tetralogue is a discussion among four, an important point, he says, for they must understand what they are about. Quoting from Cicero in Latin, he says he simply follows the advice of the Roman rhetor to begin an exordium with a definition and explanation of names. The pedantic rhetoric and Peripatetic disposition is sharply contrasted with the benign but incisive mien of Theophil who quickly corrects him, saying that they will in effect have a dialogue, "for four persons as we may be, we shall be two in functioning, namely, to propose and to reply, to reason and to listen" (54).

Invoking the muses to inspire him, Theophil introduces the thesis of Copernicus and waxes eloquent in a paean to the astronomer, making it clear, however, that the Nolan does not see with the eyes of Copernicus or Ptolemy but with his own (55).

After describing the qualities in which Copernicus excelled his predecessors, Theophil remarks: "But for all that he did not move too much beyond them; being more intent on the study of mathematics than of nature, he was not able to go deep enough and penetrate beyond the point of removing from the way the stumps of inconvenient and vain principles, so as to resolve completely the difficult objections, and to free both himself and others from so many vain investigations, and to set attention firmly on things constant and certain" (57). Showing a remarkable grasp of the nature of Copernicus's contribution and the limitations of its proofs, he continues:

And though deprived of effective reasons, he seized those rejected and rusty fragments which he could have from the the hands of antiquity, and repolished, matched and cemented them to such an extent with his more mathematical than physical discourse, that there arose the argument once ridiculed, rejected and vilified, but now respected, appreciated and possessed of greater likelihood than its contrary, and certainly more convenient and useful for theory and for computational purposes. (57)

His acute understanding of the nature of Copernicus's astronomy may lead us to expect too much. Jaki notes that Bruno is "disdainful of mathematics," claiming that he is an expert in "'physical astronomy'," which he knows is the only path to a realist description of the universe. "But his version of physical astronomy or his explanation of the motion of the earth and of other celestial bodies bogs down in gross animism (to say nothing of his Hermeticism), which vitiates much of the forcefulness of his assertion of the infinity of the universe."[22]

Judged then by a modern historian of science, Bruno contributed nothing to Copernican astronomy. But he evidently had a different conception of his mission; he did not seek to advance the Copernican project but to build on it, and what he built was a temple to ancient Egyptian knowledge. This is revealed as Theolphil extends his account of the Nolan's encomium for Copernicus to speak of Bruno himself: "Now, what shall I say of the Nolan? Would it perhaps be improper that I should praise him, just because he is as close to me as I am to myself?" Waxing grandiloquent, he rises to a climax:

22. Ibid., 57, n. 21.

Now here is he who has pierced the air, penetrated the sky, toured the realm of stars, traversed the boundaries of the world, dissipated the ficticious walls of the first, eighth, ninth, tenth spheres, and whatever else might have been attached to these by the devices of vain mathematicians and by the blind vision of popular philosophers. Thus aided by the fullness of sense and reason, he opened with the key of most industrious inquiry those enclosures of truth that can be opened to us at all, by presenting naked the shrouded and veiled nature; he gave eyes to moles, illumined the blind who cannot fix their eyes and admire their own images in so many mirrors which surround them from every side. (61)

This is not the discourse of an astronomer; it is that of the magus. Thinking that Copernicus has eliminated the spheres, Bruno sees himself as piercing through them and entering the celestial regions to participate in divine knowledge.[23] "Thus we know as many planets, as many stars, as many deities, which are those hundreds of thousands that assist in the service and contemplation of the first, universal and eternal efficient" (61).

The "vain mathematicians" he speaks of in the passage are probably Ptolemy, John of Sacrobosco, and Campanus of Navarre. In referring to the "blind vision of popular philosophers" Bruno would seem to include Aristotle, Aquinas, and the scholastics, and he does mention Aristotle specifically with contempt as a rigid thinker. But as his thought is developed in the dialogue, we see that in fact he follows Aristotle's teachings on the elements, motion, and logic. He ridicules grammarians in Prudenzio, and the philological philosophers in the Doctors, who examine classical texts for answers to everything.

With regard to Thomas Aquinas, like Campanella Bruno seems to have deeply admired his thought, viewing him and Albert the Great as Christian Magi. Frances Yates notes that a diary kept by the librarian of the Abbey of St. Victor records that Bruno voiced his admiration for Aquinas but disapproved of the subtle distinctions of the scholastics concerning the Sacraments. St. Peter and St. Paul knew nothing of such subtleties, he said. Were these eliminated, the present religious strife could be dissolved.[24]

By the end of the first dialogue Theophil has made it clear that truth in this forthcoming debate can only be learned at the feet of the Nolan. He urges his listeners: "Now watch how vigorous is his philosophy to maintain and defend itself, to unmask vanity and to lay open the fallacies of the Sophists, the blindness of the crowd, and the vulgar philosophy" (71). In the last exchange of the first dialogue Bruno promises through Theophil that

23. Yates (*Bruno*, 239–40) explains that the first line quoted here is the same in the original Italian as the description in Cornelius Agrippa's *De occulta philosophia* of the experience the magus must undergo in order to work his magic in the heavens.

24. Yates, *Bruno*, 230–31.

more of the Nolan's wisdom will be revealed on the next day. The fictional
Theophil's relationship to the Nolan bears a remarkable resemblance to
Salviati's deference to "our friend, the Lincean Academician"—Galileo—in
the *Dialogue*. But the esteem of Bruno for his own wisdom far outstrips
Galileo's.

The Second Dialogue

Theophil begins with a lengthy monologue in which he recounts the circumstances of Sir Fulke Greville's invitation to the Nolan to attend an Ash
Wednesday supper at his home, and the Nolan's subsequent journey there.
Before he agrees to accept the invitation, the Nolan sets forth the rules of the
debate. Although Greville asks the sage to state the reasons for his advocacy
of heliocentrism at the outset, he demurs, saying that he prefers his opponent to first set forth his arguments. Then the rebuttal will focus on the
strength and weakness of the arguments. Explaining the strategy behind
this procedure, Theophil says of the Nolan

that since he takes delight in showing the imbecility of contrary positions on the
basis of the very same principles with which they seem to be proven, it would be of
no small pleasure to him to find persons who would be qualified for such a procedure, and that he would always be prepared and prompt to reply. In such a way one
might so much better see the excellence of the foundations of this philosophy as contrasted with the accepted one, the greater are the opportunities presented for giving
a reply and a clarification. (73)

Surprisingly, Greville professes pleasure at the prospect.

Galileo was to use a similar technique in drawing out Simplicio to give
the Peripatetic's side of an issue. He does so in a much more even-handed
manner, however. Bruno has the Oxford doctors begin with banal quib-
bling, in this way showing them to the worst advantage.

After the arrangements are completed, Theophil reflects on the English
populace. Most of his observations are uncomplimentary, except for a few
short encomia to the queen and some prominent figures of the day: Robert
Dudley, Earl of Leicester, Sir Francis Walsingham, and Sir Philip Sidney.

With lavish detail Theophil then describes the rude and boorish treatment he received at the hands of the Englishmen he encountered on the way
to the dinner and the great difficulties he had in getting there because of the
deplorable condition of the streets and the chicanery of the boatmen who
contracted to deliver them. Poisoning the well has rarely been more imag-
inatively accomplished.

The Later Dialogues

The description of the debate itself begins in the third dialogue. Bruno
frames the discourse with headings giving the number of the proposition in

contention, but the exchanges are simply caricatures of formal dialectics. For example, the first proposition of Nundinio is "Do you understand sir, what we said?" The Oxford doctor is asking whether Bruno understands English. The Nolan does, but disdains to use it.

The crux of the issue is introduced by Nundinio in his second proposition when he claims that Copernicus developed the system simply for "ease of calculation." The Nolan is quick to inform him that he bases his proposition on the falsehoods averred in the preface, which were written by a "presumptuous jackass." Jaki points out that Bruno was ignorant of the identity of the writer, although Osiander was known to be its author long before Kepler published the name in his *Astronomia nova*.[25] Nevertheless, without knowing the name of the "jackass," Bruno had penetrated to the truth of the matter. He cites Copernicus's dedicatory letter to the pope and other passages in the text to refute Nundinio's claim.

Novel Philosophies

The subsequent propositions and their rebuttals give place to a confusing discussion of the infinity of the universe and the animistic force moving the heavenly orbs. At intervals Bruno provides diagrams which are neither adequately drawn nor marked by symbols and so convey little information. Frances Yates suggests these are meant to convey Hermetic truths.[26]

Bruno also uses verbal illustrations, some of which anticipate in content those of Galileo, but again the details are not always clear. For example, Smith describes a thought experiment that adumbrates the ship's mast experiment in the *Dialogue*. Addressing Theophil, he takes up Aristotle's claim in *De caelo* that, if the earth were in motion, a stone thrown up in the air would not come down in a straight line. He continues:

Therefore, given this projection [back] into the earth, it is necessary that with its motion there should come a change in all relations of straightness and obliquity; just as there is a difference between the motion of the ship and the motion of those things that are on the ship which if not true it would follow that when the ship moves across the sea one could never draw something along a straight line from one of its corners to the other, and that it would not be possible for one to make a jump and return with his feet to the point from where he took off. . . . Thus, if from the point D to the point E someone who is inside the ship would throw a stone straight [up], it would return to the bottom along the same line however far the ship moved, provided it was not subject to any pitch and roll. (121–22)

The figure that accompanies the text fails to communicate the intent since the letters referred to are missing.

25. Bruno, *Ash Wednesday Supper*, 95, n. 2.
26. Yates, *Bruno*, 241.

Theophil carries the idea further, explaining the motion of a thrown object on a ship and on land:

From that difference we cannot draw any other explanation except that the things which are affixed to the ship, and belong to it in some such way, move with it; and one of the stones carries with itself the virtue [impetus] of the mover which moves with the ship. The other does not have the said participation. From this it can evidently be seen that the ability to go straight comes not from the point of motion where one starts, nor from the point where one ends, nor from the medium through which one moves, but from the efficiency of the originally impressed virtue [impetus], on which depends the whole difference. (123–24)[27]

This notion of impressed virtue is similar to Galileo's, but the level of heavy contemptuous humor and insult that pervades the dialogues is very different.

Theophil describes the appearance of Nundinio: "it is here that he begins to show his teeth, broaden his jaw, blink his eyes, frown his eyebrows, widen his nostrils and send the croaking of a capon through the pipe of his lungs, so that due to this laughter others may think that he understood it well, that he was right, and that this other said ridiculous things" (114–15). Frulla replies that the Nolan was laughing too. Theophil responds, "This happens to the one who gives sweet-meat to pigs" (115).

Scriptural Texts

The major concern of the fourth dialogue is the biblical passages asserting the opposite of what Copernicus claims. The basic principles advanced by Bruno to reconcile the two are like those followed by the commentators we have already treated, which will again be used by Galileo. The similarities in their hermeneutical methods are not surprising because the authors were all familiar with the principles developed by the Church Fathers. Zuñiga, Foscarini, Kepler, Campanella, and Bruno were each trained in theology, and all but Kepler were in religious orders, where careful instruction in philosophy and theology prepared members to protect the faith.

In his defense of heliocentrism Theophil first offers the argument from irrelevancy, saying that the Scriptures were never intended to furnish us with "demonstrations and speculations about natural things" but instead treat of "the practice concerning moral actions set by laws" (126). Next he draws upon the principle of accommodation to explain scriptural references to celestial things.

Moving to a related point, Bruno notes the importance of genres in interpreting Scripture, something not discussed explicitly by the other exegetes,

27. The bracketed explanatory words throughout these quotations are Jaki's.

although it too is treated by St. Augustine—from whom Bruno seems to have drawn his other two principles. Bruno's expression of it is ingenious:

Smith. It is certainly an appropriate way, when someone intends to present history and to give laws, to speak according to the general understanding and not to be solicitous about points that are irrelevant. He would be a stupid historian who, in treating his material, wanted to introduce words that are considered new and to reform the old ones, so that the reader would be forced to consider and interpret him more as a grammarian than to understand him as a historian. (126)

He goes on to supply an example:

Many times, therefore . . . it is stupid and ignorant to refer to things according to [intrinsic] truth rather than according to occasion and convenience. Just as if the wise man [i.e., the author of Ecclesiastes, who] said "The sun rises and sets, orbits through half a day and moves towards Aquila," would [instead] have said: the earth revolves [rotates] towards the east and leaves behind the setting sun, moves along the two tropics, from that of Cancer toward south and from that of Capricorn toward Aquila [north]; would [not] the listeners have been startled and asked, in what sense does he state the earth to move? what novelties are these? they would in the end have held him for a fool, and he would indeed have been a fool. (127)

Bruno carries the discussion to the creation of the two great lights in Genesis, explaining that the text was not meant to be taken literally. In his consideration of the equality of the earth with these great lights, the text surprisingly adumbrates the passage in *Sidereus nuncius* on the same theme, which also quotes Scripture in that context: "What is lacking to the earth that it should not be a luminary more beautiful and larger than the moon, and that by receiving in the same manner in the body of the Ocean and other mediterranean seas the great splendor of the sun, might it not match as a most shining body the other worlds, called stars, no less than those appear to us as so many lamp-like torches" (128).

In a tactic later to be employed by Galileo, Bruno moves from his arguments for the Scripture's irrelevancy in treating natural philosophy to its use as a proof for the Copernican system. The passage selected by Bruno is different from Galileo's, however, and he interprets it with a mix of Aristotelian and animistic principles. Theophil states that in Job we find the heliocentric philosophy "much favored and preferable." Thus:

In that book of God, one of the personalities, in wishing to describe the provident power of God, says that he keeps the peace among his eminent ones, that is, sublime sons, that are the stars, the Gods, of which some are fire, some are water (as we say some are suns, some are earths), and that these are in harmony; for however contrary they may be, nevertheless one lives, feeds and grows through the other; meanwhile, they do not mix confusedly together, but one moves around the other at certain distances. Thus, the universe becomes differentiated into fire and water, which are subject to two primary, formal and active principles, the cold and the hot. The bodies that breathe hot are the suns, because they are luminous and warm in themselves; the bodies that breath cold are the earths. (129)

Bruno expresses his exasperation with the conservative theologians who oppose Copernicus on scriptural grounds, and does so much more freely than Galileo was to do. Smith, the intellectual layman, serves as his spokesman:

I am, for sure, very much moved by the authority of the Book of Job and of Moses, and I can readily acquiesce in these real[istic] sentiments [views] much rather than in metaphorical or abstract ones, were it not for some parrots of Aristotle, of Plato, and of Averroes,—from whose philosophy they were promoted to the rank of theologians,—who say that these meanings are metaphorical, and thus, with the aid or their metaphors, they let these meanings signify anything they want, through their zeal for that philosophy in which they were brought up. (130)

Theophil acknowledges the truth of Smith's observation, saying that the practice of metaphorical interpretation used by Jews and Mohammadens and sectarians breeds "innumerable other most contrary and different sects, all of which know how to find in the Scripture that proposition which pleases them and suits them better."

Concluding this discussion of the Scriptures, Theophil calls for humane criticism:

One is not to fear the censure of honorable minds, of truly religious and also naturally well-meaning men, who are friends of courteous conversation and of good doctrine. For upon having these things well considered they find that this philosophy not only contains the truth, but also favors the [true] religion to a greater degree than does any other kind of philosophy: like those philosophies which posit the world as finite, the effect and efficiency of divine power as finite, the intelligences and intelligent natures as being merely eight or ten; the substance of things as corruptible, the soul as mortal, as if it rather consisted of an accidental disposition, of an effect of complexion and of dissolvable temperament and harmony, the execution of divine justice over human actions as per consequence nothing, the knowledge of particular things as being removed from the primary and universal cause. (131)

In the last part of the passage Bruno inveighs against an Averroist version of Aristotle that would remove the possibility of immortality and of moral responsibility.[28] Averroism is particularly repugnant to Bruno also because it assumes the finitude of the world and this would limit his own pantheistic God who fills the world infinitely.

28. Jaki, *Ash Wednesday Supper*, 131, n. 18.

Bruno's Cosmology

In the remaining part of the fourth dialogue, Theophil describes the Nolan's cosmology. He proposes that the universe is one and infinite, containing one heaven, an "immense ethereal region." The globes in it are some great luminous "animals," and tens of thousands of other animals roam throughout space. These communicate with each other by taking part in their movements. As Yates points out, precedents have been found for his ideas in Nicholas of Cusa, and it has even been suggested that the work of Thomas Digges may have influenced him. Lovejoy has convinced us, she notes, that Bruno's belief stems from the "principle of plenitude, that an infinite cause, God, must have an infinite effect and there can be no limit to his creative power."[29]

Although many have credited Bruno with unusual scientific insight, Yates finds the major source of his inspiration in Hermeticism. She writes that "Bruno's acceptance of Copernican earth movement was based on magical and vitalistic grounds, and that, not only the planets, but also the innumerable worlds of his infinite universe move through space like great animals, animated by the divine life."[30]

While the Nolan was discoursing, Doctor Torquato shouted "*Ad rem, ad rem, ad rem*," that is, get to the point. The Nolan, who always remains the epitome of courtesy in the debate, replies that he does not wish to argue with him, but that these are in fact the points: "*ista sunt res, res, res.*"

The remarks of the interlocutors who follow amply illustrate Bruno's vitriolic wit, and also the reason his dialogue was so offensive to the English:

> *Smi.* Thinking that he was in the midst of blockheads and idiots, that jackass [Torquato] thought that they would let this ad rem of his pass for an argument, and for a proof, and that he would satisfy the whole gathering with a mere tinkle of his golden chain.
>
> *The.* Listen further. While all stayed there waiting for that coveted argument, Doctor Torquato, now turning to his dining companions, draws from the depth of his self-sufficiency and throws in their faces the Erasmian adage *Anticiram navigat* [He is sailing toward Anticyra].
>
> *Smi.* A jackass could not have spoken better, and one cannot [indeed] hear other words when busy with jackasses. (135)

The fifth and final dialogue finds the Nolan explaining the reality behind the appearances of the celestial regions. His explanation is a curious mixture of

---

29. Yates, *Bruno*, 244. She cites A. O. Lovejoy, *The Great Chain of Being* (Cambridge, Mass.: Harvard University Press, 1942).

30. Ibid.

Aristotelian physics, Hermeticism, and cabalism. In the course of this discussion he interprets the word *ethera,* finding its Greek etymological roots in the term "runners." Then he takes that origin to illuminate the movement of the bodies coursing through space. Again he notes their animation, saying that they move by their natures, not through external forces. He goes on to give an Aristotelian account of the elements, gravity and levity, and the concept of natural place. Smith carries on these ideas to argue that celestial globes in their natural place do not have gravity or levity, so that the earth is not heavier than is the sun in its natural place. For this reason, he says, the teaching that the earth's immobility causes it to be "heavy, dense and cold" is erroneous (153–54).

After this there ensues a discussion of the "true cause" of the range and changes in temperature on earth; the earth's circular motion around the sun is proposed, in contrast to Aristotle's ambiguous explanation. Theophil then assigns four motions to the earth, explaining that these emanate from final causes, among them the necessity of giving light and darkness by diurnal revolution, providing life to things on the earth by inclining its center uppermost, renewing the hemispheres by tilting, and giving different attitudes to the poles. These gyrations and twists are appropriate to the animation of the creature. In expanding upon them he points out that these motions are not obvious to those who have worked so tirelessly upon them, for they are not "regular and capable of [being worked upon] by the geometrical file" (165). Obviously, though, they do yield to those who like Bruno have privileged insight.

Prudenzio is ordered to devise a fitting epilogue to the "tetralogue." Accordingly he summons the divine spirits to guard the Nolan from "vile, ignoble, barbarous and unworthy conversations" that might make him appear "as a satirical Momus among the gods, and as a misanthrop Timon among men" (168). He turns to advise the Doctors Nundinio and Torquato to "recall that boorish and impolite teacher of yours . . . who did your training, and that other chief jackass and ignoramus who taught you how to dispute so that they may repay you the futile expenditures . . . and the brain they let you lose" (169).

In reflecting on the *Ash Wednesday Supper,* one cannot but conclude, as those earlier Oxford witnesses did, that the real intent of the discourse was to publicize Bruno's new Gospel of Hermeticism.[31] But one also has to conclude that at least Bruno was convinced that the Copernican system provided evidence for what he thought was the larger picture, a picture that thus far had escaped scholars. That picture extended the universe far beyond

31. For the place of the *Cena* in the missionary effort for hermeticism see Yates, *Bruno,* ch. 13.

Copernicus's conception of it, and it included a metaphysical analysis foreign to the case the astronomer presented. It did not conflict with Scripture if it were interpreted according to proper principles.

The reception of his discourse at Oxford, where it is said that he quoted long passages from Ficino, passing them off as his own, was so contemptuously hostile that his pride demanded a devastating retaliation.[32] The coarse repetitive sarcasm with which he garnishes his accounts, however, worked to his disadvantage, for it fell far below the brilliance of his vivid descriptions and the insights he occasionally offered.

In another work Bruno describes his conception of the cosmos within the larger Hermetic frame.[33] He sees it evolving to a purer state, where the animated virtuous bodies will replace the signs of vicious ones that exist with them in space. Through ethical magic, the beast images can be driven out of heaven. Unfortunately, the Hermetic temple he labored so copiously to erect soon fell in ruins and too few workmen came to refurbish it.

The *Ash Wednesday Supper* is a strange conglomerate of dialectics, rhetoric, philosophy, and Hermeticism. The reasoning is sharp enough when it remains within the dialectical mode. The rhetoric is by contrast clumsy and unpersuasive. The content itself, especially in its hermeneutics, is as penetrating and insightful as any of the apologetical works we have studied.

Did Galileo read Bruno's work and find some inspiration for his *Dialogue* in it? The evidence would suggest that he did. The characters as spokesmen for similar views, the interplay among them, the chronological division of the dialogue by days, occasional similarities in passages, the use of diagrams and thought experiments, all remind us of the *Dialogue*—but in science as in literature, parallels do not necessitate a common source.

32. Robert McNulty discovered an eye-witness account by George Abbot, Archbishop of Canterbury, who was at Balliol at the time of Bruno's visit, and describes Bruno's appearance, the effect on Bruno's audience of his discourse, and his use of Ficino's *De vita coelitus comparanda;* see his "Bruno at Oxford," *Renaissance News* 13 (1968): 300–305. Frances Yates quotes McNulty's discovery in *Bruno,* 208–9.

33. Giordano Bruno, *Spaccio della bestia trionfante* (1584).

CHAPTER SEVEN

Galileo's Appeal to the Church

Of all the exegetical works in this part of our study, Galileo's is by far the most impressive. Not only does it offer valid hermeneutical principles for reinterpreting Scripture in order to reconcile its apparent contradictions of the assumptions of heliocentrism, it provides firsthand physical evidence that there is a need to apply those principles. The theological content of the *Letter to Madame Christina of Lorraine, Grand Duchess of Tuscany,* which Galileo completed in 1615, resembles closely that of the writings of the theologians and religious philosophers we have discussed.

The letter differs from the other pieces in its more extensive and more skillful use of rhetorical strategies. Except for Bruno, whose *Cena* properly belongs to a literary genre, the other exegetes argue dialectically for the most part, while Galileo argues rhetorically more than dialectically, and he does so in a unique manner. He touches upon the physical evidence needed for a complete demonstration, speaks often of the importance of necessary demonstrations, combines these assertions with dialectical arguments and rhetorical appeals to create a moving petition for ecclesiastical tolerance of the new philosophical system. It is obvious from the polished text and the care with which he developed it—writing it took more than a year—that he intended to publish the letter; however, the climate of opinion prevented its appearance in print until it was published in Strasbourg in a small edition in 1636. Thomas Salusbury's translation, published in 1661, made it available to an English public.[1]

The Historical Background

The *Letter* grew out of a conversation at breakfast at the court of the Medici in December 1613, when the mother of Galileo's patron Cosimo II de' Medici voiced her concern to Cosimo and others present about the implications for the meaning of Scriptures posed by the Copernican system and Galileo's recent discoveries with the telescope.[2] Among the guests that

---

1. Galileo may have been unaware that the translation was being prepared by Elio Diodati in France. Stillman Drake discusses the circumstances of its publication, *Discoveries,* 171, n. 33.

2. Castelli's letter is in *Opere* 11:605–6. Drake translates part of it in *Discoveries,* 151–52.

morning was Benedetto Castelli, the Benedictine friend of Galileo who now held Galileo's former postion as professor of mathematics at the University of Pisa. The monk was an esteemed figure at the court and the Grand Duchess's confessor. In the course of describing the observations he had made with his own telescope, Galileo's discovery of the orbits of the planets of Jupiter, and the inferences that could be made from these, Castelli was privately and quietly contradicted by another guest, Cosimo Biscaglia, also a Pisan professor, who had the ear of the Grand Duchess. He told her that although the reality of what was seen with the telescope could not be questioned, the movement of the earth was not proved thereby and, moreover, was in conflict with the Scriptures. After breakfast the Grand Duchess questioned Castelli about the scriptural difficulties; she was particularly concerned about the text from Joshua where the prophet commands the sun to stand still. The Benedictine tried to answer that and other problems with what he thought were convincing theological arguments. Nevertheless, the Duchess was not reassured, although the Grand Duke and others of the family appeared to be. Castelli wrote to Galileo a few days later, giving him an account of the discussion and the arguments he had advanced.[3]

Upon reading his friend's letter, Galileo decided to take on the issue himself and he quickly drafted a well-argued reply within a week, addressing his response to Castelli. The letter was the germ of his more elaborate *Letter to Christina*, which expanded the original arguments and retained much of its language.

Castelli was greatly pleased by the letter and circulated it among his friends. In this way it came into the hands of a Dominican, Father Niccolò Lorini, who, troubled by its contents, sent it on to the Holy Office for further investigation.

Lorini also seems to have shown a copy of the letter to another Dominican, Tommaso Caccini, who, on 21 December 1614—one year after Galileo's letter—preached a sermon in which he condemned Galileo and mathematicians in general, whether as a response to the letter or not is unclear. The passage set for explication that day was particularly appropriate to the Copernican controversy; it was the problematic episode in Joshua. As a priest at the imposing church attached to the Convent of Santa Maria Novella in Florence, Caccini had developed a series of sermons on the book of Joshua. The miracle in which Joshua commands the sun to stand still natu-

---

3. See Drake's account of the events surrounding the composition of the two letters in *Galileo at Work*, 222–39 and *Discoveries*, 148–71. James Langford supplies another view in *Galileo, Science and the Church* (Ann Arbor: University of Michigan Press, 1971), 50–78. An earlier and very careful recapitulation of the events leading up to the trial is found in Karl von Gebler, *Galileo Galilei and the Roman Curia*, trans. Mrs. George Sturge (London, 1879).

rally became a moot text. Legend has it that Caccini used as a theme the text from Acts: "Ye men of Galilee [with an obvious pun on Galileo], why stand you gazing up into heaven?" Drake suggests that the sermon was the beginning of a vigorous open campaign against Galileo in Florence.[4]

The opposition to Galileo expressed by these two priests did not reflect the opinion of all Dominicans. A Preacher-General of the Order, Luigi Maraffi, wrote to Galileo apologizing for the excesses of Caccini.[5] When Galileo heard about Lorini's complaint to the Holy Office, he asked Castelli to return the letter. In case the text had been doctored in Lorini's copy, he could send an accurate copy to Rome and ask his friend Piero Dini to circulate it. Dini, a nephew of Cardinal Bandini, held an office at the Vatican and eventually became an archbishop in 1621. Galileo sent him a fair copy in mid-February 1615.

The discussions of hermeneutics were now no longer academic, and they soon became crucial as the deliberations about the legitimacy of the Copernican thesis grew more heated during the year. Two weeks after Galileo sent his recopied letter to Dini, another friend, Giovanni Ciampoli, wrote to Galileo of a conversation with Cardinal Maffeo Barberini. Dini and Ciampoli had been working on Galileo's behalf to calm the suspicions being raised in Rome about him. Ciampoli mentioned at the beginning of his account that Barberini always expressed great affection for him. But, he said, the cardinal voiced the hope he would exercise great caution in arguing about the Copernican system so as not to exceed the disciplinary boundaries of mathematics and physics and move into that of theology. Theologians are jealous of their prerogative of interpreting the Scripture, Barberini warned, and would not look kindly on the introduction of "novelties" even by a man of "admirable ingenuity."[6]

Then on 12 March Castelli wrote to Galileo telling him of a visit from the Archbishop of Pisa, who had come to Castelli hoping to obtain the original of the letter Galileo had written to him regarding science and the Scriptures. But the monk explained to the prelate that he no longer had it, having sent it back to Galileo.[7] In the course of his visit the archbishop warned Castelli to give up the opinions he had expressed about the motion of the earth, saying that these "in addition to being foolish, were dangerous, scandalous, and rash."[8] The archbishop eventually became angry and told Castelli that a public condemnation of the opinions would soon be made. On leaving he urged Castelli to

4. Drake, *Galileo at Work*, 238–39.
5. Maraffi to Galileo in *Opere* 12:127–28.
6. *Opere* 12:146.52–54.
7. Ibid. 153–54.
8. Ibid. 154.8.

ask Galileo for a copy of the letter. Castelli did so, pleading with Galileo to revise the letter as he had said he wanted to do and to send it quickly in order to quell some of the storm that had begun to rage about the issues.

Meanwhile, Dini wrote to Galileo on 7 March 1615 to describe an interview he had had with Cardinal Bellarmine. The cardinal told him that there was no question of banning Copernicus's book, indicating that the only action being contemplated was the introduction of glosses in the manuscript to the effect that the work offered only an hypothesis. He also mentioned that if Galileo had written something about the interpretation of Scripture in light of the Copernican thesis he would be happy to look at it.

In the interim Foscarini's *Letter* was received by Cardinal Bellarmine. He replied to the Carmelite on 12 April in the letter discussed in chapter 5. Foscarini's book had also been sent to Galileo, relayed to him by Prince Cesi, the founder of the Lincean Academy. Fortunately Bellarmine's response to Foscarini was conveyed to Galileo by his friend Dini, who reported in a letter of March 7 that the explanation for Bellarmine's congratulatory comment in his letter to Foscarini, namely, that he was glad to see that both he and Galileo had limited themselves to speaking *ex suppositione e non assolutamente,* was probably the result of his (Dini's) having given him a copy of the controversial letter to Castelli.

Dini was pleased with what he perceived to be the effect of both Foscarini's and Galileo's letters, and on 16 May 1615 he wrote reassuringly about the atmosphere in Rome. He urged Galileo at the same time to polish his revision of the Castelli letter in order to bring the best arguments to bear on the subject, arguments well-founded on Scripture and mathematics. Galileo responded that ill-health had prevented him from further progress on the promised letter. He notes with chagrin that he would like to write against those philosophers who have impugned the teachings of Copernicus, but he says that he has been ordered not to enter into discussion of the Scriptures. The implication here is that although he could not publicly introduce his own interpretations, the private letter he has been planning to finish would not come under that ban if it were not published.

Galileo goes on to say that he thinks the best way of showing that "the position of Copernicus is not contrary to Scripture would be to show by a thousand proofs that it is true, that the contrary cannot be supported in any manner; since it is impossible that two truths can be contradictory, then the Scripture must be ultimately in accord."[9] Yet he wonders how it would be possible even then to convince those philosophers who "show themselves incapable of the simplest and easiest arguments" and put their trust in

9. Ibid. 184.20–23.

"meaningless propositions."[10] Nevertheless, he says he would try even that if he had the opportunity to use "his tongue instead of his pen." If he regains his health he hopes to come to Rome and demonstrate his love for the Church. He thinks that it would be wrong to maintain that Copernicus had only an hypothesis in mind, and that he would not want eminent men to think that he agrees with this view.

The Methodological Background

About this same time Galileo wrote a series of notes, entitled by Favaro "Considerations on the Opinion of Copernicus," which reveal the basic philosophical foundation of the arguments he makes in support of *De revolutionibus* and for a new interpretation of the contradictory passages of Scripture. The "Considerations" is in three parts and undated, apparently composed over a number of months.[11] The first part of the notes concern the prefaces and text of *De revolutionibus* and the general question of whether Copernicus was arguing ex suppositione merely to "save the appearances" or whether he believed his thesis dealt with physical reality. The last two parts take up the difficulties of scriptural interpretation Galileo also discusses in the *Letter to Christina;* the final section seems to be composed in direct response to the letter Bellarmine sent to Foscarini.

The "Considerations" is important to an understanding of the argument of the *Letter to Christina* for it takes up the question of proof in detail. Moreover, what Galileo has to say on the subject highlights well the issues in our discussion of the nature of scientific and rhetorical arguments. In the first part of the notes, where Galileo is concerned with the nature of Copernicus's thesis, he speaks of the proofs of astronomers in general:

Astronomers have so far made two sorts of suppositions [*supposizioni*]: some are primary and pertain to the absolute truth of nature; others are secondary and are imagined in order to account for the appearances of stellar motions, which appearances seem not to agree with the primary and true assumptions. For example, before trying to account for the appearances, acting not as a pure astronomer but as a pure philosopher, Ptolemy supposes [*suppone*], indeed he takes from philosophers, that celestial movements are all circular and regular, namely uniform; that heaven has a spherical shape; that the earth is at the center of the celestial sphere, is spherical, motionless, etc. (75)

In other words, Ptolemy has argued ex suppositione, that is, from the suppositions of circular, regular, and uniform motions, and from other sup-

10. Ibid. 184.26–28.
11. The text is included in *Opere* 5:351–76. Maurice A. Finocchiaro has translated the notes in full in *Galileo Affair*, 70–86. References in my discussion refer to this text. Drake includes a partial translation in *Discoveries*, 167–71.

positions accepted by philosophers as first principles in astronomy. Ptolemy has argued from these primary suppositions and looks for the causes of the observed regularities.

Galileo continues to trace Ptolemy's chain of reasoning:

Turning then to the inequalities we see in planetary movements and distances, which seem to clash with the primary physical suppositions [*supposizioni naturali*] already established, he goes on to another sort of supposition [*altra sorte di supposizioni*]; these aim to identify the reasons [i.e., to supply the causes], without changing the primary ones, why there is such a clear and sensible inequality in the movements of planets and in their approaching and their moving away from the earth. To do this he introduces some motions that are still circular, but around centers other than the earth's, tracing eccentric and epicyclic circles. This secondary supposition is the one of which it could be said that the astronomer supposes it to facilitate his computations, without committing himself to maintaining that it is true in reality and in nature. (75–76)

Although Galileo does not say so here, the secondary type of supposition is that which would be also termed a hypothesis in dialectical reasoning, the sort explained in the first chapter. In this sense it is what stands under the thesis, the primary supposition as Galileo calls it in this context, since it is just that, a suppositional thesis. The hypothesis, treating as it does of causes of the observed phenomena, the epicycles and eccentrics, then introduces the middle term for his argument. The nature of this second kind of supposition permits it to be held in the mind without being itself proved at this point. In continuing the chain of syllogisms, that hypothesis could be entertained as a major premise in an argument. Its use as a supposition in this way would simply be an argument to "save the appearances." Wallace has analyzed Galileo's recognition of the distinctive and complex uses of ex suppositione and notes that "he was unequivocally distinguishing between *suppositiones* that are fictive and merely calculational and others that are true in nature and applicable to the physical world."[12]

Galileo goes on to contrast Ptlolemy's supposition with what Copernicus would supply as a "hypothesis" (*ipotesi*):

Let us now see in what kind of hypothesis Copernicus places the earth's motion and sun's stability. There is no doubt whatever, if we reflect carefully, that he places them among the primary and necessary suppositions about nature [*le posizioni prime e necessarie in natura*]. . . . In fact, to say that he makes this supposition [*questa supposizione*] to facilitate astronomical calculations is so false that instead we can see him, when he comes to these calculations, leaving this supposition [*questa posizione*] and returning to the old one, the latter being more readily and easily understood and still very quick even in computations. (76)

---

12. Wallace, *Galileo and his Sources*, 340–347.

Thus Copernicus posits a primary supposition different from Ptolemy's, but it has the same status: it is taken for granted. The later astronomer argues from the earth's motion and the sun's stability, whereas Ptolemy argues from circular, regular, uniform motion of the heavenly bodies and the earth's stability. This supposition of Copernicus, like the first kind Ptolemy uses, is not a hypothesis in the scholastic sense but a primary supposition, a thesis. One would conclude then from Galileo's interpretation that Copernicus did not argue hypothetically. He argued from the primary supposition that the physical reality was the earth's movement, etc. By this means he removed most of the inequalities seen and thus solved the problem presented by the apparent irregularities in the heavenly appearances. He had little need to posit the elaborate secondary kind of supposition or hypothesis of epicycles posited by Ptolemy. (Copernicus did make some exceptions, however, when forced by his assumption of circular motion to explain several phenomena by eccentrics and epicycles.)

Another aspect of the problem of scientific proof implicit in the argument is important also, one related to Bellarmine's criteria regarding demonstrations and scriptural interpretation. In the passage noted above, Galileo does not mention necessary demonstration, although the terminology he uses concerning "primary and necessary suppositions" that treat of the "absolute truth in nature" would indicate that these are necessary demonstrations. When he speaks of secondary suppositions, on the other hand, he hints at dialectical argument. Either dialectical reasoning or necessary demonstrations presumably lie behind the basic suppositions of astronomy. Ptolemy's primary suppositions, which Galileo refers to here, were taken from philosophy and thus were the result of elaborate arguments accepted as true. He says that Copernicus's primary supposition was like them but different, namely, that the movement of the earth is real. The source of this supposition is not noted by Galileo in this context, but it has been supplied in an earlier passage in the "Considerations," as follows.

Galileo explains that Copernicus arrived at this supposition, first, by extrapolating from a "large number of physically true and real observations of the motions of the stars (and without this knowledge it is wholly impossible to solve the problem)" (74). Kepler has made this point also, as we have noted, in his *Mysterium cosmographicum*. In addition, Copernicus "through long observations, favorable results, and very firm demonstrations [*fermissime dimostrazioni*] . . . found it so consonant with the harmony of the world that he became completely certain of its truth." He adds: "Hence this position is not introduced to satisfy the pure astronomer, but to satisfy the necessity of nature" (74).

Galileo did not outline in the "Considerations" the physical proofs he says stand behind this supposition. One can entertain several different explanations for his silence: because he knew the proofs were many and complex; because he was aware that they were insufficient to prove the case completely; or because he was simply working through the problem and was not altogether sure himself.

The ambiguity persists throughout the letter he wrote to the Grand Duchess. The omission there seems not to have been engendered by uncertainty regarding the canons of proof needed, but rather by the extent of the proofs available to him at the time.

In the "Considerations" Galileo next turns to the exegetical problems of the scriptural texts that seem to contradict the heliocentric view (80–83). He first speaks of the impossibility of two truths contradicting each other, a point he developed at greater length in both the letter to Castelli and that to the Grand Duchess. He emphasizes that nature's truths are based on "sense experiences, accurate observations, and necessary demonstrations" and notes that if scriptural truth seems to say the opposite we must have misunderstood its meaning. Galileo does deal in the "Considerations" with the difficulty of the testimony of the Church Fathers on the meaning of the Scriptures, and he repeats this in the *Letter to Christina*. The Fathers are not mentioned in Galileo's letter to Castelli, which indicates that the "Considerations" were written after the Castelli letter and before the letter to the Grand Duchess. His concern was prompted, no doubt, by Bellarmine's response to Foscarini.

In the last section of the letter to the Grand Duchess Galileo considers the very points Bellarmine made in his letter to Foscarini, sometimes using the same language as Bellarmine, obviously intending a direct rebuttal of Bellarmine's letter.

With respect to Bellarmine's belief that Copernicus spoke "hypothetically," Galileo notes first that the preface to *De revolutionibus* was certainly not written by Copernicus, and he attributes it to the bookseller who wanted to promote sales. In this Galileo unaccountably omits what Kepler and others have said about its composition by Osiander.

He comments first that Copernicus gave up the Ptolemaic system not simply because of the epicycles and eccentrics mentioned by the cardinal, but because of other absurdities (83). Next he observes that true philosophers should not "get irritated" in finding they are wrong, nor should theologians. "Moreover certain theologians who are not astronomers should be careful about falsifying Scripture by wanting to interpret it as opposed to propositions which may be true and demonstrable" (83). It is significant that this sentiment is repeated more strongly in the following note, made years later in Galileo's hand on the fly-leaf of his own copy of the *Dialogue*: "Take care, theologians, that in wishing to make matters of faith of the propositions attendant on the motion and stability of the sun and the earth, in time you probably risk the danger of condemning for heresy those who assert the earth stands firm and the sun moves; in time, I say, when sensately and necessarily it will be demonstrated that the earth moves and the sun stands still."[13]

In the third point he suggests that "our own ignorance" may cause our difficulties in determining the true meaning of Scripture and not the fact that these are at odds with "demonstrated truth."

The fourth and fifth notes argue from the same principles expressed in the letter to the Grand Duchess: irrelevancy and accommodation. Galileo answers Bellarmine regarding the directive of the Council of Trent to the effect that nothing should be accepted that contradicts the consensus of the holy Fathers, but that the Council had in mind matters of faith and morals, not those of science.

His answer in the sixth point to Bellarmine's assertion that a demonstration is the prerequisite for reinterpretation is important for the illumination it sheds on the stance Galileo takes in the final version of his *Letter*. He agrees that it is "prudent" not to believe that a proof of the earth's motion exists unless one has been given, "nor do we ask that anyone believe such a thing without a demonstration." But, he points out, the Church should be evenhanded. If the reasons given "are not more than ninety percent right, they may be dismissed; but if all that is produced by philosophers and astronomers on the opposite side is shown to be mostly false and wholly inconsequential," then what the supporters offer should not be "disparaged, nor deemed paradoxical, so as to think that it could never be clearly proved" (85).

He concedes in the next argument that to prove that the earth really moves is different from proving that the motion "saves the appearances." But he says it is also true that the received system cannot supply the reasons for these appearances.

The two notes that follow take up Bellarmine's concern that the new system may cause the authoritative teaching of the Fathers to be doubted and along with it the literal meaning of the writings of Solomon and Moses as well. Galileo says that there is no desire to overthrow the Fathers but only to investigate the truth, and he reiterates the point that these writers accommodate their language to the common man.

Galileo's final response to Bellarmine's letter concerns the cardinal's rebuttal of one of Foscarini's arguments. In effect Bellarmine says to

13. This is recorded in my "The Rhetoric of Proof in Galileo's Writings," in W. A. Wallace, ed., *Reinterpreting Galileo* (Washington, D.C.: The Catholic University of America Press, 1986), 203. As I stated there, other comments in the margins that look like planning for revisions are also in Galileo's hand. The copy is in the Bibliotheca Seminarii in Padua. Drake has included the note on the fly leaf of his translation of the *Dialogue*.

Foscarini: if you maintain that Solomon only spoke in terms of the appearances and argue that we are also doing so in saying the sun moves, and if we then compare it to our saying aboard a ship that the shoreline recedes from us when in effect we are leaving it, you have not made a valid analogy. We can correct our knowledge by seeing that this is really not true; but we have no such correction to our experience that the sun moves. Galileo, in an ingenious reply, says that it is true that our observations concerning the shore can be corrected and that this is done by the observation of others who see that the ship moves away. In the same way, if we were able to stand on the sun or another planet and then on the earth, we could verify that the earth moves. But when we must look only from the earth to the sun, we will always think that the sun moves. He says the best analogy would be to compare the earth and the sun to two ships that pass on the ocean. When we are on one and look at the other, we think that it moves and we are still (86).

Many of the points made here are retained and expanded in the *Letter to Christina*, making it obvious that Galileo wished not only to revise his earlier letter to Castelli but also to respond Cardinal Bellarmine. Since Bellarmine discounts the arguments offered in support of Copernicus, and, seemingly, the inferences made by Galileo in the *Sidereus nuncius,* and since he makes demonstration the sole grounds for accepting the thesis, Galileo knew he would have to address that problem head on. A tentative defense of the Copernican thesis was not officially constrained at this time, even though it might overturn principles of the reigning philosophy. Constraints would only be imposed if what was conjectured was interdicted by the authority of Scripture or by theology. Bellarmine, as a Jesuit and a former teacher of natural philosophy and astronomy himself, recognized the value of rational argument as an avenue to truth, but as a prelate and a cardinal he also had to accord primacy to the Scriptures as revelation of the truth. A new interpretation of Scripture, especially during the troubled times of the Reformation, was a delicate art, and only those trained by the Church had full permission to attempt it. Were revisions deemed necessary, a lengthy process of intensive review would even then be needed to substantiate changes.

The *Letter to Madame Christina of Lorraine, Grand Duchess of Tuscany*

The *Letter* itself is a formal composition, nearly a treatise, not quite within the genre of *litterae familiares* as were Galileo's *Letters on Sunspots,* but more closely resembling the formal letter.[14] It is written in the vernacular, the lan-

14. I have previously analyzed the letter in "Galileo's *Letter to Christina:* Some Rhetorical Considerations," *Renaissance Quarterly* 36:2 (Winter 1983): 547–76, and in the "Rhetoric of

guage Galileo preferred and a suitable choice for the Grand Duchess. She was a laywoman, not a scholar, but certainly in 1615 skilled in the language of her husband, having married Ferdinando I de' Medici in 1589. After his death in 1609 she remained at court, retaining her title and her interest in court life. She seems to have been fond of Galileo, probably getting to know him when he became her son's tutor in mathematics in 1605. It was to her son, Cosimo II, that Galileo dedicated the *Siderius nuncius*.

The vernacular was the best choice also for reaching the audience beyond the Grand Duchess, from whom Galileo hoped to gain, if not approval, at least sufferance for Copernicus's book. Since Cardinals Bellarmine and Barberini had warned against invading the province of theologians, a treatise in Latin might have been perceived as a formal entry into the field. And a letter, too, was a good choice. A private letter to his patroness, in Italian on a topic of conversation she had raised, would not look as if he had deliberately ignored that warning. His decision not to publish the letter at that time and simply to circulate it in manuscript form must also have been prompted by the admonitions.

The Audience of the Letter

Galileo seems to have arrived at the decision to address the letter to the Grand Duchess in stages. The earlier version was addressed to Castelli, and the revised version seems also to have been addressed to him, for it begins "Paternità." Modifications in Galileo's hand direct it instead to "Her Most Serene Highness" (*Sua Altezza Serenissima*).[15] The decision to shift the titular audience from Castelli to Christina might have been another way of emphasizing the informal nature of the discourse, so that it could be "overheard," as it were, by the primary audience in the shadows. From a rhetorical standpoint the titular audience was particularly apropos. The Grand Duchess had shown herself to be interested in the topic and desirous of enlightenment on subjects beyond her ken: philosophy, mathematics, and theology. She was also devout. In addressing such a personage Galileo would not have to be embarrassed at starting at ground level to build his argument.

Proof." Much of the substance is repeated here. I have used the translation of Drake in *Discoveries*, 175–216, as the basis for my paraphrase, but have emended the wording in the quotations when his may diverge from some of the terminology used in the original; citations appearing in my discussion of the letter without other appellation are to the emended translation. The instances in which I use the original are indicated by citations to *Opere*, where the text is that reprinted in *Opere* 5:367–70. A more recent translation is in Finocchiaro, *Galileo Affair*, 87–118; while it may be more precise in places, I have preferred Drake's rendering because I think he more nearly captures Galileo's style.

15. Favaro discusses the earlier draft of the *Letter* found in Codex Volpicelliano, *Opere* 5:274–75.

He need not suppose a reader more familiar with theology than himself, as he would were he to address Dini or Castelli. In this way too he might hope to reach a much wider audience than if he were to direct his discourse to either of them.

Certainly there is no doubt that Galileo intended a much larger audience for this work. The entreaties of Castelli and Dini to hurry and complete it show that on its receipt they expected to circulate it to an audience of influential prelates and scholars. Its underlying purpose was to dissuade the religious authorities from condemning Copernicanism. They were actually his primary audience, and he appeals to them in the implicit *petitiones* he interjects at various points in the letter. But by directing his discourse to the Grand Duchess, he could address also a general lay public, a kind of secondary audience that contained persons like Marc Welser, or Prince Cesi, people of some political power, and secular mathematicians and philosophers like himself. That these considerations are important to him is evidenced by the thrust of his arguments, which seem to be directed to a larger appreciative readership of friends. In this regard, of course, Galileo followed the general practice of humanists of his day, who almost always had more than the titular audience in mind for their elaborate letters. Interestingly enough, an even larger audience of modern readers find this letter quite convincing, as is shown by the reprinting of the letter in anthologies and by the enthusiastic response to it by Galilleo scholars such as Stillman Drake and Giorgio de Santillana.[16]

Perhaps one of the reasons the letter failed in its purpose to prevent the ban lies in Galileo's focus on Christina as the titular audience and on a secondary audience of laymen sympathetic to him, instead of the primary shadow audience he really needed to move. In that audience were powerful Church authorities who were or had as their consultants scientists and theologians. Galileo could not really address the theologians themselves, for he had been ordered not to try. The scientists he needed to convince were, of course, those who did not agree with him. In effect, he compounded an already difficult task by attempting to persuade a public so different in terms of familiarity with the subject and so disparate in their attitudes towards the matter he discusses.

The letter is quite long, thirty-nine pages of printed text in the National Edition, as compared with seven pages in the original letter to Castelli in the same collection. The traditional captatio benevolentiae and narratio found

16. Drake's attitude is apparent in his introduction to the letter in *Discoveries*, where he presents it as a valiant and uncompromising effort to describe the "proper relation of science to religion" (145, cf. 165). Santillana places the letter on a plane with Milton's *Aeropagitica* in his well-known work on the trial, *The Crime of Galileo* (Chicago: University of Chicago Press, 1955), 96–98.

in the letter exhibit, appropriately, the ethos of the writer. Galileo is straight-forward and logical, a style that suits well the image of an earnest, devout, yet embattled philosopher. He projects himself as a man of good will who seeks only to disclose the truth. Still, the tone of the emotional appeals he introduces on occasion seems to undercut the ethical appeal, at least for his shadow audience. In the captatio benevolentiae, where one would expect him to seek the sympathy of the audience, he mentions that he has been un-fairly treated by "no small number of professors" (175). These men, he says, appear to be upset because what he has discovered in the heavens has con-tradicted traditional views. It is as if they believe "I had placed these things in the sky with my own hands in order to upset nature and overturn the sciences." He goes on to relate that "the increase of known truths stimulates the investigation, establishment, and growth of the arts; not their diminu-tion or destruction" (175). The edge of ridicule and impatience in his voice establishes at once the stance he is to maintain throughout. This tone might be expected to arouse a sympathetic response in the Duchess, who would not want to see her resident philosopher insulted, and also from philoso-phers with views similar to his, but he could not expect his opponents to be placed in a receptive mood for what was to follow. And what of the primary audience whose minds were not yet made up? Cardinal Bellarmine and other theologians were conservative in the original sense of the term. They were primarily interested in conserving the teachings of the Church, and these new theories were indeed revolutionary. Moreover their implications threatened traditional wisdom regarding the cosmos and man's place in it.

Galileo's Adversaries

With a deprecating tone Galileo effectively marks off a group of philoso-phers and theologians as adversaries whose faults he proceeds to define in the narration. They are, he says, men determined in "hypocritical zeal" to preserve at all costs what they believe, rather than admit what is obvious to their eyes (179). Instead they go about invoking the Bible to disprove argu-ments on physical matters "they do not understand." On the other hand, those who are well versed in physical science and astronomy are quite able to see the truth of his discoveries (175–76).

Ethos and pathos commingle as he adds that his enemies prefer to "cast against me imputations of crimes, which must be and are more abhorrent to me than death itself" (176). The reference undoubtedly is to the allegations of Colombe and the Dominicans Lorini and Caccini, among others, that Galileo's views were opposed to the reigning theological opinion. Here his titular and secondary audience of friends with whom he could freely give vent to frustration may have distracted him, causing him to overlook the necessity of appealing to those who might sympathize with his opponents.

At this point Galileo's critics were scattered and did not yet present an organized or powerful opposition. He also had many admirers among clerics and the scholarly world in general following the publication of his *Sidereus nuncius*. In retrospect, we can see that this was a crucial period. Whatever Galileo wrote or said was to be extraordinarily magnified.

The author's castigation of his adversaries for their stupidity and hypocrisy is repeated often throughout the letter. In this, Galileo departs from advice offered by classical rhetoricians and the *dictatores* not to antagonize the audience or readers through arrogance. The astronomer's rivals were vituperative, and invective had been polished to a fine art in Renaissance Italy, but one wonders why Galileo, sensitive to his audience as he was in the *Letters on Sunspots,* fanned the flames by responding with equal invective. The answer seems not to lie in any innate maliciousness: rather it appears that Galileo was very sensitive to criticism. Evidence of this trait occurs in the memoranda for his *De motu,* written as a young man, long before his writing had become known and provoked controversy. He conjectures even then that many on reading his writings will "turn their minds not to reflecting on whether what I have written is true, but solely to seeking how they can, justly or unjustly, undermine my arguments."[17] The same defensiveness is evident in Galileo's references in the *Letter* to the professors opposed to his discoveries. He says that a few of these men have been persuaded, but others "now take refuge in obstinate silence" and in their exasperation "divert their thoughts to other fancies and seek new ways to damage me" (176). Two paragraphs later he maintains that they are "persisting in their original resolve to destroy me and everything mine by any means they can think of " (177). Whether or not such motives existed, Galileo was evidently convinced they did.

One of the reasons some academicians were wary of Galileo's claims regarding his telescopic discoveries was their fear of the erosion of their discipline.[18] This is behind Galileo's statement that his opponents are afraid he will "overturn the sciences," something Osiander was determined to show Copernicus had not done. Astronomy had been viewed as a mixed science, a

17. *Opere* 1:412 (m. 17), English translation by Drake in *Mechanics in Sixteenth Century Italy* (Madison: University of Wisconsin Press, 1969), 382. The same sentiment is sounded in Galileo's *Dialogue on Motion* of c. 1586–87, *Opere* 1:398, also in *Mechanics,* 364–65.

18. Robert Westman discusses the disciplinary rivalries in "The Copernicans and the Churches," 93–94. In his Oberlin lecture on Aristotelianism Paul Kristeller makes a similar point, emphasizing the fact that Galileo's new conception of a physics based on mathematics was thought to be an intrusion by a mathematician and astronomer into natural philosophy, a field that had previously been separate from mathematics and astronomy (*Renaissance Thought,* 48–49). See also the discussion of such rivalries among Florentine humanists and philosophers before Galileo's day in Jerrold E. Siegel, *Rhetoric and Philosophy in Renaissance Humanism,* ch. 7.

discipline inferior to mathematics and natural philosophy; the latter two, as speculative or pure sciences, were consequently of a higher order than the applied sciences of astronomy or mechanics. As we have remarked earlier, many scholars believed these lower studies very probably could not achieve the goal of certainty through knowledge of causes expected of a science.

The Theme of the *Letter*

Following his initial reference to the intentions of his enemies, Galileo introduces a quotation from St. Augustine that becomes a theme of the letter. In this part of his commentary on the book of Genesis, *De Genesi ad litteram* 2.18, Augustine is considering what can be said with certainty about the heavenly bodies in view of the Scriptures: "Now keeping always our respect for moderation in grave piety, we ought not to believe anything inadvisedly on a dubious point, lest in favor to our error we conceive a prejudice against something that truth hereafter may reveal to be not contrary in any way to the sacred books of either the Old or the New Testament" (175–76). The quotation furnishes a perfect transition to Galileo's description of the circumstances behind the current controversy over the Copernican system. He repeats the motif at several places in the letter, using it as the context from which to issue his petition to the ecclesiastical authorities for freedom of thought. (Ancient and medieval rhetors often urged their students to preface the heart of an oration or a letter with a *sententia* or *proverbium* that would set the stage for what followed.) St. Augustine thus becomes the most frequently cited source throughout the letter.

The references to Augustine as well as those to Tertullian, St. Thomas, and other authorities were not in the original version of the letter. Galileo has taken great pains to include these citations from authorities, attempting thereby to establish common ground with the primary audience. Much of the research for these seems to have been done by a contact of Castelli's, a Barnabite priest. In a letter of January 1615 Castelli to Galileo mentions that he will send on some opinions of St. Augustine and other recognized authorities.[19] Galileo may also have received a copy of Foscarini's reply to the anonymous critic of his *Lettera*, although the friar cites many more authorities. Several of these are given by Galileo and in the same order as in Foscarini's letter, and a few quotations are identical, including the theme

19. Castelli reports that the passages confirm Galileo's preferred interpretation of Joshua developed in his earlier letter to Castelli, *Opere* 12:126–27. François Russo conjectures that Galileo used St. Augustine's commentary on Genesis so frequently because it was the source most sympathetic to his views; see his "Lettre à Christine de Lorraine Grande-Duchesse de Toscane (1615)," *Revue d'histoire des sciences* 17 (1964): 337.

from St. Augustine.[20] The elaborate argument Galileo develops in his letter rests initially upon the previously noted assumption that his opponents are seeking to discredit him, and it is against them that he directs his refutation. But he adds an ethical appeal to his statement of purpose: "I hope to show that I proceed with much greater piety than they do, when I argue not against condemning this book, but against condemning it in the way they suggest—that is, without understanding it, weighing it, or so much as reading it" (179).

His stated aim is to "justify myself in the eyes of men whose judgments in matters of religion and reputation I hold in great esteem" (179). Thus, from the beginning the letter has a twofold purpose: to seek toleration for Copernicus's book and to defend himself for his support of it. The defense Galileo develops here he hopes might aid the Church, but if his effort is not viewed as constructive he vows to "renounce any errors" he might make concerning religious questions. He does not "desire in these matters to engage in disputes with anyone, even on points that are disputable." "And if not [constructive]," he adds, "let my writing be torn and burnt, as I neither intend nor pretend to gain any fruit that is not pious and Catholic" (180–81).[21] These words are movingly prophetic of the events of the trial that was to follow sixteen years later. Here Galileo probably was referring to the letter itself, not, as one might be tempted to conjecture, to the *Dialogue* on which he was already at work.

Galileo explains that in condemning the twofold claim that "the earth rotates on its axis and revolves around the sun," his detractors would also suppress any discussion of other related observations and physical statements. This view of the planetary system, he finds it necessary to point out, was really not original with him but was that of Copernicus too, a fact that his enemies have attempted to hide from the "common people." The academic philosophers "pretend not to know" that Copernicus was "not only a Catholic, but a priest and a canon"; yet the work of this esteemed scholar, *De revolutionibus,* "has been read and studied by everyone without the faintest hint of any objection ever being conceived against its doctrines" (178–79). Galileo evidently was unaware that Copernicus was not a priest. Regardless, the fact that he was a canon is enough to buttress the point that he had the respect of the Church.

---

20. For example, both cite St. Augustine and Pererius, using the same language regarding manifest experience and demonstration. And both refer to St. Jerome and St. Thomas concerning the customary practice of scriptural historians to defer to common opinion; see Finocchiaro's translation of the *Lettera,* 96, 107, and Blackwell's translation of Foscarini's *Defense,* 259–60.

21. *Opere* 5:315, lines 2–3. I have substituted "my writing" for Drake's "my book." The original reads "*mia scrittura.*"

Surprisingly, Galileo then claims that only the campaign to discredit him has prompted this effort to have Copernicus's book condemned (178–79). Although one is tempted to see his statements as a rhetorical strategy intended to shift the center of the oppositions' attack to himself and then to discredit it by pointing out the physical evidence he has found supportive of Copernicus, he actually seems to have believed that his own enemies are responsible for the attacks made on Copernicus and to have been unaware that the Dominican theologian and astronomer Tolosani had called for the work's condemnation.[22] In this view, Galileo greatly underestimated the strength of the commitment to conservative theology by those in positions of authority. Certainly the opposition of the General of the Jesuit Order, Acquaviva, to Copernicus was not aimed at Galileo, nor was Bellarmine's.

In *The Sleepwalkers* Arthur Koestler agrees that Galileo is the principal cause of the growing sentiment against Copernicus, but for reasons different from those Galileo gives. Rather than seeing increasing opposition to Galileo's earlier teachings as responsible for the disapproval of Copernicus, Koestler sees the *Letter to Christina* as a "theological atom bomb" and "the principal cause of the prohibition of Copernicus and Galileo's downfall." He adds that its "radioactive fallout is still being felt."[23]

In turning again to the work, one notes that in developing the main body of his argument against his opponents the author's tone is not as querulous as in the introductory parts, although the text is still interlaced with incisive, scornful comments at strategic points. Since the exordium was added to the second version of the letter, completed almost two years after the orginal letter to Castelli, it reflects the antagonism to his views that Galileo saw growing around him in Florence. Thus the pretext for writing the letter has subtly shifted in the exordium from the questions voiced by Madame Christina about scriptural problems to a defense of himself. In the body of the letter Galileo proceeds in the manner of a philosopher-scientist who is also skilled in rhetoric. Because of this it is especially important to note the

22. Galileo voiced a similar sentiment in a letter to Dini of 16 February 1615, where he says his opponents "have opened a new front to tear me to pieces" and bring about "the condemnation of Copernicus's book, opinion, and doctrine"; some even maintain this work was "my own" (Finocchiaro, *Galileo Affair*, 55–57).

23. Koestler, *The Sleepwalkers*, 433–34. Koestler's treatment of Galileo's *Letter to the Grand Duchess* and the character of the astronomer is harsh, and his book has been countered in reviews by Santillana, Drake, and others. Koestler does not distinguish carefully between the earlier version of the letter written to Castelli and the later one to Christina. On the other hand, in my view Santillana attributes too much to the effect on Cardinal Barberini of the letter to Christina, basing his conclusions on a conversation with Galileo recorded by Giovanfrancesco Buonamici in the latter's diary. My reading of Buonamici's diary does not yield the interpretation that the cardinal was persuaded by the letter to counsel the Holy Office against accusing Galileo of heresy in 1616; cf. Santillana, *Crime of Galileo*, 203, 289, and *Opere* 15:111.

precise terminology he uses when advancing an argument, and particularly when characterizing it as a "necessary demonstration."[24]

The Problem of Necessary Demonstration

Galileo was fully cognizant of the Aristotelian canons of demonstration, as has been noted in our discussion of his earlier work and in his annotations of Bellarmine's response to Foscarini. His study of the *Posterior Analytics*, recorded in the manuscript commentary that he copied out while he was teaching at the University of Pisa, we have described earlier.[25] Viewing the letter in relation to the contents of this text permits a fuller appreciation of the interlacement of rhetorical and scientific arguments.

Galileo's argument is conceived as a *refutatio*. It begins immediately following the declaration of his aims. He states the principal issue in a provocative proposition: "The reason produced for condemning the opinion that the earth moves and the sun stands still is that in many places in the Bible one can read that the sun moves and the earth stands still. Since the Bible cannot err, it follows as a necessary consequent that anyone takes an erroneous and heretical position who maintains that the sun is inherently motionless and the earth movable" (181).

That he presents the issue in this way after showing his own ideas to be identical with Copernicus's is a direct and unprecedented challenge for an avowedly devout Catholic to most of his primary audience. It also signals his decision to pursue the issue on theological grounds to persuade his audience that he is not a heretic. Even as empathic a commentator as Stillman Drake sees that decision as a daring move. He remarks that Galileo was proceeding against "advice from his friends at Rome [Prince] Cesi, [Monsignor] Ciampoli, and [Cardinal] Barberini to keep the battle on general grounds."[26] They, like Bellarmine, had said that as long as Galileo spoke as a mathematician and regarded the Copernican system as an hypothesis there would be no problem. But to venture into theological arguments and to maintain that the theory was demonstrable would be foolhardy. Galileo recognized this much earlier in remarking to Bishop Dini that "no astronomer nor natural

---

24. Finocchiaro discounts the significance of the term "necessary demonstration" in his discussion of Galileo's *Dialogue* in *Galileo and the Art of Reasoning: Rhetorical Foundations of Logic and Scientific Method*, Boston Studies in the Philosophy of Science, vol. 61 (Dordrecht: Reidel, 1980), and in his "The Methodological Background to Galileo's Trial" in *Reinterpreting Galileo*. Finocchiaro thinks Galileo employs a more modern fallibilist methodology. But Wallace has convincingly argued for Galileo's having followed Aristotelian canons of apodictic proof.

25. See Wallace's English translation of the *Tractatio de praecognitionibus et praecognitis*, along with its interpretative volume, as mentioned in the Preface, n. 4.

26. Drake, *Galileo at Work*, 245, and *Discoveries*, 167.

philosopher who stayed within his boundaries has ever entered into such things."[27] Why did he do so? Drake thinks that reading Foscarini's work, which cited Galileo's discoveries, led him to write the longer letter to the Grand Duchess.

The magnitude of the task Galileo has set for himself in light of the opposition and the options open to him now becomes clearer. To argue for the Copernican system without offering demonstrative proof in view of its contradiction of Scripture would be to defy traditional procedures in the eyes of his principal audience.

The manner in which Galileo handles this critical dilemma, as we shall see, is simply to presume at the outset that such proofs exist. When he remarks on the Church's supposed prior acceptance of Copernicus's book, Galileo says flatly that he finds it difficult to believe that people would see the statements therein as heretical, "now that manifest experiences and necessary demonstrations have shown them to be well grounded" (179). The Italian reads "*ben fondata sopra manifeste esperienze e necessarie dimostrazioni*" [*Opere* 5:312.27–28).[28] In view of Galileo's understanding of the expression "necessarie dimostrazioni," there is an ambiguity in this statement that will be exploited throughout the *Letter*. Does he mean that the system is actually demonstrated on the basis of sense experience, or that it is merely a plausible hypothesis that can be supported partially by observation and by strict mathematical reasoning, the reasoning Ptolemy employed, as Galileo had pointed out in the "Considerations"? The first meaning is the impression Galileo intends to convey, as can be seen by the rest of the text; the second sense would be consonant with Bellarmine's assessment of Copernicus's thesis, but if this were Galileo's meaning he would not have sufficient grounds for demanding a reinterpretation of the Scriptures. Bellarmine had said that the authority of the Bible "ought to be preferred over that of all human writings which are supported only by bare assertions and probable arguments, and not set forth in a demonstrative way" (183; *Opere* 5:317.21–24).

The most surprising thing about the letter is that Galileo never presents a confirmation of systematic inductive or deductive proofs drawn from astronomy for his position, but instead relies upon a refutation of deductive arguments from theology to counter his opponents' contentions. For the Grand Duchess, and other unsophisticated readers, he evidently assumes that he need only state that demonstrations exist and then in a relaxed dialectical manner take up the theological difficulties. As for his opponents, he

27. *Opere* 12:183.9–184.1.

28. For further details on Galileo's use of the term "necessary demonstration," including a fuller exposition of the original texts and their relation to the rhetorical proofs of the argument in the *Letter*, see my "The Rhetoric of Proof in Galileo's Writings."

simply lumps them together as Peripatetics, those academicians who look only to the text of Aristotle for proof of a proposition. They would not be expected to listen to arguments, whatever the physical evidence offered or however cogently proofs were presented, if corroboration could not be found in Aristotle's works. In this characterization of his opponents Galileo fails to consider that among his audience for the letter were others, opposed or unconvinced, who were progressive Aristotelians like himself, such as Bishop Dini, Cardinals Bellarmine and Barberini, and the Jesuit astronomers at the Collegio Romano.[29] They, unlike the conservative Peripatetics, would have been responsive to a scientific demonstration.

Accommodation and Irrelevancy

To explain to the Grand Duchess how one could possibly entertain a view that is at variance with the literal meaning of the Scripture, Galileo employs arguments he had first raised in his letter to Castelli: the arguments from "accommodation" and from "irrelevancy." He tells her that at times Scripture speaks in the language that men understand, using analogies and popular wisdom, for the Bible is concerned with spiritual truths and not in teaching man about the complexities of the universe. Scripture is not meant to be taken literally in every sentence. Granted that this is sometimes the case, he suggests that when contradictions seem to occur we should not begin to resolve the difficulty through the authority of scriptural passages but "from sensate experiences and from necessary demonstrations" (*dalle sensate esperienze e dalle dimostrazioni necessarie*) (182; *Opere* 5:316.24).

This approach is commended, he explains, because two truths cannot contradict one another; Nature like Scripture cannot be false because both

---

29. Galileo describes himself as an Aristotelian in his scientific reasoning in a letter of 14 September 1640 to Fortunio Liceti, *Opere* 18:248. The progressive Aristotelianism of Galileo in his logical methodology is analyzed by Wallace in his "Aristotelian Influences on Galileo's Thought," in Luigi Olivieri, ed., *Aristotelismo Veneto e Scienza Moderna*, 2 vols. (Padua: Antenore, 1983), 1:349–378, reprinted along with other essays in his *Galileo, the Jesuits and the Medieval Aristotle* (Aldershot: Variorum, 1991). This is not to deny that Galileo was also influenced by Plato, as Kristeller has noted in *Renaissance Thought*, 64 and nn. 47, 48 (269–70), and also urged by Alexandre Koyré in *Metaphysics and Measurement: Essays in the Scientific Revolution* (Cambridge, Mass.: Harvard University Press, 1968), 16–43. During Galileo's days at Pisa the oppositions between Aristotelianism and Platonism were not as clearly noted as they are in our times; both Jacopo Mazzoni and Cosimo Boscaglia taught Aristotle and Plato at the university there, and Mazzoni even attempted a complete reconciliation of the two philosophers. Galileo studied with Mazzoni in 1590, as he records in his letter to his father on November 15th of that year (*Opere* 5:44–45), and seems to have been particularly impressed with the way in which his father's friend used mathematics to remove *impedimenti* to man's knowledge of the physical world. For more details see Frederick Purnell, "Jacopo Mazzoni and Galileo," *Physis* 3 (1972): 273–94.

have their origin in the Holy Spirit. Nature is what our senses and necessary demonstrations show her to be. Therefore, since Nature cannot be other than she is, while Scripture can be and sometimes is interpreted differently than the strict meaning of its words, Nature should not be called into question because of particular biblical passages.

Galileo concludes this line of reasoning by quoting Tertullian: "God is known first through Nature, and then again, more specifically by doctrine; by Nature in his works and by doctrine in his revealed word" (183).

The support from "accommodation" that Galileo has used so adroitly in the foundation of his argument above is returned to several times, and he buttresses it by attributing it to St. Jerome and to St. Thomas Aquinas (200–201). Nevertheless, the effect that the argument would have on his primary audience, regardless of such appeals to authority, is predictable. In their eyes it would be acceptable to apply the accommodation principle to selected texts if one were a theologian, but it would be improper, even presumptuous, for a nontheologian to advance it without permission. As he reported to Galileo, Dini had discussed that particular argument in a conversation with Cardinal Bellarmine, probably because of the Florentine's use of it in the original letter to Castelli, but the prelate had warned against it.[30] No doubt Bellarmine feared that some theologians would be incensed at a mathematician deciding which texts do not say what they patently mean.

Concerning the "irrelevancy" argument, there is no similar admonition recorded. Galileo had used it in the letter to Castelli, and he touches on it in the "Considerations." In developing this particular argument for the Grand Duchess he quotes St. Augustine: "Hence let it be said briefly, touching the form of heaven, that our authors knew the truth, but the Holy Spirit did not desire that men should learn things that are useful to no one for salvation" (185). Galileo adds the inescapable conclusion that since the Holy Spirit did not give us knowledge about the heavens because it is "irrelevant to our salvation," then belief about celestial bodies should not be made obligatory to faith. He inquires: "Can an opinion be heretical and yet have no concern with the salvation of souls?" Lightening the tone, he quotes the words of Cardinal Baronius: "the intention of the Holy Ghost is to teach us how one goes to heaven, not how heaven goes" (185–86).

As we know, these are common arguments, offered in the earlier exegetical defenses of Copernicus. Interestingly, both are in accord with Pope Leo XIII's encyclical of 1893, *Providentissimus Deus*, which outlines how Scripture should be interpreted. Yet most theologians and philosophers of the

---

30. Drake notes Dini's concern, *Galileo at Work*, 245.

seventeenth century who were opposed to the Copernican thesis were not persuaded by these proofs. When Galileo advanced them in the letter, many must have resented his arrogant tone and his presumption in speaking about theological matters, when they knew his proper discipline was mathematical astronomy. The most important reason, however, was the one first mentioned by Cardinal Bellarmine. For the Church to relinquish an authoritative theological position that might have vast repercussions on the faith of the people, a necessary demonstration of the physical realities would have to be presented.

The Seduction of Partial Proof

After speaking as if demonstrations are available at the beginning of the letter, Galileo leaves the matter undeveloped. He then leads his readers through the argument just reviewed regarding the twin truths of the Holy Spirit: Nature and Scripture, and how these may be harmonized if one only understands that the Holy Spirit suits its language to common parlance and is really unconcerned with science. His sceptical readers might then expect him to return to the reason for offering the argument in the first place: the physical evidence that eliminates the scriptural difficulties.

Instead, using the rhetorical strategy of repetition, he continues to mention sense experience and necessary demonstration or demonstration obliquely more than forty times. It almost becomes a mesmerizing litany, and yet he does not specifically provide the evidence. Generally the terms are introduced in the context of the need for new scriptural interpretations.

Galileo does introduce some sense observations later in the letter, which he says should accompany necessary demonstration: his sightings of the great variations in position of the orbits of Venus and Mars relative to earth and the changes he saw in the appearance of Venus. But these he does not attempt to incorporate in a demonstration. He claims, however, that these and other observations "can never be reconciled with the Ptolemaic system in any way, but are very strong arguments for the Copernican" (196). In this, of course, he was right, for the sightings do offer partial support. He must have meant that evidence of Venus's orbit of the sun is manifest from his telescopic observations of the planet's crescent phases and its change in size as it recedes or processes. Thus, it could be argued that the planet rotates around the sun. And the belief that all heavenly bodies rotate around the earth could have been challenged by his observations of the movement of the satellites around Jupiter. He probably thought that the argument from the tides he developed in the *Dialogue* could substantiate the earth's rotation and that the observations of the sunspots might be used as the foundation for the supposition of the earth's annual revolution, but these were not "manifest" to other observers nor were the arguments from them "necessary" for his peers.[31]

With the same problem before him, Foscarini had seen fit to explain the probable arguments that inclined him to think that the Copernican system was the best proposed. He explained in some detail the inferences that could be drawn from the sense evidence provided by the telescope. Regardless of that precedent, Galileo does not present any of those arguments here.

Nor does he mention that his observations could also be used as support for the Tychonian system. Unfortunately, many of his opponents had embraced that solution to the dilemma. Galileo must have known the evidence could be so assessed, but he does not mention Tycho's view here or try to refute it, as he might have had he written a disputation instead of a letter. But, on the other hand, Campanella did not take up Tycho's system, nor did Foscarini in his more extended astronomical discussion.

Galileo does develop a lengthy defense of demonstration in general following his discussion of hermeneutical principles. He does so to reassure the Grand Duchess that the method is the key to determining the true sense of the Scripture when it touches on disputed matters not related to faith and morals. Having declared earlier that he cannot believe that "God who has endowed us with senses, reason, and intellect has intended to forgo their use and by some other means to give us knowledge which we can attain by them" (183), he goes on to show that the demonstrative method was supported by the most highly regarded authorities. The testimony he offers he says he has selected from more than a hundred attestations of scholars and sainted theologians. Since these are important to the argument of the book, the texts are worth examining in some detail.

The first is from Benedictus Pererius's work, *In Genesim*. Pererius was a Jesuit and a professor who lectured on natural philosophy at the Collegio Romano in the 1560s. As a priest and philosopher, his opinion might be expected to carry some weight with Galileo's clerical audience. The passage Galileo chooses is Pererius's warning that in treating the teaching of Moses we should be wary and avoid "affirming and asserting whatever is refuted by manifest proof and philosophical reasoning (*manifestis experimentis et ra-*

---

31. Nevertheless, one of the consultants to the Inquisition, Melchior Inchofer, regarded the *Letter to Christina* as prime evidence at the trial for Galileo's heretical teachings (*Opere* 19:349). Olaf Pedersen, who has described Galileo's personal religious views, thinks that he was truly sincere in wishing to follow the teaching of the Church, but that he was also convinced of the truth of Copernicus's system and in this period was bent upon establishing physical proof for it. He likewise wished to establish the "irrelevance of Biblical 'proofs' to the contrary as he did by the perfectly sound theological reasons in the Letter"; see Pedersen's "Galileo's Religion," *Galileo Affair*, 80–83.

tionibus philosophiae) . . . because with truth other truths are congruent" (5:320.2–3). Pererius continues that it is not possible for "the truth of Sacred Writings to be contrary to true reasoning and proofs (*veris rationibus et experimentis*) of human science" (5:320.4–6).

Following this citation of a contemporary philosopher, Galileo strengthens his argument with the opinion of St. Augustine, who also emphasizes the crucial role of demonstration: "If manifest and certain reasoning (*manifestae certaeque rationi*) are set up against the authority of Holy Scripture, whoever does this is not aware of what he is doing" (5:320.67). Augustine explains that such an act opposes only what is supposed to be the sense of the Scripture, whose truth has not been fully penetrated.

We might note in passing that Galileo departs from the Italian to use the Latin in all his quotations from the texts of the saints and scholars. The shift from the more informal Italian to the language of scholars has the effect of magnifying the importance of the quoted passages, and it also elevates the letter above the genre of familiar epistles.

Drawing the obvious conclusion from his citations, Galileo offers advice to the theologians. He says that since "two truths cannot contradict each other, then wise expositors should toil diligently to penetrate the true meaning of scriptural passages, which indubitably will be in accord with the natural conclusions that manifest sense or necessary demonstrations (*il senso manifesto o le dimostrazioni necessarie*) have first made certain and secure" (5:320.13–16). Driving his point home, he adds: "I believe that it would be more prudent not to permit anyone to teach passages of the Scripture and in a certain manner force the meaning in order to sustain as true this or that natural conclusion, for which one day (*una volta*) the senses and demonstrative and necessary reasoning (*il senso e le ragioni dimostrative e necessarie*) would be able to show the contrary" (5:320.22–25).

The presumptiveness of the advice is excused by the theme of the letter—which he previously quoted from St. Augustine—and by Galileo's mention here of the proof he thinks is impending "one day." The difficulty for those who want to discover Galileo's meaning in regard to the existence of such demonstrations is that it is not clear whether he intended the present passage to be taken as a general principle, an echo of St. Augustine's mention of what "truth hereafter may reveal," or whether he intended to imply that proofs were still not available. The ambiguity is never clarified.

Galileo adds to these admonitory sentences two rhetorical questions: "Who indeed will set bounds to human ingenuity? Who will assert that everything in the universe capable of being perceived is already discovered and known?" (187). The line of thought that Galileo is intent on conveying for this section of the letter, beginning with the passage about two truths, is retained verbatim from the orginal letter to Castelli. Even though he knows that the Castelli letter was disturbing to some who saw it and prompted warnings not to "irritate" theologians, he evidently thought he had now provided enough foundation for his claims to risk requiring submission to the evidence.

He then observes that since the Scripture itself declares that God has given the world over to disputations and that it is difficult to discern the work of his hands from beginning to end (Eccles. 3.11), then no one should bar man's inquiries—an adroit riposte to that supposed limitation. Many distinguished philosophers throughout history have believed in the stability of the sun and the mobility of the earth, he says, but the opinion was "amplified and confirmed with many observations and demonstrations" only by Copernicus (188). Although there is an ambiguity in the words "amplified and confirmed," the tone conveys the impression that there can be no question about Copernicus's accomplishment.

Theology and Astronomy

Galileo next turns his attention to the delicate problem of the relations between theology and astronomy. He describes the academic theologians as obstinate in their desire to preserve their domain. They maintain that "theology is the queen of the sciences" and therefore that she does not need to adjust herself to the findings of "less worthy sciences." He next considers in what sense theology should be termed a queen, whether for the reason that her study contains the fruits of all the other sciences or because her subject matter "excels in dignity" and is "divulged in more sublime ways?" The latter explanation is his guess, and he suggests that if theology does not deign to descend to the "humbler speculations of the subordinate sciences" it should behoove her professors not to make pronouncements on subjects they have "neither studied nor practiced" (191–93).

The major problem with these professors lies in their demand that astronomers retract their proofs as fallacious. But, he says, "this would amount to commanding that they not see what they see and not understand what they know, and that in searching they find the opposite of what they actually discover" (193). Although the passage is a stirring defense of intellectual freedom, it is actually a misrepresentation of the Church's position as Bellarmine presented it. His letter had asked only that until proof was at hand astronomers refrain from making strong truth claims and present their results merely hypothetically.

Burdens of Proof

Following these assertions, Galileo performs the most remarkable rhetorical feat of the letter. Almost imperceptibly he turns the tables on the theologians and ends by maintaining that they must offer proof that the astronomers are wrong. First he makes the distinction between truths that are merely stated and those that are demonstrated, echoing Bellarmine's words to Foscarini. Next he argues that if "truly demonstrated physical conclusions" do not have to be modified in light of the Bible but instead the Scripture must be reinterpreted, then before authorities condemn a physical proposition "it must be shown to be not rigorously demonstrated" (194). Now he demands that a physical proposition be accepted even if it conflicts with Scripture unless it can be proved false! The most startling point follows: the proposition (not to say its demonstration) must be disproved "by those who judge it to be false" (195). In support of this demand he reiterates the theme of his exordium, the words of St. Augustine, which he now quotes at even greater length (196). He returns to the same point a few pages later in a crowning passage from *De Genesi ad litteram*, which he presents in the following way: "And later it is added, to teach us that no proposition can be contrary to the faith unless it has first been proven to be false: 'A thing is not forever contrary to the faith until disproved by most certain truth. When that happen, it was not holy Scripture that ever affirmed it, but human ignorance that imagined it'" (206).

Near the close of the letter, continuing in the same vein, Galileo says "these men are wasting their time clamoring for condemnation of the motion of the earth and stability of the sun which they have not yet demonstrated to be impossible or false" (210–11). Now the burden of proof is on the theologians. Bellarmine's cautions about irritating the theologians were obviously ignored when Galileo hit upon this superb reversal. Campanella had come near to adopting that ploy when he demanded that theologians must have knowledge of astronomy if they are to judge the Copernican issue correctly, but he did not advance to the next step that they must prove Copernicus wrong.[32]

Having summarily disposed of the pretensions of the theologians, Galileo turns to an objection that Cardinal Bellarmine raised against the new astronomy: the necessity of following the consensus of the Fathers, as mentioned by the Council of Trent. Contradicting Bellarmine, Galileo contends that the Fathers were not in agreement; in fact, he says, they never treated the issue because it had not been raised. On the other hand, he adds, some theologians have lately begun to consider that the mobility of the earth is compatible with the Scriptures. Evidently unaware that Zuiñga had been vigorously reprimanded by Juan de Piñeda, the well-known Jesuit theologian, Galileo cites a passage from Zuñiga's *Commentaries on Job*,

32. See chapter 5, Campanella's discussion of the prerequisites for judges.

where the author quotes Job's "Who moveth the earth from its place" as a confirming test for the earth's motion (203).

Galileo next takes issue with the application of the ruling of the Council to the case of physical matters:

Besides I question the truth of the statement that the Church commands us to hold as matters of faith all physical conclusions bearing the stamp of harmonious interpretations by all the Fathers. I think this may be an arbitrary simplification of various council decrees by certain people to favor their own opinion. So far as I can find, all that is really prohibited is the "perverting into senses contrary to that of the holy Church or that of the concurrent agreement of the Fathers those passages, and those alone, which pertain to faith or morals, or which concern the edification of Christian doctrine." (203)

One wonders how this direct challenge to the current interpretation of the Council's ruling and the hortatory passage that follows must have sounded to prelates and theologians in the heat of that controversy. "Hence it remains the office of grave and wise theologians to interpret the passages according to their true meaning" (203). He adds that they should do so after first "hearing the experiences, observations, and proofs of philosophers and astronomers on both sides" (205).

Freedom of Thought

Galileo concludes this refutation with an implicit petition for liberty of thought directed to the ecclesiastical authorities. The passage is especially moving in its prophetic irony:

Anyone can see that dignity is most desired and best secured by those who submit themselves absolutely to the holy Church and do not demand that one opinion or another be prohibited, but merely ask the right to propose things for consideration which may the better guarantee the soundest decision—not by those who, driven by personal interest or stimulated by malicious hints, preach that the Church should flash her sword without delay simply because she has the power to do so. Such men fail to realize that it is not always profitable to do everything that lies within one's power. (206–7)

In spite of the tone and content of this implicit refutation of Bellarmine's position, the cardinal seems not to have allowed it to govern his treatment of its author. In his audience with Galileo concerning the ruling that the Holy Office made in March 1616, he remained courteous and even protective of the astronomer's reputation. According to the latter's testimony at the trial years later, Bellarmine furnished him with a letter after the audience that attested to his good standing in the Church. One may well imagine the effect that Galileo's letter to the Grand Duchess had on more irascible readers. Perhaps it is not going too far to suggest that much of the animosity exhibited during the trial may have been fired by the letter's rhetoric.

The fears that haunted academic theologians and the ecclesiastical hierarchy were raised in the same paragraph as the preceding plea to the Church for liberty of thought, even though Galileo intended the point to augment his theme. In attempting to show the negligible effect that new interpretations permitted by this freedom would have upon infidels trained in the sciences, he points out that the infidels know more astronomy than those who would keep the literal meaning of the disputed texts. They will simply laugh at the ignorant and discount the Bible: "And why should the Bible be believed concerning the resurrection of the dead, the hope of eternal life, and the Kingdom of Heaven, when it is considered to be erroneously written as to points which admit of direct demonstration or unquestionable reasoning?" (208). Not believing that such demonstrations exist, the conservative theologians feared that even to suggest the reinterpretations of texts would encourage doubts in the minds of the faithful concerning other passages, as Protestant practice had already shown.

Inadequacies of Interpretations

The passages that follow continue to amplify Galileo's thesis that in the face of necessary demonstrations Scripture must be reexamined. In one of these, where he repeats his castigation of the practices of some theologians, Galileo again retained the sentiments uttered in the letter to Castelli, with similar dubious judgment. This time, however, he tempers the original by directly addressing the Grand Duchess, as if the criticism were for her ears alone. He says that he is sure she can see "how irregularly those persons proceed who in physical disputes arrange scriptural passages (and often those ill-understood by them) in the front rank of their arguments" (209). He grants that if they really know that what they are saying is absolutely true, they will also have physical proofs and those who argue against them would simply be guilty of employing sophisms and such. But if the conditions of scientific inquiry are truly met and the canons of philosophy are preserved, "why do they, in the thick of battle, betake themselves to a dreadful weapon which cannot be turned aside, and seek to vanquish the opponent by merely exhibiting it?" He supplies the same answer in both version of the letters: "If I may speak frankly, I believe they have themselves been vanquished, and feeling unable to stand up against the assaults of the adversary, they seek ways of holding him off. To that end they would forbid him the use of reason, divine gift of Providence, and would abuse the just authority of holy Scripture—which, in the general opinion of theologians, can never oppose manifest experiences and necessary demonstrations when rightly understood and applied" (209).[33]

Galileo then returns to the suggestion made earlier: those who would argue that the Copernican view is false should occupy themselves "in demonstrating its falsity" (*Opere* 5:342.24–25). He concludes this line of argument: "In summary, if a conclusion must not be declared heretical when it might be true, then vain are the efforts of those who would aspire to condemn the mobility of the earth and the stability of the sun, if they have not first demonstrated it to be impossible and false" (*Opere* 5:343.6–15).

Galileo's Exegesis

The major part of Galileo's petition is at an end. The remaining portion of the letter is curious. Expanding on an argument used in the earlier letter to Castelli, Galileo seeks to prove that the text from Joshua in which he commands the sun to stand still accords better with the Copernican system than with the Ptolemaic. In developing this line of thought not only does he use Scripture to hallow a physical conclusion, a practice he has just condemned, but he structures his support in a thoroughly medieval way: he appeals to the authority of Dionysius the Areopagite, a sixth-century Neoplatonist whose opinions on astronomy he would not ordinarily entertain.[34] Galileo says that Dionysius spoke of the "admirable power and energy of the sun," whose energy is in turn the cause of the planets' motion, and he suggested that at Joshua's command the primum mobile "stood still" and all the other orbs stopped because of that. Galileo adds the endorsement of St. Augustine and the Bishop of Avila to this explanation. Then he explains that under the Copernican system if God willed the sun to stand still all the other motions of the planets would cease as well, since they are dependent upon it, whereas in the Ptolemaic system the text would make no sense at all (211–16).

Introducing next one of his own observations in support of Dionysius, Galileo says that in his *Letters on Sunspots* he has shown that the sun rotates

---

33. Compare Galileo's letter to Castelli, translated by Finocchiaro, *Galileo Affair* (52) and his translation of the *Letter* to the Grand Duchess (113). See *Opere* 5:285 and 5:342.

34. Dionysius is generally termed the Pseudo-Dionysius to differentiate him from the Dionysius converted by St. Paul, Acts 17.4. The works of Pseudo-Dionysius that were translated by Ficino had been written in the early 6th century by a Neoplatonic philosopher who adopted the famous name. Lorenzo Valla questioned the ascription of the works to the first century convert, but not until the time of Galileo's *Letter to Christina* was the authority seriously challenged. The writings were cited by responsible Church scholars through the Middle Ages and the Renaissance. The argument Galileo develops here about the sun's energy was prefigured in more detail in his letter to Dini of 23 March 1615.

on its axis, completing a revolution in about a month. He adds a rhetorical slant to the remaining proof:

If we consider the nobility of the sun, and the fact that it is the font of light which . . . illuminates not only the moon and the earth but all the other planets, which are inherently dark, then I believe that it will not be entirely unphilosophical to say that the sun, as the chief minister of Nature and in a certain sense the heart and soul of the universe, infuses by its own rotation not only light but also motion into other bodies which surround it. And just as if the motion of the heart should cease in an animal, all other motions of its members would also cease, so if the rotation of the sun were to stop, the rotations of all the planets would stop too. (212–13)

The appeal from pathos in his reference to "the nobility of the sun" and the analogical comparison of the sun to the heart and soul would have had slight effect upon his primary audience, although Christina may have been reassured. Other scriptural texts contrary to the Copernican view, Galileo asserts, he is sure can be similarly harmonized with the physical realities. He ends with a hymn of St. Ambrose to the brilliant splendor of the heavens and a quotation from Proverbs: "He had not yet made the earth, the rivers, and the hinges of the terrestial orb" (8.26). Galileo comments that "hinges" would be an inappropriate term if the earth did not need them to turn upon.

In our discussion of the *Letter to Christina* we have noted the dialectical arguments and scholastic terminology used by Galileo to imply that complete proof of the Copernican system exists. The rhetoric of his treatment lies not only in the capital made of his claim to have proof, but it figures prominently also in the passages in which he implies he has not, and bases his request for reinterpretation on the probability of proof becoming available. In the seventeenth century, *scientia* implied certainties, not suasion. Thus his use of the passage from St. Augustine as a theme serves him well. The saint was of course speaking of situations concerning heavenly bodies where proof was not yet forthcoming but could be in time. Since this might one day come about, one should not take a firm position on scriptural interpretations so as to prejudice oneself against the new sense of the Word. This is a good dialectical argument.

The manner in which Galileo handles the fuller explication of the Augustinian theme allows for an explicit admission that a complete demonstration is still to be effected. He could then have mounted a persuasive argument from probabilities—that, since particular evidence exists that supports the thesis, more will be forthcoming. But he never does explicitly make the admission that only partial proof is available. He comes near to it when he remarks that his arguments against Ptolemy and Aristotle "relate to physical effects whose causes can perhaps be assigned in no other way" (177), and that his discoveries "plainly confute the Ptolemaic system while admirably agreeing with and confirming the contrary hypothesis" (177). But he chooses instead to base the bulk of his arguments on the supposition that the demonstration has been effected. The letter then in essence provides a formally sound dialectical argument, but the major premise is materially equivocal. This notwithstanding, if the Tychonian explanation had not offered an attractive alternative to his opponents, perhaps his arguments would have found a receptive audience among those who saw the need to reject the Ptolemaic view. By 1615, however, Galileo's supporters among the Jesuits had to rein in their inclinations to accept innovative opinions concerning the heavens. Acquaviva had admonished all to maintain a uniform Aristotelian doctrine in philosophy and to follow Ptolemy in astronomy.[35] The most they dared advance was the Tychonian view. As the vanguard of the Counter-Reformation, their views were impressive to the educated public. Their dilemma will become clearer in the discussion of Grassi's debate with Galileo.

From a rhetorical standpoint the letter is a very moving piece. It must have been particularly persuasive for those friends and admirers who could enjoy Galileo's elegant audacity and the erudition with which he argued for a reinterpretation of Scripture but who did not know what kind of proof was needed for a demonstration. It is even more moving to us in light of the events that we know were to follow in Galileo's life. And it is convincing now because we are aware that his claims have since been vindicated. He was right. The Church's treatment of him following the publication of the *Dialogue* seems unnecessarily cruel in its humiliation of a brave scientist who may have been imprudent, but certainly should have been permitted to advance arguments on the evidence he already had. The worldview of his primary audience, however, was very different. Science and religion were not then separate spheres where autonomy in each could be respected and demanded, and claims in science were expected to be validated by rules of classical logic. Rhetoric was accorded no place in the proofs of that process.

35. Acquaviva wrote two letters outlining the uniformity expected of Jesuits. The second called for those who taught "novelties" to be removed from teaching positions; see Blackwell, *Galileo, Bellarmine, and the Bible,* 140, 148–57.

PART THREE

THE TRIUMPH
OF RHETORIC

CHAPTER EIGHT

The Delicate Balance:
Galileo versus Grassi

The *Letter to Christina* failed of its purpose, as did Galileo's visit to Rome in the winter of 1615-16 to argue for the cause in person. During the previous year the Congregation of the Holy Office had been investigating the orthodoxy of the Copernican thesis. Some of the details of their inquiries enable us to glimpse the range of the opposition, those who made up a powerful segment of the readers of the writings we are considering.

Galileo's earlier letter to Castelli, brought to the Congregation's attention by the Dominican Niccolò Lorini, had been read and examined by the fathers, but aside from some inappropriate connotations in the diction employed, the chosen consultor did not find it dangerous to the faith. The Dominican priest who preached against Galileo in Florence, Father Caccini, had asked for a hearing with the Inquisitors. In his testimony in March 1615 he reported much hearsay about the Paduan professor and his followers, claiming that he corresponded with Germans (a reference to the correspondence with Welser), that he interpreted Scripture to agree with his own opinion, and that he had many followers in Florence who were convinced of his opinion that the earth moves and the heavens stand still.[1] But Caccini had to admit that he had never met Galileo and did not even know what he looked like.

Galileo under Suspicion

In November 1615 another Dominican, Father Ferdinando Ximenes, also swore to more damaging knowledge of Galileo and the Galileists. He alleged that he had heard one of Galileo's followers declare that not only does the earth move but that "God is an accident; that things do not have a substance or continuous quantity, but rather discrete quantity and empty

1. The documents from these proceedings are translated by Finocchiaro, *Galileo Affair*, 134–53; Caccini's deposition, 136–41.

spaces; and that God is sensuous and subject to laughter and crying."[2] He remarked that he was not sure that the disciple of Galileo understood all that he said but surmised that he was merely repeating Galileo's opinion more than his own.

The disciple, Gianozzo Attavanti, a nobleman in minor orders, testified shortly afterwards and explained that he was not a student of Galileo but knew him well and often discussed scientific matters with him. He had heard him support Copernicus's teaching and even heard him argue that the passage from Joshua where the sun is said to stand still could be differently interpreted. Regarding the dangerous opinions about God's laughing and crying, he explained that these were brought up in the process of his taking instruction from Father Ximenes, under whom he had been studying. As he participated in a disputation with Ximenes he took the position described, but the determination of the argument was made in accord with St. Thomas that God is not sensuous, and so on. Attavanti thought that perhaps Caccini, who was staying in the priory in a room nearby, may have overheard the disputation and thought that he was espousing the opinions in earnest as a disciple of Galileo. This possibility had occurred to him because on his later visit to Father Ximenes, Caccini overheard the two discussing the earth's motion, came into the room, maintained that the opinion was heretical, and said that he would preach against it.[3]

These statements vividly reveal the mounting suspicion of Galileo and the role that rumor played in it. Regardless of the support shown by Campanella and the Preacher General of the Order, who had apologized to Galileo for Caccini's sermon, the testimony before the Inquisition shows that opposition among Dominicans in Florence was growing. The "black and white hounds of the Holy Office," as Lorini phrased it, took very seriously the Order's responsibility of guarding the faith. (Dominicans had early in the history of the Order been punned as *Domini canes*, "hounds of the Lord.") Some of their zeal may be attributed to their concern to carry out the directives of the Council of Trent. Fortunately, the Inquisitors showed themselves skeptical about the testimony they heard and sharply questioned the witnesses.

The Edict against Copernicus

During the ensuing months Galileo continued to press for his views. The campaign came to an end near the end of February 1616. On the 24th, a special panel of "Father Theologians" appointed by the pope to examine the opinions of Copernicus reported on their conclusions. The eleven Fathers

2. Ibid., Ximenes's deposition, 141–42.
3. Ibid., Attavanti's deposition, 143–46.

on the panel, five of whom were Dominicans, agreed that the idea that the sun is at the center and motionless is "foolish and absurd in philosophy, and formally heretical since it explicitly contradicts in many places the sense of Holy Scripture, according to the literal meaning of the words and . . . common interpretation and understanding of the Holy Fathers and the doctors of theology."[4] The proposition that the earth is not the center of the universe but rotates and moves around the sun they thought was likewise "foolish and absurd" as far as philosophy is concerned, and for theology the opinion was "at least erroneous in faith."[5]

The minutes of the Inquisition record that the next day the secretary of the Holy Office told the members of the Congregation that on hearing the assessment of the commission of theologians, the pope had ordered Cardinal Bellarmine to call in Galileo and warn him to "abandon these opinions." If he should not so agree he was to be served with an injunction "to abstain completely from teaching or defending this doctrine and opinion or from discussing it." If not he would face imprisonment.[6]

Accordingly, on 26 February 1616, Cardinal Bellarmine informed Galileo of the decision of the papal panel. Since Galileo acquiesced to the directive, there was no need to serve him with the injunction. At Galileo's request Bellarmine did give him a certificate stating that rumors that Galileo had been forced to abjure and forced to do penance were untrue. The cardinal declared that Galileo had simply been notified about the decision of the fathers that deemed the opinion of Copernicus to be contrary to the Scripture and "therefore cannot be defended or held."[7] The difference in phrasing in the Inquisition's minutes concerning what Galileo was to be told and what Bellarmine wrote in the certificate was to figure largely in the trial. Did he leave Rome thinking he could explain Copernicus's views without seeming to defend them, or was he expressly forbidden to say anything at all about Copernicus?

Galileo expressed satisfaction with the outcome of the proceedings in a letter to the Tuscan Secretary of State, Curzio Picchena. He noted that in spite of the efforts of those who complained to the Inquisition about his opinions, their allegations did not prevail. The Church did not condemn his writings, but the Congregation of the Index did suppress Foscarini's book and order the opinion of Zuñiga corrected. In the case of *De revolutionibus,* ten lines in the preface were to be removed, those in which Copernicus tries to reconcile his opinion with the Scripture, along with a number of other

4. Ibid., the consultants' report, 146–47.
5. Ibid.
6. Ibid., the minutes of the meeting of the Inquisition, 25 February 1616, 147.
7. Ibid., 153.

minor corrections. Concerning his own behavior Galileo reassured the court that regardless of what malicious tongues might report, he had comported himself with "calmness and moderation."[8]

The decision was not nearly so harsh as it might have been, and Galileo must have felt himself to have been partially effective in countering the passionate opposition ranged against the Copernican opinion and himself. Certainly he had the good will of Bellarmine and, from his own report, that of the pope as well. He recounted to Picchena an audience with the pope in which he told him that he had come to Rome to defend himself against slander and that he had told his patrons that he would not avail himself of their protection in facing questions concerning his faith or morals. The pope praised him, saying that he made the right decision and was indeed well regarded by himself and the Congregation. He told Galileo to be at peace, for he need not fear such slander as long as he (the pope) lived.[9]

Galileo had at least held his own against the attacks of the Dominicans, and for the time being they seemed to be held in check by the pope. Not long after that battle was over, his attention was diverted from the Dominicans to the Jesuits as a fresh controversy emerged.

The Problem of the Comets

Following the decision concerning the Copernican thesis, many more astronomers turned to the theory of Tycho Brahe in order to answer the difficulties posed by recent observations. As we have seen, Brahe maintained that the planets did move around the sun, and that they and the sun circled the earth. His theory could be used to explain the movements of the planets revealed by the telescope, which the Ptolemaic system could not, and, most importantly, the Tychonian system preserved the scriptural texts.

Brahe had also made careful observations of the comet of 1577. On the strength of his measurements and computations, he had claimed that the comet was real and lay beyond the moon. He also conjectured that it orbited the sun. This was a controversial point, for it implied that change could take place in the heavens, contrary to what Aristotle had taught. Many accepted Tycho's opinion on celestial alteration, Galileo among them, as we have noted in discussing his letters to Welser concerning the sunspots.

Thus in 1618, when three comets were seen in the heavens, there was an immediate outcry for an explanation of the phenomena and how they related to the rival explanations of the movements of the universe. An anonymous disputation, published in 1619 and entitled *On the Three Comets of the Year 1618: An Astronomical Disputation*, attempted to explain the comets in

---

8. Ibid., Galileo to the Tuscan Secretary, 6 March 1616, 150–51.
9. Ibid., Galileo to the Tuscan Secretary, 12 March 1616, 151–53.

accord with Tycho's system.[10] The disputation, described on the title page as having been delivered publicly at the Collegio Romano, opened what was to become a heated exchange between Galileo and its author, Orazio Grassi, who held the chair in mathematics at the College. Grassi, like Scheiner, also observed the Jesuit stricture not to publicize his identity when engaged in disputes of dubious outcome. Once more, then, Galileo became involved in a controversy with a Jesuit, and as was the case with Scheiner, the identity of his opponent was soon discovered. But this time the man involved was speaking from the "platform" of the Collegio Romano itself, and giving his disputation as a part of the many academic occasions celebrated there. The filial spirit at the Collegio and the respect accorded the premier studium of the Society was to surround the controversy with more than casual interest. The debate also had the potential of arousing the collective ire of all the Jesuits should they feel their Order slighted.

The Collegio Romano

The mission of the Collegio was to further orthodoxy and to support the pope in his battles against heresy. Its founder and the founder of the Order, Ignatius Loyola, recognized the invaluable role education would play in these battles and accorded central attention to the establishment of institutions for the training of its members and other intellectually promising young men, whether they intended to join the Order or not. Ludovico Carbone, who so carefully preserved their teachings, was one of the innumerable externs trained at the Roman College. That college, as we noted in describing Carbone's texts on dialectic and rhetoric in chapter 1, was a model for the rest of the vast network the Jesuits established throughout Europe and in their foreign missions.

Founded in 1551, the Collegio had inscribed over the door "School of Grammar, Humanities and Christian Doctrine, gratis."[11] Its aim was to fashion a well-rounded scholar who was learned in the arts and sciences of the day, skilled in Latin, Greek, and Hebrew, trained in Aristotelian philosophy and the theology of St. Thomas Aquinas, but who valued above all the life of the spirit. The training of its members in persuasion was thought to

---

10. The disputation, as are all of the publications in the debate, is translated by Stillman Drake and C. D. O'Malley in *The Controversy on the Comets of 1618* (Philadelphia: University of Pennsylvania Press, 1960). Citations or quotations in my discussion refer to this work or to the original reproduced in the National Edition, *Opere* 6. When both the translation and the *Opere* are cited, the first page numbers are references to the translation.

11. Riccardo G. Villoslada, *Storia del Collegio Romano* (Rome: Gregorian University, 1954) describes the history of the College and the place of the rhetoric course in the curricula (67–68). He also describes its commitment to humanism and its growth (19, 58). The rhetoric course is described in detail in my "The Rhetoric Course at the Collegio Romano," 137–51.

be an important part of their formation. Ten years after its founding, the student population exceeded 700 and in another decade 900. The *ratio studiorum* or program of studies adopted at the college was instituted at the other Jesuit colleges established throughout Europe and the lands in which Jesuits had missions. The number of their colleges and seminaries increased to 700 by the end of the seventeenth century. But this was not by any means the limit of their influence. Students went on to teach in other schools and universities and to influence those beyond them through example and through publications. They were thus able to dominate the teaching of philosophy and theology in some of the major universities.[12] At the time of Galileo's encounter with Grassi the college was little more than sixty years old, yet it dominated the academic landscape in Italy in the teaching of philosophy and mathematics.

Grassi's *Disputation*

The Jesuits evidently thought that the public disputation of their chief mathematician merited a wider audience than those who heard it in person. Accordingly they published it, introducing the author only as "one of the fathers of the Society."[13] The book is elaborately packaged. The Latin text of the disputation proper is preceded by two introductory poems: one magnifies the comets' mysterious and "fearsome" appearance while maintaining the beneficent import, and the other, a supplication, asks that the comets whose light so disturbs the heavens will yield knowledge of the stars. Following the poems, a dramatic panel of engravings depicting the comets' appearance in the heavens fills two facing pages. The illustrations accompanied by brief explanatory legends offer a remarkable pedagogical supplement to the disputation. Perhaps enlarged versions of these illustrations served Grassi as proto-visual aids when he delivered his disputation. This kind of display would accord with historical records of elegant celebrations at the Collegio Romano during the period. Printed programs and other published miscellany disclose that disputations were often part of a longer exhibition, featuring dramas, ballets, musical interludes, poetry readings, and orations.[14]

12. Stephen Harris states that the Jesuits actually "controlled" the faculties of theology and philosophy "of several leading universities"; see his "Transposing the Merton Thesis: Apostolic Spirituality and the Establishment of the Jesuit Scientific Tradition," *Science in Context* 3 (Spring 1989): 54.

13. A facsimile of the original is in *Opere* 6:20–35.

14. The Clementine Collection at the Catholic University of America contains a number of published programs and assorted orations and disputations delivered at the Collegio around this time.

The audience for the productions included not only students and professors of the Collegio but distinguished guests: cardinals, other Church officials, civil authorities, prominent political and social figures. The illustrations may well have been printed and distributed for the occasion.[15]

The panel distills the import of the lecture, illustrating the arguments succinctly and gracefully. Two drawings, one above the other, appear on the left page and two explanatory geometrical figures below these. The uppermost figure on the left is the Great Bear, with a bearded comet shown both at its rear and front feet. A legend headed "From Rome" states that this comet first appeared on 29 August 1618 under the hind feet of the Great Bear, Ursa Major, and that by 2 September it was seen near the forelegs. The second illustration depicts another comet as it appeared in Hydra on 18 November 1618. Below the constellations are two geometrical figures detailing the points of parallax obtained for the comet on 18 November 1618 and the relation of these points to the heavens.

On the facing page is a very large drawing depicting the path through the constellations of the third and brightest comet, which first appeared on 29 November 1618. A table of its sightings is provided beside the drawing. The engraving shows the constellations of both Ursa Minor and Ursa Major, flanked by Boötes, Libra, Virgo, and Leo. Superimposed on the figures are optical sight lines indicating parallax. A tabulation at the left gives sightings for the comet in specific constellations as observed from the Jesuit colleges in Rome, Parma, and Antwerp, and from Innsbruck as reported by Johannes Remo.

Galileo's Interest

That Remo is cited is interesting because he was a friend of Galileo's and probably was once his student at Padua. He corresponded with Galileo about the comets in 1619.[16] It may be that the reference to Remo's observations and the reputation of the Collegio Romano's mathematicians stimulated Galileo's interest in Grassi's little book. Modern commentators have been puzzled about his concern when there were so many other works on comets to which he did not reply.[17]

Galileo's interest, however, may have been drawn by Grassi's subtle intimation that the renowned sidereal observer had somehow missed the

15. These are reproduced in *Opere* and are printed also in Drake and O'Malley's translation, where they are considerably reduced.

16. The reference to Remo's observations is probably to those contained in a little tract published in early 1619, *Observationes et descriptiones duorum cometarum, qui anno Domini 1618* . . . Remo discusses his observations in a letter to Galileo of 24 August 1619, *Opere* 12:484.

17. Drake and O'Malley comment to this effect, *Controversy on the Comets*, xv, as does Shea, *Galileo's Intellectual Revolution*, 75.

comets and by his suspicion that the lecture was designed to taunt him. The prologue of the disputation contains some oblique references to the discoveries of Galileo and to the text of the *Sidereus nuncius*. Grassi begins: "The human mind, Most Illustrious Ones, is so desirous of novelties that occasionally it grows weary of the long continuance of things which are good and desires to improve upon the situation." He then recounts the new knowledge of the heavens that astronomical observation has yielded: the disfigurement of the moon's surface, the orbits of Venus, Mercury, and Mars, and the satellites of Jupiter and Saturn. These observations had by this time been confirmed by Jesuit astronomers. The comment concerning the ingenuity of the human mind echoes a commonplace of the preface of the *Sidereus nuncius,* and "novelties" recalls the remarkable character of the discoveries recounted in that text. But at the same time "novelties" hints at a kind of dangerous fascination. The Church repeatedly warned its congregations against the pursuit of novelties at this time, just as it was to do in the early twentieth century, in the encyclicals outlining the errors of modernism. And we must not forget that the Reformers were called the *Novatores.* While the admonitory overtone may have been unintended, it could have riled Galileo.

Grassi then notes that only comets have evaded the notice of these "lynx-eyes" (6). The latter reference to the phenomenally-sighted animal would be recognized as an allusion to Galileo, who was the most illustrious member of the famed Lincean Academy begun by Prince Federico Cesi. The indirect reference grants a kind of grudging recognition to Galileo's discoveries even as it calls attention to his oversight. The remark might be viewed as either playful, or sarcastic, depending upon the disposition of the reader.

Content of the Treatise

Grassi remarks that the sudden appearance of three comets in the year 1618, and one of these an exceedingly bright one, have turned men's attention to that phenomena and to the problem of their location—whether they are in the celestial or the sublunar regions. This is the major problem of the tract and one that Grassi says he will try to solve with mathematics and without entering into the popular controversies concerning the portents these are thought to embody. Although his avowed purpose is scientific, in typically humanistic fashion Grassi exhibits his literary and rhetorical erudition in the process of unfolding his proofs. Departing from the practice of the scholastics, Grassi frames his disputation with a prologue and postscript and maintains a stately, erudite manner throughout. As in an oration, Grassi employs a narration and partition to introduce the proofs adduced to solve the problem he poses. The proofs themselves are presented in the fashion of a disputation, with the arguments for the position carefully outlined and the opposing view aired and refuted in an orderly manner. If the disputation had not been given as an entertainment and had been intended primarily for classroom instruction, Grassi would probably have imposed more scholastic rigor on the material. He would have included the arguments opposed to the position of the debater along with an exhaustive airing of relevant authorities and other evidence, a statement of the proponent's side—again with convincing support, followed by a response to the opposing opinions, and finally a response to the author's opinions.[18] All of these we have seen in Campanella's disputation.

In its humanistic form, the disputation becomes a medium for rhetorical as well as dialectical display. Grassi's disputation permits a close look at the unusual prominence rhetorical artifice plays in the amplification of the dialectical and demonstrative proofs.

The Jesuit's description of the comets in the disputation proper contains many metaphors taken from the figures of the constellations. He speaks of the first comet "licking the hind feet of the Great Bear," and in describing the third separates the traditional loci of epideictic discourse from the dialectical discussion that follows: "Since I believe that in this duty I ought not deviate from the masters of eloquence, in accordance with their practice, taking the first argument of my discourse from the comet's birth, I have sought its native land and parentage, and I have opened a pathway for myself through the illustrious circle of its subsequently famous life to the far from obscure character of its death" (8).[19] Grassi then observes that since the sun and Mercury were "lodged together in Scorpio," they necessitated "a very elegant and splendid feast to be prepared for the guests and, as well, that a very bright torch be kindled" (9). He calls the newly arrived comet "a foetus" remarking that its appearance shows it to be the "offspring of Mercury" (10).

The description is a part of the narration preceding the first demonstration supporting his theory that the comets are located beyond the moon, between it and the sun, and that they describe a circular orbit. In the mode of the disputation he begins his discussion with a statement of the problem: "To investigate the nearly true distance of the comet from the earth" (7). In the partition that follows he says he will explain what occurred and then describe the times and motions of each of the comets before treating their locations.

Grassi develops his argument carefully in the remaining five and a half

18. Charles Lohr sums up the basic form and notes the great concern at the Collegio for clarity and order in the disputation. For an outline of the parts generally used, see his "Jesuit Aristotelians," 9–10.

19. Epideictic oratory invokes the topoi of ancestry, native land, education, deeds, character, etc.; see *Ad Herennium* 3.7.13.

pages. He offers the sightings of astronomers in other parts of Europe to ground his conclusion, those listed in the prefatory illustration. He uses these and his own observations as the basis for establishing the geometric points and measurements in the demonstrations regarding parallax. If no parallax were found, Grassi and his observers at points some distance from each other would have seen the comet at the same place against the backdrop of the "fixed" stars. Lines drawn from these points on earth to the comet and to the stars behind would show no angle of difference. That would mean the comet was as distant as the stars. Small degrees of difference, small parallax, would mean that the comet was not that far away, but certainly beyond the moon. A greater parallax would, of course, mean the comet was close to earth, sublunar. The geometric drawings preceding the disputation offer important illustrations of the evidence of his arguments.

Focusing on the third comet in the discussion that follows, Grassi first asserts that the comet should not be placed in the earth's atmosphere (12; *Opere* 6:29.32–34). Using the observations from Rome and Antwerp, he provides the first mathematical demonstration of small parallax, proving that although the comet was not at an infinite distance, as are the stars, it was not within the earth's atmosphere. He next makes a second assertion that it was, in fact, not in the sublunar region, the region of fire, at all (13; *Opere* 6:30.30–33). This is supported by another mathematical demonstration based on sightings from Antwerp and Parma over two days that revealed no parallax.

A third supportive argument is furnished by the Innsbruck observations of Remo, which, when compared with those in Rome, showed minimum parallax (14; *Opere* 6:31.5–9). He remarks that instruments must be very large to make even more exact readings (without the telescope), like those of Tycho Brahe. But even though such were not available to them, the observers still made a significant discovery. In both Rome and Cologne the comet almost covered the tenth star in the constellation Bootes. This would mean "that our comet was not sublunar but clearly celestial" (14; *Opere* 6:31.28–29).

The argument, Grassi claims, is a demonstration. The major premise states a principle accepted by all: whatever is sighted in the heavens that exhibits no parallax or a small parallax must be at a distance beyond the moon. The minor premise declares that the comet showed only minimum parallax, which Grassi supports by observations. If all the premises are granted the conclusion is a necessary one. It could remain only probable to critics, however, for two main reasons. Some could challenge it because they did not think that mathematics could properly be applied to physical phenomena: the dilemma of the mixed sciences. Others might accept the application of the measurements to the problem but challenge the accuracy of the observations. While Grassi feels obliged to offer refutations of possible objections, he does not consider the more basic difficulty of the mixed sciences, for he believed that was resolved to the Jesuits' satisfaction by Clavius. But the observations and their significance with regard to the domains of the five elements of the universe engage his attention.

Answers to Difficulties

Grassi abbreviates the dialectical process and refutes arguments as he takes them up, instead of presenting the pro and contra side separately and then refuting each, as Campanella did in his more formal scholastic disputation. As he begins his consideration of objections, Grassi employs the topos of opposites. Even if it were granted that the comet was sublunar, he says, it would have to be very close to the moon, given the very small degree of parallax found. He then refutes this possibility by arguing that if the comet were sublunar it would have to be in the region above the upper air, that is, the region of fire. This would require the comet to be propelled by "exhalations from the earth, but, good God, how great an amount of fuel would be consumed by such an immense fire over so long a time" (15; *Opere* 6:32.5–6). Probable reasoning supports his dialectical counter-proof.

In another refutatory argument Grassi's major premise states that the movement of fiery eruptions is very erratic; the minor holds that the comet in fact follows a planet-like regular motion. This premise is supported by observations of the comet's path, showing that it made a great circle as do the planets. He calls upon a rhetorical figure to amplify the conclusion:

Now what is that which the poets have said, that by their motion and pace the usual gods are recognized, so that he who moved in the manner of the gods was considered as a god? Thus, according to Virgil, Aeneas recognized his mother as Venus. Hence does not this light by its venerable and august pace also reveal a goddess? that is, not arisen from the dross of this earth into the air but granted a seat among the celestial lights, where in a manner certainly not unworthy of the sky it shone with brief and transitory brilliance—yet so long as it lived in nowise did it display itself unworthy of the sky from which it imbibed its celestial quality. (17)

On the basis of these movements, Grassi says he "shows," or "declares," (*ostendo*)—not "demonstrates" (*demonstro*), as Drake and O'Malley translate the passage—that the comet is a heavenly body and not merely an eruption of fire close to the earth.

A third argument against the sublunar position is drawn from observations with the telescope. Grassi says: "I am convinced of the same thing by the fact that when the comet was observed through a telescope, it suffered scarcely any enlargement." He adds that "it has been discovered by long experience and proved by optical reasons" that although objects are generally enlarged by the telescope, those very distant show little magnification. Thus, "fixed stars, the most remote of all from us, receive no perceptible magnification from the telescope" (17). Since the comet was enlarged very little, it would have to be very far from the moon. The moon appears much larger when viewed through the telescope. Grassi adds that this argument may not be very convincing to some, "but perhaps they have accorded little moment to the principles of optics," for these "are necessary to give maximum persuasive force to what we are considering" (my translation of *sed hi fortasse parum opticae principia perpendunt, ex quibus necesse est huic eidem maximam inesse vim ad hoc quod agimus persuadendum*) (*Opere* 6:33.27–29). The dialectical nature of the argument is underscored by his phrasing here.

This passage annoyed Galileo, who seems to have had a proprietary view about conclusions based on telescopic evidence, although he devoted little attention to the principles of optics in his discussion of the telescope in *Sidereus nuncius*. As we know, Kepler took him to task for this omission in his "conversation" with that work.

Grassi concludes the disputation with the following proposition: "Thus, in order that we may now determine almost the true place of the comet, let us say that it can probably be placed between the sun and the moon." Grassi says "probably" (*dicimus probabiliter Solem inter ac Lunam illum statui posse*) in order not to claim absolute accuracy for his assumption (17; *Opere* 6:33.30–31). In support he offers as a major premise the optical principle that the more slowly lights move when "excited by particular motions" the higher they are, with the implied conclusion that the motion of "our comet" requires us to place it midway between the sun and the moon (17–18). Grassi then offers computations for the size of the comet based on a comparison of observations of the comet on a particular day with its distance from the center of the earth. Without the tail, it is 19,361,555 cubic miles in volume (18).

At the end of the disputation, he takes up a problem that he says has worried many people: the fact that the comet appeared to be near the arctic circle, which means that it should not have disappeared from sight. Yet the comet did disappear and did not become visible until late at night. Arguing from similarities, Grassi contends that this is because its light is not brilliant enough to be seen and that vapors on the horizon tend to conceal lights. The same is true of stars in the constellation of the Great Bear that seem to disappear when they near the horizon.

The disputation ends with an elegant postscript summarizing its content: "I have believed that the comet, shining on all directly from the same place and appearing the same from all sides, must be considered as worthy of the heavens and very near to the stars." He expresses the hope that his audience of "distinguished gentlemen" will grant the motives of his reasoning to be lofty, and, as a final touch, appends a verse from Horace, "With my head exalted I shall touch the stars" (18–19).[20]

This little book enables us to see first hand that the academic exercise of disputation has spawned a hybrid, a kind of *declamatio philosophiae naturalis*, a cross between an oration in epideictic mode and the formal, tightly ordered, closely reasoned disputation (of Campanella's type) that its title implies. This is certainly not Grassi's innovation, but his *Disputation* shows that a notable change is taking place in this period, one that enables an author to interject rhetoric more liberally into subjects in natural philosophy.

The original audience for the disputation, as we have noted, was composed of Grassi's peers, students, and others who would have understood and been persuaded by his mathematical opus and enjoyed the embellishments with which he presented it. Among those who listened or read the disputation were those who were not mathematicians and would not have been able to follow the proofs; some of these might have been persuaded by the rhetoric of its printed illustrations and by the respect engendered by the Collegio itself. As the audience for scientific discourse widened, so did the opportunities for the effects of the rhetoric employed in it. To experts in the field, the mathematicians, the part of the discourse that counted in creating the science was that of the actual demonstrations and the dialectical proofs. The rest was what made the production entertaining, delighting the hearer and the reader.

The *Discourse on the Comets*

The reply to Grassi's book, published soon afterwards, included a similar mix of science and rhetoric, and it also appealed to an audience of experts and interested amateurs. Galileo did not respond immediately with an opinion about the comets, but he did so indirectly through one of his disciples, Mario Guiducci, consul of the Florentine Academy. As consul, Guiducci was obliged to give occasional public lectures, and topics of current interest were naturally preferred. Probably urged by Galileo, he decided to focus on the comets, delivering two lectures in all, both of which were printed in 1619 under the title *Discourse on the Comets*.

Perhaps to reflect the tone of the Florentine Academy as a center of humanistic studies, Guiducci's lectures as well as the published text were in Italian. More than likely, however, the choice of Italian rather than Latin and of "Discourse" rather than "Disputation" was also meant to convey a

20. Drake (362, n. 16) attributes the verse to Horace *Carmina* 1.1.36.

more sophisticated refutation, and at the same time subtly insinuate a kind of elitist opprobrium toward the pedantic scholasticism of Grassi. It appeals to a knowing "in-group."

The author explains in the preface that he dedicates the work to Leopold Archduke of Austria because of his interest in literature and astronomy and because of the attention he paid to Galileo when passing through Florence. For his part Guiducci feels emboldened to take up the subject only because that renowned astronomer has provided opinions on the comets that form the "foundation of this essay." Guiducci's acknowledgement is significant given the controversy that followed over Galileo's contribution to the text. Although Galileo staunchly maintains in the *Assayer* that Guiducci was the real author of the *Discourse*, he may have privately thought of "author" in an honorific or qualified sense. For the editor of the National Edition of Galileo's *Opere*, who studied the manuscripts from which the text of the published version derived, there is far more evidence to hold that Galileo was the real author and Guiducci the honorary author. Favaro's conclusion is that the most important parts of the *Discourse* are Galileo's work and that Guiducci simply began the composition, probably while Galileo was re-covering from the illness that had prevented him from making systematic observations of the comets. Even these "were retouched and improved by his [Galileo's] additions, so that the entire *Discourse* may be said to be essen-tially his work."[21]

The dedication to the Archduke underscores the international character of the audience for whom Guiducci and Grassi assume they are writing, even though their immediate audiences were assembled within their hear-ing. The discourse, as published, bears few traces of the original lectures. Aside from occasional direct addresses to the audience, the work has much more in common with a disputation. It is certainly not as "humanistic" as Grassi's hybrid declamation-disputation. Except for the introduction and conclusion, it is marked by few rhetorical flourishes other than those in-tended to diminish the opponent's credibility, found more often in rhetori-cal disputes. Like Grassi's *Disputation*, the *Discourse* has an introduction, a partition, and a narration. It also resembles a scholastic disputation in that it takes up authorities and refutes opposing positions before turning to its thesis and confirmation.

Preliminary Considerations

In the introduction Guiducci's tone is sometimes harsh; he is sarcastic in his references to the Jesuits and to the author. Even though Grassi had made no

---

21. Favaro's comments are in the *Avvertimento*, 8–9, in *Opere* 6. Shea accepts this view in his discussion of the writings on the comets, *Revolution*, 75–76. Drake also shares Favaro's opinion, summarizing it and translating the conclusion, *Controversy*, xvi–xvii.

direct reference to Galileo or to his views in his disputation, except for his allusion to the lynx-eyed, Guiducci's opening remarks indicate that Galileo must have taken offense at that comment and, perhaps, to what he perceived as other references as well.

Addressing his audience of "learned Academicians," Guiducci notes that although everyone who wishes can gaze upon the great universe exposed to their view, only a few are able to penetrate deeply into it to ascertain "the seat of government for all this beautifully contrived expanse" (23). That vision is open only to those acquainted with philosophy, and even they can be dazzled by it. When some few do manage to discover something of great value, they should be especially honored. Then begins a curious indirect defense of Galileo, designed to counter Grassi's indirect allusions. The shadow-boxing of this the first exchange of the debate becomes direct combat in later pieces. Guiducci says that because only a short time has elapsed since the comets' appearance, no great blame should be attached to those philosophically enlightened elite who have not been permitted a prolonged opportunity to probe the causes of the phenomena. He turns then to novelties, saying that when these evoke more wonder than ordinary things they should also spark a desire to learn their causes.

That observation leads effortlessly to Guiducci's partition, which indicates he will follow the time-honored practice of scholastic disputations. To fulfill the desire to discover causes, he proposes first to review what ancient authorities have said and then to turn to the conjectures of "your Academician Galileo." These he says are proposed not "positively but merely probably and with reservations" (*affermativamente, ma solo probabilmente e dubitativamente*) (24; *Opere* 6:47.10–11).

Again Guiducci acknowledges Galileo's part in this work, saying that he hopes to explain his views well. He adds that he would rather be praised for "having been a good imitator" than attempt to pass himself off as the originator, as do "those who have attempted to make themselves the inventors of views that are really his [Galileo's], pretending to be Apelleses when with poorly colored and worse designed pictures they have aspired to be artists, though they could not compare in skill with even the most mediocre painters" (24). Even today the accusation is jarringly out of place given the context of the remarks. Scheiner and his confreres would surely be as much angered by the accusation of plagiarism as by the disdainful assessment of their efforts. Guiducci, like Galileo (or with Galileo), seems to be addressing an audience of intimates in the Florentine audience, who might be expected to snicker appreciatively at the allusion.

In the course of his review of the authorities Guiducci refutes some false opinions and fallacious arguments of Aristotle in the *Meteorologica*. One of these is the view that comets have their origins in the hot and dry exhalations that collect and condense in the upper air. Aristotle conjectured that this material is warmed by the turning of the spheres, which breezes from below fan and feed, until it becomes ignited, creating a comet. With ingenious arguments Guiducci dismisses the opinion, showing it to contain unprovable assumptions. The conclusion to this part of the argument would agree with Grassi's that an enormous amount of fuel would be needed to propel the comet for so long.

Discounting Tycho and Grassi

In the next section of the discussion Guiducci takes up the opinions of Tycho, with whom the "Professor of Mathematics at the Collegio Romano" agrees and for whom he even provides additional support.

Guiducci addresses the topic of parallax first. He remarks that this is a well-accepted method of determining distance, but one must determine above all whether one is focusing on what is real or on an appearance. He mentions a number of illusory phenomena created by the sun's shining upon and through clouds, mists, and vapors. He concludes that, given these difficulties, parallax can only be accepted if it can be proved that the comets are not just reflections but "unique, fixed, real, and permanent objects." He adds, "I do not know of anything which more exactly resembles a comet than those projections of rays through holes in the clouds" (39). He notes that some Pythagoreans thought that comets were simply reflections of the sun on vapors. Observations of parallax could thus be obtained by the refraction of light across the vapors, which would seem to be issuing from the same point.

After discounting the proofs from parallax, Guiducci attacks two more of Grassi's arguments. That based on observations of the comet's movements he judges to be very weak; even weaker is the argument based upon the small degree of enlargement of the comet the telescope offers. The scorn with which that argument is treated leaves little doubt that this was written by Galileo. Favaro's printing of the parallel texts lends corroboration: the published versions appear above with the manuscript fragments below. The manuscript of this portion is in Galileo's hand and the printed version follows it almost verbatim. Consequently we may infer that it is Galileo who termed the argument "incorrect and false," and who went on to remark:

I did not believe it had gained assent except from persons of so little authority that it would not be worthwhile to consider it. But recently, I have seen in the discourse on this matter delivered at the Collegio Romano how mathematicians there had so high a regard for these arguments as not only to approve them, but to criticize those who had deprecated them, calling them little skilled in the principles of perspective and in the telescopic effects which they themselves had understood and observed by virtue of long experience and from theorems of optics. (41)

In the persona of Guiducci the author then mentions that he will present the views of "the Academician" who had long ago "contradicted that reasoning and deemed it worthless." Moreover, the Academician had done so "more positively and more publicly than anyone else." (Interestingly, Galileo's handwritten text also provides the deferential allusions to the "Academician.")

The reason for Galileo's perception that Grassi's statements were insulting to him is made explicit now. Guiducci hopes that his audience and "those very learned geometers too" may profit by the corrections he will offer. The original in Galileo's hand adds a direct reference to "the Reverend fathers" that is excised in the printed version; perhaps on reflection it seemed pointedly disrespectful.

In support of the counterargument Guiducci-Galileo maintains that the telescope does enlarge distant objects, that it has in fact made visible stars that before were invisible to the naked eye. The Medicean stars are a case in point. Furthermore the differences in magnitude of stars is made clearer by the instrument. They introduce a number of experiments to provide additional evidence.[22] The conclusion of this counterargument states that "all objects are enlarged by the telescope in the same ratio, and, if very close objects seem to be more enlarged, this arises from the use of a longer instrument" (47).

In the opinion of William Shea, it was Grassi, not Galileo, who was in the right on this point. Grassi, too, had "experimental evidence for his position: stars observed with a telescope appear little larger than when observed unaided." Shea states that Galileo knew less about optical principles than Grassi. In addition, as Pietro Redondi reminds us, Grassi was the author of a scholarly work on optics.[23]

Since Tycho's cosmography undergirds Grassi's contentions regarding the comet's motions, a discussion of the inadequacies of Tycho's observations and conclusions follows. Guiducci first takes up the assumption that the comets are embedded in spheres as are the planets, saying that the concept of the spheres has been discarded by most scholars and implying by this that Tycho believed in them (which he did not). Since the comets of 1577

22. Shea discusses the validity of these experiments, *Revolution*, 79–106.
23. For Shea's assessment, see *Revolution*, 80. Redondi mentions Grassi's *Disputatio optica de iride* (1618) in his *Galileo Heretic* (Princeton: Princeton University Press, 1987), 43. Grassi's research on the optics of the rainbow would prepare him particularly well for a refutation of Galileo's notion of the comets being a reflection of light from mists.

and 1618 are declared to have different orbits, they would each have had to have different spheres to carry them around. He proceeds:

Now this multiplicity of spheres, forever idly waiting in the hope that a comet may come along, God knows when, to turn round in them for a brief time through a small part of its circle, I cannot reconcile with the extreme neatness which nature maintains in all other works by retaining nothing which is superfluous or idle. Tycho says in effect that such an arrangement of the heavens suffices for such pranks of Nature and playthings of the true stars, for though infirm they have a natural inclination to follow every manner and custom of the skies. This savors much more of poetic grace than of soundness and rigor, and deserves no consideration from you whatever, as Nature takes no delight in poetry. (49–50)

Continuing to pour contempt upon the Danish astronomer, he says that "Tycho tries his best to reduce this [motion] to uniformity" (50). Brahe "pretends not to see" that his solution makes the comet move in a different direction from the planets. The contempt spills over on Grassi because of his acceptance of Brahe's assertion that the comets move along a great circle. The demonstration Grassi offered is further held to have omitted "important points," and so he is guilty of "defective logic" (50).

Addressing the audience directly, Guiducci summarizes his opponent's methodology: "Academicians, you may note for yourselves the other kinds of absurdity lightly passed over by those who desire too anxiously that physical things shall correspond and accommodate themselves to ideas which they have formed casually" (52).

Remarking that it is "a great source of wonder to me," Guiducci-Galileo asks how the "Fathers of the Collegio" could possibly "have been persuaded to call the comet the offspring of heaven; being in effect a triple goddess, it would have to be made an inhabitant of the heavens, of the elemental regions, and also of hell" (52–53). This excursus into figurative amplification is supported by a geometrical proof based on the path of the comet, delivered in condescending tones: "a smattering of geometry suffices to show that if its orb encircles the sun" and does not orbit the earth, it would have to plunge eventually "into the infernal bowels of the earth." The author then remarks pointedly that to support his contention of the circular orbit, the Jesuit would have to propose the comet's rotation around the earth, or around the sun, either of which would get him into more difficulties (53).

Shea explains that Galileo felt obliged to refute Brahe's position because the Danish astronomer regarded the uniformity of the motion of the comet of 1577 as evidence that the earth was stationery. The Dane reasoned that, if the earth were moving, comets would follow the same retrogressions as the planets; but comets do not display retrograde motion. His new system with its stationary earth he thought better accommodated to comets' movements. Tycho did not realize that the comet's path could actually be a parabolic curve, and hence different from the closed orbits of the planets. Galileo criticized Brahe for his inconsistencies, but "he glossed over the break-down of his own attempt to explain the progress of the comet by rectilinear motion."[24]

The explanatory hypothesis Galileo proposes through Guiducci is far more difficult to sustain, as many commentators have noted. He argues that a comet is an optical illusion, caused by the sun striking vapors that arise from the earth, in much the same way as it illuminates the aurora borealis. Using a geometrical figure to illustrate his argument, Galileo contends that the motions of comets are simple and follow a straight line, maintaining that the four main observations of Grassi support this. Yet the Tuscan's argument is inadequately supported by observations, and its implications place him in a logical quandary; for if the comets are illusions, to what does he ascribe a path? He admits his bafflement, but in Galilean fashion dismisses the problem and turns to another topic.

At the end of the *Discourse* Guiducci apologizes, somewhat awkwardly, for providing his auditors with conjectures instead of finished sound arguments, but he prefers this, he says, to offering nothing. (The ending seems to have been prepared by Guiducci; no parallel manuscript by Galileo is recorded by Favaro.) Not to be outdone by the erudite author of the *Disputation*, Guiducci offers two quotations from the classics, the first from Euripides:

> Being poor, I do not wish to offer gifts
> To you rich, lest you deride the giver
> And think me by my giving to be begging. (65)

He then expresses a final hope that what he has said will take root in the fertile soil of the brains of the "gentle Academicians" and yield "positive demonstrations from which we may come to a full knowledge of that truth," as Dante notes, "Which rolls away the clouds from all our minds" (65).

The problem posed by the comets was complex. In sum, the *Discourse,* which is actually a counter-disputation, does not pretend to offer a demonstration, as Guiducci-Galileo admit. It posits only dialectical arguments to counter Grassi's contentions and seeks to force him to prove that the comets are real.

24. Shea, *Revolution*, 87.

The Astronomical and Philosophical Balance

The *Discourse* generated a reply from Grassi, who used the pseudonym of Lothario Sarsi of Sigensana, an anagram of Orazio Grassi Savonensi.[25] The author directs his response to Galileo himself, dismissing Guiducci as an amanuensis. Sarsi explains that he attributes the authorship to Galileo because he has heard that the astronomer himself claimed to be responsible for the ideas of the treatise and because Guiducci stated in the *Discourse* that he was relaying the opinions of the philosopher. Parodying the approach of the *Discourse*, Sarsi says he will present the opinions of his master, Horatio (in the English translation) Grassi (70). We shall retain the fiction and refer to the author as Sarsi.

As the title page informs the reader, *The Astronomical and Philosophical Balance* (1619) purports to *weigh* the "opinions of Galileo Galilei regarding the Comets . . . as well as those presented in the Florentine Academy by Mario Guiducio."

Introduced with the same humanistic flourishes as the initial piece in the controversy, the work bears a foreword to the reader in verse that sustains the metaphors about the comet and its fearful light in Grassi's initial work. The comet "brandishes a fiery torch in the frozen Bear" by which light the "balanced scales" of Libra appear (68). The title page displays a striking engraving of the comet illuminating Libra in the sky above an observatory. Drake's prose translation cannot capture the craft of the original, but the play on words in the last lines is conveyed. Speaking of the comet, the final lines say:

> I recognize the silent dominion of a new light. It orders
> its light to be weighed on one scale and the other, and
> on those scales its tail is tested. We also shall
> weigh our remarks on them.

> *Agnosco tacitum lucis imperium novae:*
> *Hac illa trutina lumen expendi suum*
> *His et probari lancibus comam iubet;*
> *His nostra nos et dicta pendamus licet.*

A detailed examination of this and the remaining works in the comet debate would require far too much ink. Although the later pieces are more important, they are also quite lengthy. Since the foundation of the dispute has been laid out in the analysis of the first two pieces, perhaps it is enough to summarize the rest and to see how the themes of the earlier pieces are taken up and amplified with rhetoric.

25. Ibid., 83.

The prose in this reply is far less literary than in Grassi's earlier work. It contains no poetic embellishments or figurative flights, but confines its rhetoric to occasional exclamations at Galileo's replies. The first piece is an intellectual entertainment; the second is a scholarly refutation.

Grassi's Rebuttal

The introduction of the *Balance,* echoing Guiducci-Galileo's theme of the need to search for causes, declares that the Gregoriana (the Collegio Romano) has been implored to find explanations of these appearances. Its conclusions, the author says, were not deemed to be disappointing except by Galileo, who "disapproved of our explanation, and rather sharply." This fact was at first distressing, "but afterward we were consoled to find that Aristotle himself, Tycho, and others were not considered much more gently by him in this dispute" (69).

Sarsi implies that Galileo has broken the code of polite discourse appropriate to a refutation of a festive public disputation. He returns to the point, saying somewhat self-righteously that in his reply to the arguments he will "constantly abstain from words which are more indicative of an exasperated and angry spirit than of knowledge, although I readily grant that method of reply to others if they desire it" (70).

The author says he decided to continue the debate in order to show that the arguments of his teacher were not fallacious and to consider carefully the confutation presented by the other side. For his part, Sarsi says, he cannot understand why Galileo claims to have been unfairly treated in the *Disputation* when the Collegio has always shown great respect for him, honoring him and praising him for his discoveries with the telescope. Sarsi, taking a page from Galileo, appeals to his audience's sympathy when he reveals the dismay Galileo's response caused the Collegio: "Therefore I do not know what reason he has for vilifying the good name of this Collegio Romano so that he calls its teachers unskilled in logic and does not hesitate to pronounce our position regarding comets as worthless and supported by false arguments" (71).

As for Guiducci-Galileo's accusation that Sarsi's master is content to follow Tycho's lead in these matters, Sarsi says that there is little in his *Disputation* that would warrant that conclusion. He then reverses the issue: "But consider, let it be granted that my master adhered to Tycho. How much of a crime is that? Whom instead might he follow? Ptolemy? . . . Or Copernicus? But he who is dutiful will rather call everyone away from him and will equally reject and spurn his recently condemned hypothesis. Therefore, Tycho remains as the only one whom we may approve as our leader among the unknown courses of the stars" (71). In tit-for-tat, Sarsi offers here a gratuitous admonition to the best-known supporter of the Copernican hypothesis, in retaliation for the taunt by Galileo regarding the implications of his espousal of Brahe's opinion on circular orbits of the comets. He also clearly reveals the constraints regarding speculation demanded of Jesuits.

The Weighings

The *Balance* contains three "weighings" or general headings under which the major propositions in the Guiducci-Galileo *Discourse* are considered. Sarsi first summarizes Galileo's objections to the positions of the *Disputation,* wherein the testimony and evidence of respected authorities furnished the ground for the original arguments. He counters Galileo's ascription of rectilinear motion to comets with the assertion that Kepler had discarded this possibility and had suggested that it could be explained in conjunction with the earth's motion. But Sarsi says this solution is not open to Catholics. He suggests a number of other possibilities, including the possibility that its movement might be elliptical.

Sarsi then goes to the major business of the first "weighing," remarking that since Galileo complained about his master's want of logic, he will examine whether Galileo observes logical rules, "not by many examples, for we shall be content with one or two." This litotes implies there are indeed many infractions that could be taken up. Galileo's insulting tone is catching. In this section Sarsi explains at greater length why he thinks the telescope does not enlarge distant objects as much as it does nearer ones. The defense of his position is adroit and substantial, and Shea finds Grassi to be more right than Galileo in this exchange.[26] At the end of the passage Sarsi attempts to mollify Galileo by showing him how he had misinterpreted his original comment. It was aimed, he says, not at Galileo but at those who mistrust the telescope and have no knowledge of optics, for the opinion expressed there was not "opposed to the truth and, indeed, to his [Galileo's] own conclusions." Galileo would have realized this had he "considered the matter with a less agitated mind." Sarsi goes far beyond superficial politeness in his explanation, adding a rhetorical question: "Why should it ever occur to us that at some time these things which we have considered to be wholly his own would not be pleasing to him?" (86). We have noted before the incongruous composition of Galileo's nature: the vulnerability to slights or apparent insults and the propensity for delivering the same with varying degrees of grace and wallop. His irascibility was perhaps the most unfortunate of nature's endowments, causing him more grief perhaps than his ill health.

The second weighing considers Galileo's opinion concerning the "substance and motion of comets" (86). With seven arguments (emphasized sys-

26. Ibid., 80.

tematically by the glosses), Sarsi demolishes Galileo's argument that the comet is an apparition of some sort. Using optical principles and geometrical figures, he demonstrates that clouds or vapors lit by the sun or the moon would not follow the path the comets did nor would they permit the readings of parallax of Tycho and later observers.

Next, Sarsi takes up the question of whether the great comet of 1618 displayed rectilinear motion, and he refutes the view that it did on the basis of observations regarding the speed and direction of the comet. Galileo's contention would not explain how the comet could be carried toward the pole, especially since he denied the motion of the spheres. To Galileo's demurrer that he cannot give an answer to the contradictions his view poses, Sarsi cannot keep himself from remarking: "Ought one not be astonished that a straight-forward and by no means timid man is suddenly seized by such fear that he is afraid to present his argument? But I am not one who is acquainted with prophecy" (97). The intimation that the phenomenon might be explained by the earth's motion is met immediately by an exaggerated *praeteritio* or *paralipsis*, the trope by which one pretends to pass over something in silence:

But at this point I hear something or other softly and timidly whispered in my ear about the motion of the earth. Away with the word dissonant to the truth and harsh to pious ears! It had better be whispered with lowered voice. . . . For if the earth is not moved, this straight motion does not agree with the observations of the comet; but it is certain that among Catholics the earth is not moved, and therefore it will be equally certain that this straight motion by no means agrees with the observations of the comet, and therefore must be judged inept for our purpose. Nor do I believe that this had ever come into the mind of Galileo whom I have always known as pious and religious. (98)

But Sarsi then proceeds to take up the opposite, unthinkable explanation, concluding that even if the earth moved the conjecture of straight-line motion would not be possible.

The third weighing considers a few remaining problems related to Galileo's criticisms of Aristotle. Sarsi interposes a comment before taking up the issues, saying that he desires simply to "champion the conclusions of Aristotle" (105). He does not plan to do so by explicating Aristotle, but will show that Galileo's arguments are "unsound." He adds: "and to speak plainly . . . certain propositions on which, like a foundation, the entire mass of his disputation rests display perhaps some appearance of truth, but if anyone inspects them very carefully, he will, I believe, judge them to be false" (106). The rhetoric and the dialectic of Galileo's attack on the disputation fail to provide a firm argument, Sarsi insinuates, either from probabilities or mathematical demonstrations.

In his first counterargument Sarsi introduces a number of experiments in support of Aristotle's view that air can be moved by a rotating smooth body. These become the basis of dialectical arguments to support Aristotle's contention that a comet's fire is kindled by the movement of the spheres. Galileo had denied this and said that celestial objects must have a perfect spherical shape and polished surface since they are thought to be the most noble bodies; he further claimed that experiments proved that air is not moved along by smooth surfaces. Referring to an experiment purportedly introduced by Galileo, showing that a straw on the surface of water in a dish does not move if the dish is rotated, Sarsi devises a more elaborate experiment. Before describing it he attempts to soothe Galileo by praising his ingenuity in developing "very simple demonstrations" to explain "difficult matters": "I do not wish to lessen this widespread praise of him, but as concerns the present subject, I have found each experiment entirely false—may Galileo spare me for speaking the truth" (109).

Grassi details his own experiment showing that air can be moved by a rotating surface and mentions that verification can be provided by Virginio Cesarini, who witnessed it. The witness proves to be important, for Galileo chose to address Cesarini in *The Assayer*, his response to the present work of Grassi. By the time Galileo completed that work, Cesarini was Lord Chamberlain to Pope Gregory XV. Inducted into the Lincean Academy at the age of twenty-two in 1618 and a former student of the Jesuits, he was an admirer of Galileo, as Grassi surely knew.[27] .

Sarsi's second counterargument treats the relation of motion to heat. Aristotle had taught that motion is the cause of heat and Galileo had denied it, maintaining that friction was its cause and that parts must be worn away in the process. Sarsi describes experiments Grassi has performed that show no appreciable loss of matter in the heating of copper through pounding it. In a curious mixture of proto-modern and medieval methodology, Sarsi then quotes literary authorities, perhaps because Guiducci mentioned Seneca in the *Discourse*. He introduces opinions from the poets—Ovid, Lucan, Lucretius, Virgil—that lead projectiles melt from motion through the air. He does so, he says, because the poets were "well trained in knowledge of natural phenomena," and he buttresses these views with other authorities, including Seneca's. One piece of testimony, which Galileo will turn back on him, is that of Suidas, who in *Histories* claimed that the Babylonians whirled eggs in a sling to cook them (118–19).

The remainder of the *Balance* is given over to a consideration of further arguments against Aristotle. In some cases Grassi is quite resourceful and correct in the evidence given for his propositions. In others his desire to fol-

27. Drake, *Galileo at Work*, 444.

low Aristotle takes him onto less firm ground. The space given to these arguments refuting Galileo's criticisms of Aristotle attests to the importance they have for Grassi and the Collegio. They give credence to the view that members of the Order did indeed feel bound to defend Aristotle's views since these were approved by the Constitutions.

Shea has called the opposing experiments offered in these exchanges "the war of experiments" and notes that these are always introduced to illustrate, to confirm, or to falsify the hypotheses that have previously been assumed.[28]

Guiducci's Response

After the appearance of the *Balance,* Guiducci felt it necessary to make an attempt to salvage his reputation as a scholar, desiring not to be known simply as a mask for the revered Galileo. His method, interestingly enough, was a letter, written in Italian, to his former rhetoric professor at the Collegio Romano, entitled *Letter to the Very Reverend Father Tarquinio Galluzzi of the Society of Jesus* (1620). Evidently stung by Grassi's imputations of rudeness and fearful that he may have appeared to denigrate the education he had received at the Collegio, Guiducci was at pains to set the record straight and express his veneration for its professors. Protesting that he had simply desired to provide an example of oratory for the students of the academy, he asks that his old professor of rhetoric judge whether or not he was at fault as Grassi had suggested.

At the beginning of the letter he apologizes for having written the *Disputation* in Italian instead of the Latin customary in scientific treatises. He explains that one of aims of the Florentine Academy is to perfect expression in the native tongue. From his position as Consul he hoped through his lecture to provide a suitable display for the Academy's students. "In this, I did not depart from the example of Sig. Galileo, who has also set forth his marvelous ideas in this idiom" (136).

He apologizes also for the topic of the *Discourse,* explaining that he chose it because the comets were being talked of, as was the opinion of "the most ancient and modern philosophers" and among them "the reverend Mathematician of the Collegio Romano." He declares his astonishment at finding that disagreeing with that opinion could be looked upon as an injury, "especially as I referred to it with the greatest possible honor and respect" (137).

The explanation becomes then a plea against a tyranny that would "restrict the liberty of men's intellects" (137). Bowing to Grassi's fiction, he cleverly directs his criticism to the "pupil" of the Jesuit instead of the "master" and berates him for claiming oracular wisdom for the Collegio's phi-

28. Shea, *Revolution,* 92.

losophers. Such immoderate praise he finds exceedingly inappropriate to the modest character the college sought to engender in its students.

Guiducci then registers his dismay at Sarsi's assumption that the *Discourse* was written by someone else. He takes particular offense at Sarsi's play on words when he writes that he prefers to speak to the "dictator" rather than to the "consul." He claims that he wrote the treatise simply to expose the ideas of ancient philosophers, and of Galileo among the moderns. Following Sarsi's reasoning one would have to call Clavius "an egregious copyist" because he reviewed and compiled the opinions of authorities in his commentary on Sacrobosco. This is an ingenious ploy, for Clavius is the most respected of the Collegio's mathematicians. In a crowning touch, he compares his situation to Plato's as the conveyor of Socrates' views. No one would think of calling Plato a copyist and "prefer disputing with Socrates as Dictator" (141).

Little attention is devoted by Guiducci to the scientific content of the dispute, perhaps because he knew that Galileo would attend to that himself. After spending half of the letter in an indignant response to Sarsi's slighting attitude towards himself, he turns to the passages in the *Discourse* that Sarsi found disrespectful.

Concerning the assumption by Sarsi that he wished to belittle Grassi by saying Nature takes no delight in poetry, Guiducci indicates that he meant only to criticize Tycho. Feeling at liberty to speak harshly against Sarsi, Guiducci complains: "With no greater veracity than in the foregoing, Sarsi saddles me with doctrines and conclusions that I have not held and do not hold to be true, in order then to have a field in which to vanquish them and thus expand the size of his volume" (144). Guiducci then takes up in two brief paragraphs what he does not hold in regard to the telescope and to the appearance of the comet.

The longest passage of the work is devoted to a defense of an experiment he claims he himself devised, to show that air does not follow a moving smooth surface. Through it Galileo became convinced, a man who, Guiducci says, is "open-minded" and "not excessively devoted to his own opinions" (146). But, he adds, the astronomer thought that the proof did not advance his point.

In sum, he proclaims that "Sarsi's experiments are no hindrance to what I have said, they being very fallacious and not free from suspicion of fraud" (147). In attacking the fictive Sarsi, Guiducci can present the appearance of indignation at being found impertinent, and at the same time gain the opportunity to assail the real mathematician whom he obviously does not respect.

The final attack is directed against an argument Grassi offered to support his view that luminous bodies are transparent, which explains why stars can be seen through the tail of the comet. Grassi had appealed to the passage in Scripture where Nebuchadnezzar saw Shadrach, Meshach, and Abednego walking in the middle of the fiery furnace. Guiducci counters with a text immediately following in which an angel is said to have caused a damp wind to quench the flames in the midst of the furnace, thus enabling the three young men to survive. Not presuming to offer his own exposition, Guiducci leaves it to the doctors of divinity to judge whether the three young men were seen within the fire or within the wind. He then appeals to the opinion of these doctors as to the wisdom of adapting Scripture to this purpose. For his part he is sure that stars cannot be seen through flames.

He closes with the hope that the fathers are now convinced of his loyalty to the Collegio. He advises Sarsi to cultivate the patience to endure what he has said to defend his *Discourse* "from deficiencies and defects imputed to it by him."

The exchanges of Grassi, Galileo, and Guiducci demonstrate clearly another reason for the increased presence of rhetoric in scientific discourse. Not only has it been employed because the audience is a mixed one, as in the case of the *Letter to Christina;* and because two of the occasions—the public disputation and the oration—have traditionally required ceremonial rhetoric; but also because the experiments that furnish support for arguments must seem to be credible. At this point in the development of modern science the experiments by themselves are not wholly convincing, for variables are not controlled, the conditions and the apparatuses are not comparable, and thus repetition by different people yields different results. In such cases, the character of the experimenter becomes very important. It is he who determines the point of the experiment, who testifies to what occurred, and who interprets the results. The next essay provides further evidence of this new application of ethos.

The Assayer

The most important exchange in the series, written by the master of persuasion himself, is also the lengthiest. Galileo's *The Assayer* (*Il Saggiatore*) is cleverly and rhetorically conceived as a further "weighing," this time, as the long title discloses, in "a delicate and precise scale . . . the things contained in *The Astronomical and Philosophical Balance of Lothario Sarsi of Siguenza*." Galileo, writing in Italian, casts his response in the form of a letter. The title page describes the recipient as "the Illustrious and Reverend Monsignore Don Virginio Cesarini, Lincean Academician, Lord Chamberlain to his Holiness," and its author as Signor Galileo Galilei, Lincean Academician, Gentleman of Florence, Chief Philosopher and Mathematician to the Serene Grand Duke of Tuscany. The Lincean Academy published the work in 1623; their dedicatory letter to the newly elected Pope, Urban VIII, is included in the extensive introductory matter.

Opening Endorsements

The presentation of the work is a rhetorical triumph. It more than matches Galileo's reputation against the aura of sanctity and erudition surrounding Galileo's disputant, the Reverend Father and Chief Mathematician of the Collegio Romano. Part of the credit most go to the Linceans, for as Pietro Redondi has shown, the work was the product of a Lincean conspiracy.[29]

The Roman Linceans, Cesi, Cesarini, and Ciampoli, urged Galileo to respond to Grassi and to do so by means of a letter. They also prepared the "packaging" and reviewed Galileo's text, suggesting revisions. On the reverse of the title page are two notations of the work's imprimatur, one by the Master of the Sacred Palace, who was responsible for the licensing of books, and the other by the Socius of the Palace. In between is an unusual statement by the Reverend Niccolò Riccardi, later to become Master of the Sacred Palace himself, who in this capacity later reviewed the *Dialogue Concerning the Two Chief World Systems*. Riccardi, asked by the Master to read *The Assayer* before publication, states that he found "nothing offensive to morality, nor anything which departs from the supernatural truth or our faith." He goes on to praise its contents and its author:

I have remarked in it so many fine considerations pertaining to natural philosophy that I believe our age is to be glorified by future ages not only as the heir of works of past philosophers but as the discoverer of many secrets of nature which they were unable to reveal, thanks to the deep and sound reflections of this author in whose time I count myself fortunate to be born—when the gold of truth is no longer weighed in bulk and with the steelyard, but is assayed with so delicate a balance.

Galileo could not have chosen a more suitable advertisement, nor a more apt crafting of the trope in the title. To have such praise from the censor must surely mitigate that "hint" of heresy couched in Grassi's allusion to what he dare not imagine regarding heliocentric intimations in the *Discourse*. There Grassi's praeteritio certainly might have recalled to mind statements in *Sidereus nuncius* where the astronomer revealed his conversion to Copernicanism. The enthusiastic endorsement of Riccardi may have been encouraged by the new pope, who was a great supporter of Galileo at this time. As a Dominican Riccardi must also have been happy to see a Jesuit trounced, for the Order of Preachers was smarting from recent theological battles with the Society. The dedication to the pope that follows also extols Galileo as an

29. Redondi, *Galileo Heretic,* 44–45. Redondi cites letters of Ciampoli of 18 May 18 and 2 August 1620, 15 January 1622, and 26 February 1622 (*Opere* 13:38ff, 46ff, 69, 83), and Cesarini, 7 May 1622 (*Opere* 13:89).

explorer of the heavens. An imposing full-page portrait of the mathematician and philosopher appears next, preceding six pages of laudatory poems. The first is a ceremonial introduction by Johann Faber, a Lincean and a physician. Columbus and Vespucci, he says, must yield to Galileo, who gave to mankind "the succession of the stars and the new constellations of the heavens." In verse he enumerates Galileo's discoveries concerning Jupiter, the moon, the sunspots, and now the comets. Although Aristotle "under a learned appearance . . . deceived minds," Galileo by means of the telescope "has exposed the phenomena and the meaningless nonsense believed by princes of learning."

Hyperbole prepares the way for a lengthy poem by another Lincean, extolling Galileo's piercing sight and his triumph over the Stagirite. Poetic introductions, of course, were common, but in this case they are focused with considerable care on the revolutionary nature of Galileo's universally acclaimed discoveries. Aristotle must, despite his reverend champions, yield also. The overthrow of the old philosophy was one of the chief aims of the Linceans.

In this piece there is no shadow author, but certainly there is a shadow audience beyond Cesarini himself, as publication of the work and its dedication imply. Moreover, beyond the shadow target, Grassi, is a greater one, the Jesuit order. The Linceans were determined to reduce the Jesuits' scholarly reputation. Johann Faber had written to Galileo, appealing to him to cut down to size "the pride of the Jesuits."[30] But the Academicians wanted to do so surreptitiously. On the advice of Ciampoli, Galileo addresses the fictive Sarsi in order to attack the real author with greater freedom.[31] Again, Grassi's "joke" is turned to his disadvantage.

A Victim of Injustice

Galileo begins, as he did in the *Letter to Christina,* with a captatio benevolentiae wherein he describes the injustices done to him: "I have never been able to understand, your Excellency, how it comes about that every one of my studies which, in order to please or to be of service to others, I have seen fit to place before the public has occasioned in many a certain animus to detract, steal, or deprecate that modicum of esteem to which I had thought I was entitled, if not for the work, at least for my intention" (163). He goes on to describe the outpouring of opposition that arose upon the publication of *Sidereus nuncius, Discourse on the Comets,* and the *Letters on Sunspots,* and the attempts of Simon Mayr to take credit for both the military compass and his

30. *Opere* 13:43; Faber's letter, 15 February 1620 (*Opere* 12:23). Redondi (45) describes the desire to "get" the Jesuits.
31. Redondi, *Galileo Heretic,* 44.

discoveries of the Medicean planets. Such false claims and denigrations of his work would have silenced him, he says, but for the recent importunities of his friends. Galileo's friends had indeed urged him to take up the pen in defense of his opinions, and their pleas allow him to ascribe to others opinions that are *au point:*

These gentlemen, my friends, demonstrating in no small way their approval of my ideas have managed with various reasons to change my mind about the resolutions thus made. In the first place, they tried to persuade me not to be concerned about these obstinate attacks, saying that in the end these would in fact rebound upon their authors and would render my arguments more vivid and attractive, becoming a clear proof of the uncommon nature of my compositions. They pointed out to me the common maxim that vulgarity and mediocrity receive little or no consideration and are left behind; the mind of man turns only on that which reveals marvels and transcendent things, and it is this which in turn gives rise in ill-tempered minds to envy and defamation. (168)

His own writings thus magnified and the opposition deftly minimized, he comes to the task at hand. In the narration he explains that he had thought his previous silence would quiet the fulminations against him, but his enemies have now resorted to blaming him for the writings of others. Galileo says that this false ascription shows "a spirit impassioned beyond all reason" (169). This description is an obvious retort to Grassi's remark about the "exasperated and angry spirit" displayed in the *Discourse*. The painstaking response to every critical statement and to every difference of opinion that might be perceived to imply a denigration of his own work is typical of *The Assayer*.

Galileo declares that anyone who knows Guiducci knows he is capable of writing the *Discourse*. But, in a reversal of the point, he asks why even if he himself had been the author and had wished to remain incognito, Sarsi should want to unmask him? And he says that for his part he will respect the mask of Sarsi (170). Besides relieving Galileo of the customary deference due the reverend father, this pretense excuses him from any obligation entailed by Grassi's recent overtures towards Galileo and his friends. Grassi had openly discussed his disputation with Ciampoli and spoken "respectfully" of Galileo.[32] A direct attack on the Jesuit might not only anger Grassi, it could also range the entire Society against him. Galileo's dissimulation might have actually saved the game had he not employed to such devastating effect ridicule and the poisoned epithet, thereby unmasking himself.

Galileo next bows toward Cesarini, extolling his noble qualities and im-

32. Shea, *Revolution*, 106, n. 15, citing Ciampoli's letter to Galileo, 18 October 1619 (*Opere* 12:494).

partiality, and then begins the task of refuting the *Balance* in exhaustive detail, responding to each paragraph in order.

Vanquishing Sarsi

The body of the work retains the quality of a letter by its informal tone and its occasional remarks to Cesarini. We will not attempt to offer even a summary, since it takes up Grassi's text in tedious detail. Its significant passages have been analyzed so much in the literature that we will be content to outline Galileo's attack and sample the special rhetorical qualities of the arguments that surface even in translation.

Galileo first attempts to catch Sarsi in a contradiction in the very title of *The Balance*. To produce a pun, says Galileo, Sarsi placed the comet in the constellation of Libra when he had said in his own *Disputation* that Scorpio was the comet's native land (171). It is a petty point and one not clearly supportable because Grassi had reported that the "new foetus was established in Scorpio at a longitude of about 11 1/2 degrees, between the two scales of the Balance [i.e., Libra]" (10). Galileo elaborates, turning the point into a jibe and a program for what is to follow:

Hence, much more appropriately (and more truthfully, when we consider what Sarsi has actually written), he might have entitled his work, "The astronomical and Philosophical Scorpion"—that constellation which our sovereign poet Dante called the

> . . . figure of that chilly animal
> Which pricks and stings the people with its tail;

and truly there is no lack of stings here for me. Much worse ones too than those of scorpions; for the latter, as friends of mankind, do not injure unless first offended and provoked, whereas this fellow would bite me, who never so much as thought of offending him. But as luck will have it, I know the antidote and speedy remedy for such stings, and I shall crush the scorpion and rub him on the wounds where the venom will be reabsorbed by its proper body and leave me free and sound. (172)

The motif of the scorpion is returned to later in the work when Galileo remarks that Sarsi, in attempting to refute his reasoning about telescopic enlargement, seizes on trifles. In this Sarsi's argument resembles "a serpent which, scotched and trampled, no longer has any life in it outside the tip of its tail, which still goes on twitching in order to make the passers-by believe that it is yet alive and strong" (222).

With just such care Galileo attacks the wording and implications of each of Sarsi's weighings. The reasoning is brilliant, the illustrations and analogies persuasive, and Galileo, as always, makes telling observations about the nature of science and the role of the scientist in these philosophical debates. But here, as elsewhere, he often coats his ripostes with venom. These were received with delight by those in sympathy with the letter's aim—the pope is reported to have been much amused.[33]

At first Galileo consistently maintains the fiction of Sarsi's authorship. Echoing Guiducci's letter to Galluzzi, which he had part in, he feigns disbelief that Father Grassi could have permitted him to be an intermediary: "I am sure that the said Father would never have spoken or thought or willingly have seen Sarsi write such fantastic things, far removed in every respect from the doctrine taught in Father Grassi's college, as I hope I shall make clearly recognized" (175–76).

He continues in this vein, fulminating against Sarsi's lack of judgment in pretending to have heard that Galileo had said he had composed the *Discourse*. To Sarsi's intimation that Galileo showed disrespect for the Collegio he remarks that if Sarsi's views do reflect those of the college then he is forced to conclude that they really have no regard for his opinions, since these are not approved in *The Balance:* "Nothing is to be read there except opposition, full of accusation and blame, and, if one may believe the rumors, there is in addition to what is written an open boast of power to annihilate everything of mine" (179). This seems to be the most compelling reason Galileo decided to yield to the pleas of his friends to answer Grassi. The *Balance* in itself offers far too little offense to generate such a massive counterattack. Ciampoli had reported to Galileo, however, that although Father Grassi did not speak of him harshly, in the discussions of the other fathers "the word 'annihilate'" was "quite common."[34]

In treating the matter introduced by Sarsi in the first weighing, Galileo dwells upon the fact that the arguments are the same as those advanced by Tycho. He then turns to a critique of the Dane's geometric proof in his observations of the comet of 1577. Drake comments that this criticism "is quite gratuitous," since Tycho had said that his analysis was not exact. He adds that "Galileo seldom neglected an opportunity to ridicule Tycho" (369, n. 9).

The assayer next seizes upon Sarsi's appeal to authorities. These, he says, are irrelevant to the issue at hand. He proceeds to group Sarsi with the worst of the scholastics of his day who are more convinced by arguments from authority than by sense evidence and mathematical demonstrations:

It seems to me that I discern in Sarsi a firm belief that in philosophizing it is essential to support oneself upon the opinion of some celebrated author, as if when our minds are not wedded to the reasoning of some other person they ought to remain com-

---

33. Redondi (49) mentions the description Ciampoli gave of its reception; see Ciampoli's letter to Galileo, 28 October 1623 (*Opere* 13:141).

34. Drake notes this (*Controversies,* 369, n. 8), quoting from Ciampoli's letter to Galileo, 9 December 1619 (*Opere* 12:499).

pletely barren and sterile. Possibly he thinks that philosophy is a book of fiction created by some man, like the *Iliad* or *Orlando Furioso*—books in which the least important thing is whether what is written in them is true. Well, Sig. Sarsi, that is not the way matters stand. Philosophy is written in this grand book—I mean the universe—which stands continually open to our gaze, but it cannot be understood unless one first learns to comprehend the language and interpret the characters in which it is written. It is written in the language of mathematics, and its characters are triangles, circles, and other geometrical figures, without which it is humanly impossible to understand a single word of it; without these, one is wandering about in a dark labyrinth. (183–84)

This famous passage has been extensively discussed for its reputed evidence of Platonism, but it is quoted here because it illustrates Galileo's use of rhetoric to magnify his separation from the Peripatetics' extreme reliance on authority. Yet Galileo shares with them the view that mathematical computations can provide certain proof. Whether these proofs can be applied to the realm of nature so as to yield necessary demonstrations is the problem for some of them, but such was not the case with Grassi. He was an excellent mathematician and highly respected among academicians. His *Disputation*, in fact, contains a fine example of the very method Galileo was recommending to him: the use of parallax to prove the position of the comets. Moreover, his fellow Jesuit Giuseppe Biancani had written a treatise on the application of mathematics to problems in nature in 1615, long before Galileo's statement in *The Assayer*. Biancani noted in his treatise that mathematical demonstrations are most powerful (*potissimae*) and the best means of attaining certitude in the physical sciences.[35]

Returning to Sarsi's preference for Tycho's explanation of the cosmic system, Galileo states that it cannot be compared to the "complete systems of the universe" of Ptolemy and Copernicus (185). Since Catholics are denied the Copernican answer and the Ptolemaic is untenable, they are left with none. But Galileo does not think he should be criticized with Seneca for desiring an answer regarding the "true constitution of the universe"; furthermore, he does not lament and "deplore the misery and misfortune of our age," as Sarsi maintains.

35. Biancani, like Grassi, studied under Christopher Clavius at the Collegio Romano. Clavius's course in mathematics was specifically designed to instruct students in the importance of that subject for the study of the physical world. It was he who said after the appearance of the nova of 1572 that the Peripatetics would have to revise their Aristotelian views of the matter of the heavens to account for the existence of the nova in the celestial realm, where alterations were not expected to appear. The location of the nova Clavius conjectured through mathematical demonstration; see Wallace's account of Clavius and the nova in "Galileo's Early Arguments for Geocentrism," 31–40; also Wallace's treatment of Biancani and the mathematical tradition at the Collegio in *Galileo and his Sources*, 141–48.

Optical Arguments

The conjecture of the *Discourse* that the comets are merely an appearance and that they seem to move in straight lines is defended vigorously, with more imputations of Sarsi's dishonest manipulation of the arguments. Galileo next takes up the problem of the relative enlargement of objects by the telescope. Naming Grassi directly, he says that he erred by following Tycho in his argument in the *Disputation*. (The anonymous author of the first work had been publicly revealed by Sarsi when he assumed "his master" Grassi's defense.) Galileo then turns his scorn on Sarsi for daring to impugn his logic, showing the pupil to have been greatly mistaken and guilty of egregious errors in his deductions. According to Galileo, neither the relative enlargements nor the causes of the irradiation we observe around the stars have been understood by Sarsi. Galileo continues to separate Grassi from Sarsi by commending the master for his defense of the telescope against those who claim it deceives: "In this act the Father's intention seems to me as praiseworthy and good as the choice and quality of Sarsi's defenses appear to me poor and harmful, when against the impostures of its maligners he tries to cover the true effects of the telescope by attributing to it false ones" (211). But then he goes on to extend his contempt to those who share Sarsi's views, saying that he is unable to thank Sarsi for his support: "He cannot reasonably pretend that I should increase my debt and my affection toward people who make silly and false attributions and who threaten me with loss of their friendship because I reveal their errors by speaking the truth" (211). The target is unmistakably now the larger one: the Jesuits in general.

Galileo's possessiveness regarding the telescope is evident in the passage that follows. He chides Sarsi for calling the telescope his "foster child," after praising the benefits of the instrument. He asks: "Is this rhetorically sound? I should have thought rather that on such an occasion you would have tried to make me believe it my own child even if you had been certain that it was not" (211). He rehearses his part in the telescope's history, relating his laborious construction of a superior model and demanding from Sarsi acknowledgement of his ingenuity. In reading this Grassi must have thought the passage revealed Galileo's extreme sensitivity more than any fault of his own, for he had been careful to note the acclaim given to him at the Collegio for his invention and his discoveries. Extensive discussion of Grassi's misapprehensions of the capacity of the instrument follows.

Illusory Appearances

The second weighing begins with a curious criticism of Sarsi's grasp of the nature of the opinions offered by Guiducci. The conjecture that the comets are illusions, for instance, was simply offered "for consideration by philosophers, together with such reasons and conjectures as appear suitable to convince them that such might be the case" (231). The same misinterpretation holds for his suggestion that they move in straight lines perpendicular to the earth. These opinions were simply conjectures, which Sarsi pretends were definitely held. He did so, suggests Galileo, only better to "annihilate them—which if he does, I shall hold myself obliged to him, as in the future I shall have one less opinion to consider when I set my mind to philosophizing about such matters." This is not to be the case for all of them, however, for Galileo maintains "there is still a little life left in Sig. Mario's conjectures" (232). Galileo seems to be saying that the dialectical arguments were meant to be considered but not refuted, as they might be were he really serious about them.

He refutes the first of Sarsi's arguments that the comets resemble other celestial lights too much for them to be considered illusions, using one of the ingenious comparative images for which he is famous: "I confess that I do not have such a perfect discriminatory faculty, but resemble the monkey that firmly believed he saw his mate in a mirror, and so live and real did the image seem to him that he did not discover his error until he had run behind the mirror two or three times to catch her" (232).

To illustrate further the frailty of arguments against his conjectures and at the same time to support his skeptical stance, Galileo fabricates a parable. He first gives the gist of the parable. He tells of a man who had "extraordinary curiosity and a very penetrating mind," who was quite interested in how musical sounds were produced. He raised birds whose songs he enjoyed, marvelling in their capacity to produce music. Coming one day upon a shepherd boy playing a flute, he acquired that and retired to study it; then a stringed instrument delighted him, showing him different ways of making sound. Later the hinges of a temple gate, the ringing of a goblet by drawing a finger across the rim, and insects beating their wings produced sounds in ways of which he had not dreamed. Finally, he was baffled by the source of the cicada's noise, which was not in its mouth or its wings. Thinking that the sound might come from the thorax, he decided to break its sides to find the sound. "But everything failed until, driving the needle too deep, he transfixed the creature and took away its life with its voice, so that even then he could not make sure the song had originated in those ligaments" (235–36).

The application is clarified: "Therefore I should not be denied pardon if I cannot determine precisely the manner in which comets are produced, especially as I never boasted that I could, knowing that it may occur in some way far beyond our power to imagine. The difficulty of understanding how the cicada's song is formed even when we have it singing to us right in our hands is more than enough to excuse us for not knowing how a comet is formed at such an immense distance" (236–37). The parable could be lifted from the work to live a life of its own as an eloquent mini-essay on man's noble but ineffectual efforts to understand nature's secrets. Galileo is eloquent in excusing his own lack of certainty about nature, yet he is very certain that Grassi is wrong, and just as eloquent in his desire to establish that. Having explained his skeptical position, Galileo counters Sarsi's objections with dialectical arguments and attempts to show that his earlier explanation that the comets are an illusion is feasible after all. It is a dizzying feat, but effected with such elegance that the reader cannot but be dazzled by it. The refutation runs something like this: It is very difficult to prove an illusion, but it cannot be disproved if after all one cannot penetrate nature.

After so recently amplifying his own tenuous argument, Galileo proceeds with penetrating sarcasm to explain to Cesarini how Sarsi might argue to better effect:

But he had better pay attention to his cause, and consider that to a person who wants to convince others of something which, if not false, is at least very questionable, it is a great advantage to be able to use probable arguments, conjectures, examples, analogies, and other sophisms, and to fortify himself further with unimpeachable texts, entrenching himself behind the authority of other philosophers, scientists, rhetoricians, and historians. To reduce oneself to the rigor of geometrical demonstrations is too dangerous an experiment for anyone who does not thoroughly know how to manage these, for just as there is no middle ground between truth and falsity in physical things, so in rigorous proofs one must either establish his point beyond any doubt or else beg the question inexcusably, and there is no chance of keeping one's feet by invoking limitations, distinctions, verbal distortions, or other fireworks; one must with but few words and at the first assault become Caesar or nobody. (252)

The advice is followed by a refutation of the geometrical arguments Sarsi offered concerning sunlight's reflection from the surface of water. Galileo introduces modifications that shift the application of the reference points so that he can offer different interpretations of the figures used in the argument. Grassi later objected on the grounds that the path of the comets was different from what Galileo asserted. Shea points out that none of the interpretations is really conclusive since each is dependent upon the context for its meaning.[36]

Near the end of the second weighing, while refuting a point of Sarsi's regarding the curved appearance of the tail of the comet, which Kepler and Guiducci-Galileo explained differently, Galileo continues to magnify the impression that Grassi deceitfully tried to discredit him:

Is it possible, Sig. Lothario, that you have allowed yourself to be so transported by the desire to obscure my name in the field of science, whatever it may amount to, as

36. Shea (*Revolution*, 96–98) discusses the arguments in detail. I have followed him in my discussion.

to disregard not only my reputation but even that of many of your friends? With errors and fictions, you have attempted to make them believe your teachings to be sound and sincere; by such means you have acquired their applause. But later, if they should ever happen to see this writing of mine and thereby come to understand how often and by what tricks you have treated them as simpletons, they will consider themselves to have been shabbily dealt with by you and the esteem and grace which you hold in their hearts will change its state and condition. (272)

He ends this section with an interesting explanation of the obfuscations in the *Discourse* that Sarsi complained about. He says that Grassi's original intention was to explain comets to "common people" (*al vulgo*)—Galileo so terms the audience at the Collegio—and to teach them what they could not have understood without his instruction. Guiducci, on the other hand, wrote for a better informed audience, and not to teach, but to learn. So he framed his statements "questioningly" and did not speak with "magisterial certainty" (*magistralmente ditermino*) but permitted those who were better informed to decide the issues (275; *Opere* 6:371.6–10). Galileo then tells Sarsi that the reason he found parts of the writing obscure is that he did not know enough to understand the conjectures. Other parts he did not attempt to answer because he understood only too well, but pretended he could not comprehend rather than admit their truth.

Counter-Experiments

The third weighing contains further experiments to defeat Sarsi's contention regarding the effects of motion. Galileo thinks that Sarsi distorts and misunderstands Guiducci's and Galileo's arguments in developing his refutations. To counter Sarsi's experiments introduced to discern whether air moves in a rotating vessel, Galileo invents other experiments. Again, these are not conclusive according to modern commentators, but they are remarkable for their variety.

Galileo also takes great delight in confounding Sarsi's appeals to the authority of poets and historians. If the poets cited were present at his experiments, he says, they would change their minds and admit that they were either given to hyperbole or wrong. As to Suidas and the Babylonian method of cooking eggs, Galileo's response is without peer:

To discover the truth I shall reason thus: "If we do not achieve an effect which others formerly achieved, it must be that in our operations we lack something which was the cause of this effect succeeding, and if we lack but one single thing, then this alone can be the cause. Now we do not lack eggs, or slings, or sturdy fellows to whirl them; and still they do not cook, but rather they cool down faster if hot. And since nothing is lacking to us except being Babylonians, then being Babylonians is the cause of the eggs hardening." (301)

Applying ingenious reason to Grassi's appeal to testimony, Galileo grants the conclusion, and the means, but by reductio ad absurdum effectively demolishes the whole effort to buttress a philosophical argument with such evidence.

Galileo's Atomism

One of the proofs Galileo advances in opposition to Grassi's conception of the nature of the comet was to arouse concern in his own day and much discussion in our own. The epistemological dimensions of the argument were to affect attitudes towards Galileo afterwards, and so they become important to our rhetorical considerations.

In taking up the implications of the proposition, "motion is the cause of heat," Galileo reflects upon the nature of matter and upon our perceptions of it. He begins by saying that people generally do not have a true conception of what heat is. Most believe it to be a real quality in the material itself, but actually it resides in our senses. He differentiates what are to be identified by John Locke as primary and secondary qualities. Shape, quantity, place, and movement (all primary qualities) reside in the thing, whereas taste, odor, and color (all secondary qualities) are simply names given to what we sense. "Thus, if the living creature were removed, all these qualities would be removed and annihilated" (309).

To illustrate his meaning Galileo introduces the famous example of the tickle and the feather. When a feather is brushed across the sole of the foot, we say, "It tickles." But is the tickle in the feather or in us? It belongs to us, he says. We ascribe our sensations to the material object. But if we are removed no tickling would be present.

Heat, he says, is of this character. Tiny moving particles that make up fire touch us and produce the sensation of heat. He explains that heat particles created by friction move very quickly, penetrate our bodies, and cause a burning sensation. Galileo deliberately avoids the use of the term atoms in describing these tiny particles, calling them minimal corpuscles (*corpicelli minimi*, or *minimi ignei*, etc.). Atomism was in ill repute as a doctrine, for it had unacceptable implications for theology and teachings on the Eucharist.

Galileo probably became interested in the view through his contact with the Paduan Aristotelians.[37] Agostino Nifo and Jacopo Zabarella thought

---

37. Such contact seems to have come partly through Galileo's studies at Pisa under Francesco Buonamici, partly through lecture notes he appropriated from the Roman Jesuits. On the first, see Mario Otto Helbing, *La Filosofia di Francesco Buonamici, professore di Galileo a Pisa* (Pisa: Nistri-Lischi Editori, 1989); for an illustration of the second, see W. A. Wallace, "Randall Redivivus: Galileo and the Paduan Aristotelians," *Journal of the History of Ideas* 49 (1988): 133–49, as well as his *Galileo and his Sources*.

that *minima* might be responsible for physical changes. Galileo would have learned about the distinctions between qualities in his study of philosophy, for they had been commonly taught since the Middle Ages. The *sensibilia communia*, our primary qualities, were differentiated by scholars from the *sensibilia propria*, the secondary qualities, which affect the senses directly. The sensibilia communia were ranked lower in Aristotelian philosophy, however, for knowledge of these must first come through the sensibilia propria.

For Aristotle sense perception is an interactive process, where real qualities are perceived by the action of the senses. So sensations are reliable; they do transmit knowledge of what is out there. For the atomist, perception is subjective, since it is our interpretation of what impinges on our senses. We cannot perceive the atoms that compose objects.

The problem posed for theologians by an atomistic interpretation is the following. According to Catholic teaching, a change in the underlying substance of bread and wine occurs during the consecration of the Mass. The substance is transformed into the body and blood of Christ, but the sensible species of color, odor, and taste remain the same. If the teachings of the theory proposed by Galileo were applied to the Eucharist, its detractors said, the basic elements in the bread and wine would not be thought of as natural or substantial minima but as atoms and as such would only be interpreted subjectively. The change would then take place in us and not in the elements, being the result of our interpretation.[38] Grassi was to reply to *The Assayer* and attack Galileo on these grounds. Shea sees this as an unfair argument because Galileo could not legitimately carry on a theological debate.[39] Shea's evaluation is not altogether satisfying in that Galileo chose to move into the realm of theology in his *Letter to Christina*.

Redondi, in a controversial thesis, has claimed that the imputations of atomism, supported by these and similar passages from *The Assayer*, were what really lay behind the trial of 1633. An anonymous letter to the Holy Office claimed that this doctrine could be a powerful weapon in the hands of Protestants in their renunciation of transubstantiation. Redondi thinks that Grassi was the author of the letter and that the pope was afraid that he might be identified with Galileo and his liberal views.[40] The contention that Grassi was the author and that atomism was the secret but central issue for the trial

38. For a critical examination of this argument, which was not accepted by Giovanni di Guevara, the priest (later bishop) who checked *The Assayer* for its theological orthodoxy, see W. A. Wallace, "The Problem of Apodictic Proof in Early Seventeenth-Century Mechanics: Galileo, Guevara, and the Jesuits," *Science in Context* 3 (1989): 81–84.

39. Shea, *Revolution*, 103–4.

40. Redondi, *Galileo Heretic*, 165–75.

has been sharply disputed and generally discounted by historians of science. But the rumors of Galileo's atomism were circulated, adding further ammunition to his enemies.

At the end of *The Assayer* Galileo assesses the efforts of the two parties. He says that Sarsi has implicitly stated that his teaching rests on some "mistaken experiments and defective reasoning." Further, in support of his side Sarsi has provided "a catalogue and summary of the conclusions" in the *Discourse* which he has refuted. Galileo compares this with his own effort:

I hope that my cause will not be a little favored for my having examined point by point every reason and experiment adduced by Sarsi and replied to all, whereas he has skipped over the greater number of those given by Sig. Mario, and the most conclusive ones. I had thought of reviewing all these here in exchange for Sarsi's catalogue; but though I set myself to the task, I lacked the spirit and the strength, seeing that I should have to transcribe all over again little less than Sig. Mario's entire treatise. Hence, to save both your Excellency and myself this tedium, I have decided instead to refer your excellency to a rereading of that *Discourse*. (336)

Galileo correctly gauged his public. Despite the entertainment offered by so much of his refutation, his readers must have felt relieved when he came to the end. The breadth and depth of his contentious analysis had already become tedious. Yet in his day the book was a great success; people delighted in its wit and its daring.[41]

The Significance of the Rhetoric in the Comet Debate

As we have seen, the tactics Galileo employs to refute Grassi are both dialectical and rhetorical. His experiments are confirming but not demonstrative; his mathematical arguments are sometimes demonstrative but their worth depends upon their application. His principal defense is a time-honored dodge: earthly man cannot be sure of anything.

Rhetoric embraces and permeates the whole of *The Assayer*. It pervades Galileo's dialectical counterattacks as he questions the honesty and intelligence of his opponent. Had he simply exposed his opponent's lack of logic he would have had no need of rhetoric; dialectics would have been enough. But he preferred to destroy his opponent's ethos. Rhetoric operates too in the ridicule of the testimony Grassi brings to his aid. Most importantly, rhetoric continuously emerges in his attempts to arouse indignation in the reader for the alleged injustices done by the lies, misinterpretations, and ignorance of his opponents. The intimate theater provided by the letter genre permits the insults to be ventilated more freely than they could be in a public speech.

41. Drake (*Controversy*, i, xix) mentions its popularity, as does Redondi, 51.

Part of the motivation for this marriage of negative rhetoric with dialectics is obviously Galileo's perception of ill will on the part of his opponent. This may have been what deflected him from finding a better answer to the mystery of the comets.

Grassi's scientific treatises are not free of rhetoric either. He uses rhetorical as well as dialectical arguments, but less liberally than Galileo. In his initial piece in this exchange he handled both modes in a manner appropriate to the genre. Rhetoric amplified his dialectical arguments and demonstrations, just as it had the discourse of Copernicus and Kepler. In the *Balance* negative rhetoric also entered into Grassi's discussion at various points. When he raised the possibility that further development of the views expressed by Guiducci-Galileo might verge on heresy, his intent might be read as an attempt to cast suspicion, but it might also be seen as a desire to tease and admonish at once.

Grassi does accuse Galileo of being illogical and of deriving false conclusions; as a seething Galileo expressed it, "mistaken experiments and defective reasoning" were rife. But these charges are fair play in dialectical refutations; few debates are free of logical infractions or misconceptions. In spite of his allegations, Grassi's tone comes across as genuinely respectful of his opponent. Occasionally he stoops to deliver some bitter remarks aimed at Galileo's arrogance. His motivation, plainly enough, is anger at the indignant response he received to his initial attempt to deal with what seemed to him an intriguing riddle open to any serious mathematician.

Regarding the scientific content of the dispute, we can see that Grassi was intent upon persuading his audience that the comets were real and that they were in the celestial regions, even though his assertions called into doubt Aristotelian-Ptolemaic traditions still accepted by many Peripatetic academics. On the other hand, Galileo claimed that mathematics was the only method of gaining certainty about nature's secrets. Yet he employed dialectics and rhetoric to support a surprisingly reactionary position.

The attack mounted by Galileo against Tycho was not convincing to those mathematicians who favored Tycho over Ptolemy. They would be persuaded only by orthodox methods. Certainly Galileo was aware of this, and yet, if he were, why did he advance the kind of argument he did? Shea thinks that he was greatly influenced by the Florentine humanists who were opposed to all that savored of scholasticism. Their attitude "made room for free thought and original research but it also allowed rhetoric to pass muster for rational argument."[42]

Redondi's description of the Linceans' contribution to the development of *The Assayer* provides yet another answer. Prince Cesi and the Academi-

42. Shea, *Revolution*, 88–89.

cians added their voices to the chorus urging Galileo to respond to Grassi's work. They wanted him to destroy the Jesuits through his critique of their chief mathematician.[43]

The two explanations are mutually supportive. Galileo could well have believed that if Grassi and the Jesuits were effectively dispatched, the principal voice of the Tychonian system would be silenced. *The Assayer* showed that Grassi alone was no match for Galileo's superb rhetorical skills, and in the eyes of Galileo's sympathizers he and the Jesuits were drowned out. The outcome is ironic in view of the Academy's desire to rise above rhetoric, while retaining the beauty that eloquence could bring to discourse.

Two more pieces followed *The Assayer*—a further reply by Grassi and a comment by Kepler—but these were overshadowed by Galileo's opus and need not be analyzed here.

43. Redondi (70–77) describes a 1626 Carnival celebration at the Court of the Prince Cardinal Maurizio of Savoy wherein Galileo's *Assayer* furnished the principal theme of the entertainment. The peak of the entertainment was a lecture in which the Jesuits, without being openly named, were riduculed as slavish followers of ancient authorities. Galileo was hailed as another Columbus.

CHAPTER NINE

The Final Salvo: Galileo's *Dialogue*

In 1622, while the Linceans were still reviewing *The Assayer* prior to its publication, Campanella's *Apologia pro Galileo, mathematico florentino* was printed in Germany and began to circulate in Rome.[1] The book recalled to prominence the position Galileo had taken in support of Copernicus prior to the decree of 1616. The preface to the work was written by Tobias Adami, a Lutheran, who linked Galileo to other innovators—Cardinal Cusanus, Foscarini, Bruno, Kepler, Patrizi, Telesio, and Hill.[2] The association had the effect of raising even more doubts about the illustrious Florentine among conservatives in the Church, the Jesuits included. For his part, Grassi was busy applying his mathematics to the design of a new Church, honoring the newly canonized St. Ignatius, to be built over the site of an older edifice. Its imposing architecture was meant to complement the rest of the buildings in the enclave of the Collegio Romano.

Six months after the publication of *The Assayer* in November 1623, Guiducci informed Galileo that a complaint had been lodged with the Holy Office against the atomism found in the work, but the reviewer, Father Giovanni Guevara, praised the book, declaring that whatever its stance he would not advocate repression.[3] The identity of the complainant has not yet been discerned. Stillman Drake suggests it was Scheiner, who had recently come to Rome, and Pietro Redondi claims that it was Grassi. Others have stated that the handwriting of the accuser does not match Grassi's.[4]

Grassi was said to have been much angered by Galileo's attack, however, and also by a letter from Florence claiming that the Jesuits would be at a loss

1. Bonansea, "Campanella's Defense of Galileo," 207.
2. Redondi, *Galileo Heretic*, 40. Nicholas Hill was an atomist.
3. The complaint and letter are translated in Finocchiaro, *Galileo Affair*, 202–6.
4. Drake, *Galileo at Work*, 300. Redondi (*Galileo Heretic*, ch. 6) believes that Grassi was the author. He reprints the letter in an appendix (333). Opposing his view is that of a paleographer and chief archivist of the Vatican's Secret Archives in the Vatican Library, Sergio Pagano. He says that an analysis by Edmond Lamalle, archivist of the Roman Archives of the Jesuits, who has compared the letter with other writings of Grassi written during the same period, shows that Grassi can be excluded as the author. They think his identity cannot be found at this point. See S. M. Pagano, ed., *I Documenti del Processo di Galileo Galilei* (Vatican City: Pontifical Academy of Sciences, 1984), 40–45.

to reply to Galileo. Grassi reportedly remarked that if the Jesuits could refute a hundred heretics they could surely answer one Catholic; nevertheless, he vowed he would not write his response with the same venom.[5]

If Grassi were the author of the denunciation sent to the Holy Office, his subsequent friendliness to Guiducci could only be explained by imputing to him a deceitful and crafty nature. Grassi, with at the very least an appearance of friendliness, had visited Guiducci in Rome when the latter was convalescing from an illness there in August 1624. During his visit Guiducci told him of some of Galileo's conjectures about the movements of the tides providing proof of the earth's motions, which Galileo had first worked out and described in an unpublished letter to Orsini before the Copernican ban was published. Grassi reportedly was responsive to the thesis and the Copernican hypothesis and simply mentioned what Cardinal Bellarmine had said: that if there were such a proof and it was demonstrated, then Scripture would have to be reinterpreted. Guiducci reports that on further public meetings he was embarrassed by Grassi's overtures. In an encounter in November 1624 Grassi told him that he was now forced to write against *The Assayer* and was sorry to have to do so.[6]

Perhaps Grassi's overtures to Guiducci were stimulated by the chagrin he felt at the Order's insistence that he respond and the direction they may have required his arguments to take. The uniformity in doctrine on Aristotelianism in philosophy and against Copernicanism demanded by Aquiviva, the General of the Order, in 1611 and 1613 was to ultimately close down any such free thinking on the part of Jesuit scientists, as Blackwell has argued convincingly.[7]

The publication of Grassi's work was delayed, however, until 1626, when it appeared under the title, *A Reckoning of Weights for the Balance and Small Scale*. The engraving on its title page takes up the image Galileo had applied to Grassi, the serpent, but gives it an irenic twist: a caduceus, the symbol of the messenger of the gods, featuring two serpents facing each other and entwined around an olive branch. A legend surrounds them: "Now omens of

5. Redondi, 182.

6. See the differing accounts of the meetings in Redondi, 186–89, and Drake, *Galileo at Work*, 291–95.

7. Blackwell (*Galileo, Bellarmine, and the Bible*, 156–57) points out too that Grassi never wrote another work on science although he lived until 1654.

8. The work is reprinted in *Opere* 6:376–500. Redondi (335–40) has translated Examination 48 of Sarsi, which criticizes the atomism of Galileo's treatise. Sarsi becomes quite impassioned in his denunciation: "One must therefore infer, from what Galileo says, that heat and taste do not subsist in the host. The soul experiences horror at the very thought." He goes on to refute neatly Galileo's differentiation of the secondary and primary qualities, pointing out that since motion is the cause of heat it also causes cold, taste, and odors.

peace."[8] Galileo did not bother to reply, apparently because he thought it carried no serious challenge to his scientific doctrines; furthermore, it was prudent not to do so since Grassi discussed in it the dangerous implications of atomistic thought for the doctrine of the Eucharist.

Galileo, meanwhile, was at work expanding his treatise on the *System of the World,* which he had promised in the *Sidereus Nuncius* of 1610. In a visit to Rome in the spring of 1624 he evidently discussed the possibility of publishing his views with the new pope, Urban VIII, who as Cardinal Maffeo Barberini was his old admirer and supporter. During one of his audiences with the astronomer the pope seems to have agreed that Galileo might discuss the hypothesis as long as he treated arguments on both sides. In fact, Urban seems to have suggested he reply to a work by Francesco Ingoli, written in 1616 against Copernicanism. In the course of one of their conversations on the subject, Urban reputedly expressed strong reservations about the argument from the tides, which Galileo thought was the most compelling proof for the earth's motion. The pontiff, educated by the Jesuits in his native Florence and at the Collegio Romano, was comparatively liberal in his views of science, agreeing with Copernicus at one time himself and saying to Galileo that the Church had not condemned Copernicus's views as heretical, but simply as "rash." He believed, however, that despite the arguments from the tides we still cannot know for certain the work of God's hands. In his infinite power God could cause the waters to move in some other way.[9]

On his return from Rome Galileo completed a response to Francesco Ingoli in which he presented many of the arguments for the Copernican system aired in his *Dialogue.* He sent the letter to Rome in 1624, probably to test the waters regarding the feasibility of making his views public. Ciampoli read portions of the letter to the pope, who seemed to like some of its experiments. Subsequently, rumors regarding a denunciation of parts of *The Assayer* led Guiducci not to send the letter on to Ingoli, as Galileo had requested.[10] For a time Galileo seems to have put aside the great work of the dialogue, but in October 1629 he wrote to Elia Diodati:

A month ago I took up again my *Dialogue* about the tides, put aside for three years on end, and by the grace of God have got on the right path, so that if I can keep on this winter I hope to bring the work to an end and immediately publish it. In this, besides the material on the tides, there will be inserted many other problems and a most ample confirmation of the Copernican system by showing the nullity of all that

9. Santillana (*Crime of Galileo,* ch. 8) provides convincing evidence that this was the point at which Urban voiced this view.

10. Drake, *Galileo at Work,* 296–97, 301. See also Redondi, 138–39.

had been brought by Tycho and others to the contrary. The work will be quite large and full of many novelties, which by reason of the freedom of dialogue I shall have scope to introduce without drudgery or affectation.[11]

Opening the *Dialogue*

With this brief sketch of some of the ferment that took place during the writing of the *Dialogue,* let us turn to the work itself.[12] The frontispiece transports the reader to a theater. On a stage three men stand as if in conversation; they are identified by the lettering on their gowns as Aristotle, Ptolemy, and Copernicus. Aristotle, in partial shadow with his back to the audience, stands with Ptolemy at stage left. They appear to be discussing an intricate armillary sphere which Ptolemy holds to show the orbits of the planets circling the earth. Filling the right half of the stage and standing in full light a little apart, Copernicus faces the two. He gestures with upturned right hand toward the sphere as if to say "This cannot be." At his left side he patiently holds another model, a simple sphere with the sun at its center and an earth embedded in the metal band surrounding it. Above their heads, two cherubs bear a princely crown and hold aloft the stage curtain whereon is printed amid illustrations of six circling planets: "Dialogue of Galileo Galilei, Linceo, for the Most Serene Ferdinand II, Grand Duke of Tuscany."

The stage is thus set. The three invisible figures who have inspired the text of the dialogue appear in the engraving in earthly form. Copernicus in biretta and ermine-trimmed gown clearly dominates the scene. The bold announcement of the dialogue's title and its author appear simultaneously to seek the protection of the Duke. Nowhere is there a better example of the rhetorical elements an illustration can provide, the canon of delivery of classical oratory reborn through the medium of print.

On the title page itself the single word *Dialogo* is repeated. Directly below, Galileo is again described first as Linceo, then as "Supra-ordinary Mathematician" (*Matematico Sopraordinario*) for the University of Pisa, and finally as "Philosopher and Chief Mathematician" to the Grand Duke of Tuscany. The first and last titles have been used routinely in Galileo's published writings, but the connection to the University of Pisa is not announced in other works. One must wonder why it was chosen for this one.

The university title indicates a singular honorific for Galileo in that no

---

11. The translation is from Drake, *Galileo at Work,* 310.

12. Again, I have used the translation of the *Dialogue Concerning the Two Chief World Systems* by Stillman Drake, comparing it to the original text reprinted in *Opere* 7:21–520 and to the English translation by Thomas Salusbury, which guided Drake's, in *Mathematical Collections* (London, 1661), tome 1:1–424.

duties nor residence were required, but it would also seem to magnify his responsibilities.[13] Whatever he were to do or to say would be expected to bring glory to the University. As supra-ordinary professor Galileo would probably make a rare appearance, accompanied by much ceremony, and speak in plenary sessions to the University at large.

The prominent placement of his titles underscores Galileo's status as a respected academician who speaks ex cathedra from the University of Pisa with the approval and backing of the rulers of Tuscany. Perhaps the inclusion of academic rank was meant to disarm readers at home and abroad who might see the opinions of the dialogues as the brilliant but idiosyncratic constructions of a famous mathematician, the court favorite of a wealthy and powerful family. Perhaps, too, his academic appointment is mentioned to counteract the impression made by previous work that he identified with a more perspicacious and more liberal public.

Midway down the title page, following mention of the Grand Duke, is a brief description of the contents of the book: "Presenting a congress of conversations over four days about the two Chief Systems of the World, Ptolemaic and Copernican"; and in smaller italic type, defining it still further: "Proposing indeterminately philosophical and natural arguments as much on the one, as on the other side." Here the dialogue is carefully defined as a dialectical consideration that argues for neither the Copernican nor the Ptolemaic system. This description and the rhetoric of the frontispiece that controverts it epitomize the difficulties engendered by the text itself. It is the engraving that accurately foretells the effect the text would have.

Dedication and Preface

Following the title page appears the brief dedicatory letter to the patron, the Grand Duke of Tuscany. In the first sentence of the letter Galileo links the theme of the engraving to the text of the work itself. He writes of how the extraordinary intellectual gifts possessed by certain men set them above their fellows by only a little less distance than that which separates man and beasts. The cause of that difference is their profession: they are philosophers. And among them, two, Ptolemy and Copernicus, may surely "claim extreme distinction in intellect above all mankind." Playing on a commonplace of the exordia of both the *Almagest* and *De revolutionibus*, Galileo remarks that the study of the universe elevates those who undertake it and allows them to open for us the "great book of nature." His own contribution has been illuminated by these, "the greatest minds ever to have left us

13. Drake describes the appointment at Pisa and its renewal in *Galileo at Work*, 161, 308.

such contemplations." Under the Grand Duke's patronage he hopes to offer something "from which lovers of truth can draw the fruit of greater knowledge and utility."

The preface is set in italics and bears the salutation: *"Al discreto lettore"* or "To the discerning reader" in Drake's translation. Galileo begins by declaring that he felt impelled to publish the book because "a salutary edict" of the Church has caused some to think that Rome did not have the benefit of informed advisors, that emotion and not reason prompted the decision. It was just "such carping insolence" that prompted him "to appear openly in the theater of the world as a witness of the sober truth." Ignorance did not prompt the Church to prohibit the espousal of Copernicanism. As he explains it: "It is not from failing to take count of what others have thought that we have yielded to asserting that the earth is motionless, holding the contrary a mere mathematical caprice, but (if for nothing else) for those reasons that are supplied by piety, religion, and the knowledge of Divine Omnipotence, and a consciousness of the limitations of the human mind." The ethos projected here is that of a pious Catholic who has bowed to higher religious authority, one who has recognized the frailty of human reason. These reasons are quickly tied to the need to defend his country's honor against the opprobrium of other nations. The purpose of the work, he says, is to present fully to "foreign nations that as much is understood of this matter in Italy, and particularly in Rome, as transalpine diligence can ever have imagined."

The sincerity of these sentiments has been questioned, both by the examiners for the trial and by commentators of our own day. The examiners thought that, since the preface was set in italics, what it stated was "vitiated." While it claimed simply to offer an hypothesis, the text of the dialogue itself argued quite strongly for the Copernican side and "contemned" its opponents.[14] Modern commentators point out that the very salutation with which the preface begins clearly indicates an ironic intent.[15] And we can see that one of the reasons given for acceptance of the edict in the quote above— the recognition of the "limitations of the human mind"—clashes dismally with the resounding paean to its prowess in the dedication to the Grand Duke. It is true that the preface was tacked on. Galileo states in a letter to Prince Cesi of 24 December 1629 that he has all but completed the work, remarking that what remains is "the ceremonial introduction and the arrangement of the opening of the dialogue." He adds that these are "matters

14. The charges made by the commission appointed by Pope Urban VIII are discussed at the end of this chapter. They are listed in Drake's *Dialogue,* 477, n. 103.

15. Drake's translation differs from the first English translation, that of Thomas Salusbury, who renders the preface title: "To the Judicious Reader." Salusbury's seems more to capture the sense of a defense of the Church without implying subterfuge.

more rhetorical and poetic than scientific, though I do want it to have some spirit and charm."[16] The fact that he added the preface, however, and that he was asked to make revisions to magnify the hypothetical nature of the treatise on his visit to Rome in 1630, does not mean that what he says there was foreign to his own views or that he wished to signal that fact to his readers.

The reason Galileo gives as the precipitating cause of his publication of the work—the jeers of others, especially transalpine critics—was just the kind of impetus that actually might have led him to publish the many proofs he had gathered over the years following the first announcement of his intention to reveal them in the *Sidereus nuncius*. It may also be the main reason that the pope agreed to his moving forward with the book.

Both were unwilling to let jibes pass unnoticed. Galileo had given the same reasons in his unpublished letter of 1624 to Francesco Ingoli. In explaining the delay of eight years in responding to his critique of Copernicanism, he tells Ingoli that a recent audience with Pope Urban VIII had apprised him of the fact that "it is firmly and generally believed that I have been silent because I was convinced by your demonstrations." Furthermore, he has been told that some people believe Ingoli's arguments to be necessary demonstrations and therefore "unanswerable." Galileo, of course, offers counterarguments, some of which are included in the *Dialogue*. Alluding to the work he was then composing, he notes that he has planned to treat the same topic more extensively to show "heretics" especially that Catholics are very knowledgeable but that they have greater reverence for "the writings of our Fathers" and "zeal in religion and faith." Further, "they understand how little one should rely on human reason and human wisdom."[17]

At the time Europe was embroiled in religious wars. Rumors or propaganda that could enhance the Protestant side and show them to be more perspicacious in science or more pious in religion were a source of serious concern to Catholics and especially to the pope. It is not improbable that Galileo had become caught up in a paradoxical situation. He may have truly accepted the authority of the Church to silence discussion and yet have felt obligated intellectually to show the superior reasoning of the Copernican side. The same contradictions are apparent at the trial.

Some of the discord in dedication, preface, and text also may be explained by the constraints under which Galileo labored. To clear the work for publication, the authorities of the Church provided guidelines that were imposed on him after the work was written. By early May 1630 he was in Rome with his manuscript, seeking the approval of the Dominican Niccolò Riccardi, who had now become Master of the Sacred Palace and was in

16. Drake, *Galileo at Work*, 311.
17. Finocchiaro translates the letter, "Reply to Ingoli," in *Galileo Affair*, 155–56.

charge of licensing publications. At the time Galileo entitled the work *On the Ebb and Flow of the Sea* (*De fluxu et refluxu maris*). The Pope objected to this focus in the title, believing that it gave undue weight to physical proof. He felt the book should give only mathematical proofs, proofs that save the appearances but cannot generate absolute truth. The prelates eventually seemed quite favorable to the project and Galileo returned to Florence. Shortly after Galileo learned of the death of Prince Cesi, who had wanted to publish the work at Rome under the aegis of the Lincean Academy.

Further delays ensued, and when Galileo decided to have the book published in Florence he had to seek permission also from the Inquisitor of Florence. Riccardi then began to show increasing concern about the preface and the ending. He wrote to the Florentine Inquisitor Clemente Egidi as follows:

I want to remind you that Our Master [the pope] thinks that the title and subject could not focus on the ebb and flow but absolutely on the mathematical examination of the Copernican position on the earth's motion, with the aim of proving that, if we remove divine revelation and sacred doctrine, the appearances could be saved with this supposition; one would thus be answering all the contrary indications which may be put forth by experience and by Peripatetic philosophy, so that one would never be admitting the absolute truth of this opinion, but only its hypothetical truth without the benefit of Scripture. It must also be shown that this work is written only to show that we do know all the arguments that can be advanced for this side, and that it was not for lack of knowledge that the decree was issued in Rome; this should be the gist of the book's beginning and ending, which I will send from here properly revised. With this provision the book will encounter no obstacle here in Rome.[18]

Thus, the paradox was one acknowledged by the Master of the Sacred Palace himself.

The book as a whole contained a multitude of arguments besides that of the tides, but Galileo considered it to be his strongest proof, as his letters and the notes of his friends reveal.[19] Much of the *Dialogue,* in fact, was based on the earlier work, "Discourse on the Tides," which he had decided to repress after the 1616 decree.

Rhetoric and Dialectics in the Dialogue

After this explanation of his reasons for writing the work, Galileo begins to describe the book's content. Having collected all the pertinent "specula-

18. The letter is translated in Finnochiaro, *Galileo Affair,* 212.
19. Drake examines the evidence and argues that Galileo seriously proposed the theory of the tides in "Reexamining Galileo's Dialogue" in *Reinterpreting Galileo,* 155–75; Finocchiaro views the arguments as essentially hypothetical; see his *Galileo and the Art of Reasoning,* 17; and Wallace thinks Galileo knew the proof was inconclusive; see his "Galileo's Science and the Trial of 1633," *The Wilson Quarterly* 7 (Summer 1983): 154–64.

tions," he says that he has: "taken the Copernican side in the discourse, proceeding as with a pure mathematical hypothesis by every artifice to represent it as superior to supposing the earth motionless—not indeed absolutely, but as against the arguments of some professed Peripatetics." In this sentence Galileo has announced the basic elements of the dialogue: its rhetoric— "every artifice"; and its dialectic—showing the hypothesis to be superior, "not indeed absolutely," but to combat the arguments of the Peripatetics.[20]

In the preface Galileo presents a partition outlining proofs he plans to develop: first, that all the experiments that can be performed on the earth cannot demonstrate whether the earth is moving or still; second, a practical consideration of celestial phenomena to show the validity of the Copernican hypothesis; and third, an "ingenious fantasy" regarding the tides that makes the motion of the earth seem probable (6).

The arguments Galileo promises here for the *Dialogue,* then, are not those he intimated he would provide in the *Sidereus nuncius;* that is, he will not attempt to offer demonstrations according to the canons prescribed in Aristotle's *Posterior Analytics;* he will instead offer dialectical and rhetorical proofs, the probable type of reasoning treated in Aristotle's *Topics* and *Rhetoric.*

Throughout the work dialectics is intertwined with rhetoric, such as when Galileo appeals to the emotions or uses the ethos of Salviati, or Salviati's friend the academician, to provide extra strength to the arguments, and when he undercuts the opposition by means of ridicule, sarcasm, and irony. It is present too when he develops thought experiments and introduces circumstances or elements that magnify his side and minify the other or when he uses figures of speech and allusions to amplify his arguments. His artistic genius lies in the luminescence of his illustrations, the wit, timing, and elegance with which he weaves rhetoric and dialectic together to create not only a compelling apologia for Copernicanism but a masterpiece of rhetorical literature.

The rhetorical elements reveal another problem of the dialogue. The audience for a dialectical disputation, even in its humanistic form, would presumably be an elite audience of knowledgeable people who would be expected to judge the discourse on its specialized content and regard the rhetorical elements as amusing, entertaining or, if not in agreement, as irritating, even insulting. The rhetoric would be recognized and not seen as a meaningful part of the argument as it might be in a political issue where the emotions and personal interests of the audience had to be engaged. The au-

20. Finocchiaro's version of the preface in *Galileo Affair* (215), instead of "by every artifice," reads "in every contrived way." The original is "per ogni strada artifiziosa," implying rhetorical artifice.

dience for rhetorical arguments, on the other hand, were generally not experts in the field, those who might expect to decide an issue if their emotions were touched or if they thought the arguer an ingenious, appealing fellow. In the case of this dialogue, it seems that Galileo assumed a mixed audience of both experts and those who might be swayed by what seemed on the surface to be a clever argument. As we have noticed before, he seems always to have in mind a congenial audience of friends. Consequently a less receptive audience of academics or clerics, some quite leery of the topic altogether, would find the negative rhetoric more repellent than the arguments.

Galileo's purpose in playing on the theatrical theme in the frontispiece and preface was probably to lighten the burden of the text and give it the "poetic" quality he mentions. Viewed from that angle, the rhetorical elements would merely be seen as amplificatory devices, allowable in literature. They would be meant not so much to persuade as to delight. But the topic was a real one, too serious to be taken lightly by many in the audience. This would prevent it from being viewed merely as a dramatic dialogue. The "Copernican mask" covered a philosophical face. The play was a philosophical dialogue by the nature of its content, as Galileo himself knew and demonstrates.

The philosophical dialogue engenders another problem. If it were a dialogue in Ciceronian style, one might expect voices to argue strongly for positions but not to reach a real conclusion for one side over the other. A Platonic dialogue would end with a strong conclusion about a truth on which the participants eventually agree.[21] The difficulty for Galileo is that he cannot end Platonically or he will trespass against the understanding he

21. In the course of its history the dialogue developed a dual personality. What had been intended by Plato to memorialize the Socratic method of investigating a subject—that of exposing contrary ways of looking at issues in order to arrive cooperatively at the truth—became somewhat different in Roman usage. Cicero, whom the humanists read so avidly, had made the dialogue a more managed rhetorical examination of a subject with a presiding speaker and an airing of opinions on all sides of an issue, but without one of these clearly winning. The humanists' form was even freer, exposing multiple kinds of opinions with literary style. A single answer or conclusion was not the point of a Ciceronian dialogue. Yet there were other varieties developed in the period: didactic and polemical dialogues, satirical dialogues, comic dialogues. Some of these obviously took sides. David Marsh discusses the evolution of the dialogue in *The Quattrocento Dialogue* (Cambridge: Harvard University Press, 1980), ch. 1; as does C. J. R. Armstrong in "The Dialectical Road to Truth: the Dialogue," *French Renaissance Studies* (Edinburgh: University of Edinburgh Press, 1976), 36–51. Giovanna Wyss-Morigi classifies and describes a number of different kinds of humanist dialogues, including those with didactic, polemical aims and amusing varieties, in her dissertation, Contributo allo Studio del Dialogo all'Epoca dell'Umanesimo e del Rinascimento, University of Bern, 1947. Nancy Struever discusses the use of rhetoric and invective in humanist discourse in "Lorenzo Valla: Humanist Rhetoric and the Critique of the Classical Languages of Morality," in James J. Murphy, ed., *Renaissance Eloquence* (Berkeley: University of California Press, 1983), 191–206.

has with the Holy Office. If he ends in Ciceronian style, as the title page of the book has implied, he will trespass against his own convictions and what he has stated in the preface he will do. The task is then an impossible one, but it is caused by the nature of the dialectical skeleton beneath the dialogue. Dialectical reasoning does presume that one will come to a conclusion that is more probable than that held by the other side. Indeterminate dialectics, as promised on the title page, is a contradiction in terms. One may conclude that the constraints under which Galileo was permitted to publish the work multiplied the ambiguities implicit in the genre and in the nature of the audience for which it was conceived.

Perhaps Galileo hoped that periodic statements of the anti-theme of the dedication—the limitation of human minds—would save the appearances and convince the audience that the whole was only a Ciceronian dialogue. It is easy to imagine that conscious of the power of his own human mind, which the adulation of his friends encouraged, he ignored the consequences of an inadequate assessment of the audience who would judge the import of the work.

Interlocutors of the Dialogue

The decision to write the work in the form of a dialogue was a rhetorical strategy that enabled Galileo to unfold his ideas as he wished, something harder to do in a disputation. The cast of interlocutors was carefully chosen to highlight the point of view he espoused. Although Galileo may have been inspired by Bruno's *Cena* in the organization of his work and the conception of his characters, he far surpasses the Nolan in the creation of their personas and the depth of their discussions.

The sage and witty Salviati, standing in for Galileo, is the convener and manager of the exchanges, who sets forth his opinions on the constitution of the heavens, outlining for the other two participants the evidence that has led him to his position. Sagredo, the affable and sympathetic listener, raises objections and doubts, but he is generally persuaded to see the wisdom of Salviati's reasoning. Simplicio, on the other hand, although sympathetically depicted as an alert and intelligent interlocutor, provides counterarguments and stubborn objections based on his blind adherence to the text of Aristotle. The very name, Simplicio, however, carries with it the same connotations of simpleton in the Italian as it does in English. But another reason Galileo chose the name was because it evoked the memory of Simplicius, the Greek commentator on Aristotle's *De caelo, Physica, De anima,* and *Categoriae,* and the one to whom conservative Renaissance interpreters of Aristotle turned, enamored as they were of philological methods. He, they thought, could supplying definitive answers to problems through the Greek text. More enlightened, progressive Aristotelians were not committed to a defense of all that Aristotle uttered, and updated his insights in light of new knowledge. These last have no spokesmen.

Galileo's construction of the dialogue, with two polar viewpoints aired before an observer, seems deliberately patterned on the disputation with its disputant, respondent, and impartial determiner. This underlying structure he must have hoped would signal to his readers a fair consideration of all the arguments—just as the title page forecast.

Through four days of discussion, the companions of Salviati are led to recognize the obvious superiority of the Copernican thesis. On the last day of the dialogue, in what has been called the "medicine at the end," Simplicio evaluates the deliberations, saying that although Salviati's arguments have been ingenious, man is still powerless to know for certain the real causes behind the phenomena in God's creation, returning to the anti-theme on the limitations of the human mind noted in the preface.[22]

As history has demonstrated, the presentation of the case for Copernicus was so convincing that it led to Galileo's conviction on the charge of teaching the forbidden thesis. The protestations of the preface and the testimony at the end failed to convince the audience that the author merely wished to inform without actually taking sides; the rhetoric of the whole persuaded his enemies that Galileo actually intended to persuade.

To understand the way in which this is accomplished, the exchanges must be sampled in some detail. The dialogue is so rich in dialectical and rhetorical arguments that we cannot possibly treat these in a thorough manner. Moreover, Maurice Finocchiaro has made that task unnecessary.[23] His extensive analysis, however, has guided our discussion and enabled us to narrow our focus to selected passages that illustrate well the mixture of rhetorical elements and dialectical proofs in Galileo's arguments.

22. The main problem with this "medicine" is that Simplicio expressed the pope's view. . Karl von Gebler describes an audience granted by Pope Urban VIII to Galileo in which Urban explained his opinion in *Galileo Galilei and the Roman Curia,* 116–17, 160; Santillana (160–80) reconstructs the conversation there on the basis of documents. The examiners' report is in *Opere* 19:348–60.

23. Finocchiaro, in *Galileo and the Art of Reasoning,* has made an exhaustive systematic analysis of the arguments of the work and the interrelation of rhetoric to logic and science. Finocchiaro, as a student of philosophy and particularly logic, views rhetoric as nonlogical. He states that "by rhetorical analysis is meant an examination of the non-intellectual (and non-literary-aesthetic, to be exact) content, structure, and aspects of the book" (5). His approach is ahistorical and evinces no knowledge of the discipline of rhetoric. Nevertheless, Finocchiaro's analysis uncovers the multiplicity of emotional and logical elements at work in the book and remains a valuable guide to the work. An essay by Brian Vickers focuses on the use of a literary rhetoric of praise and blame, "Epideictic Rhetoric in Galileo's *Dialogo,*" *Annali dell'Istituto e Museo di Storia della Scienza* 8 (1983): 69–102. The treatments by Finocchiaro and Vickers differ markedly from each other, particularly in their conception of rhetoric. The difference seems to stem from professional orientation. Vickers sees the literary-poetic elements of rheto-

Inadequacies of Peripatetic Philosophy

At the beginning of the first day of the dialogue, Salviati summarizes the reasons the group has decided to meet. They plan to "discuss as clearly and in as much detail as possible the character and the efficacy of those laws of nature put forth by the partisans of the Aristotelian and Ptolemaic position on the one hand and by the followers of the Copernican system on the other" (9). The burden of proof is immediately placed on the Peripatetic side when Salviati next announces that since Copernicus says the earth moves like the other planets, they will begin with the reasons why the Peripatetics hold the opposite view.

The initial duel between Salviati and Simplicio quickly establishes the characteristic modes of argumentation of each. Salviati selects for easy ridicule Aristotle's contention that there are three dimensions in nature because of the trifold nature of perfection or completeness. Simplicio points out "elegant demonstrations" that prove the omnipresence of these dimensions, quoting chapter and verse, including along the way an argument from testimony of the Pythagoreans. Everything has three parts, and three dimensions, he says, a beginning, a middle, and an end; these are necessary for perfection. Salviati counters, saying that he is not persuaded by the perfection in threes: "Therefore it would have been better for him to leave these subtleties to the rhetoricians, and to prove his point by rigorous demonstrations such as are suitable to make in the demonstrative sciences" (11). Galileo here makes a clear distinction between the rhetorical and scientific dimensions of the argument in ascribing the aesthetic balance and emotional appeals of the argument to rhetoric and asserting that science requires demonstration.

Aristotle's Spokesman

Despite the topic of his criticism, Galileo's treatment of the spokesman for the Peripatetics is decidedly rhetorical. Simplicio exhibits unswerving faith in every line of Aristotle and an amusingly pious circumspection. When Salviati's makes a derisive comment on the religious beliefs of the

---

ric as central in the Renaissance. In his recent *In Defence of Rhetoric* he describes the ways in which poetic and rhetoric interact in Renaissance discourse. The picture he paints, however, does not highlight the equally important alliance of rhetoric and dialectic; thus it does not accord rhetorical argument the prominence it possessed in Renaissance discourse, and particularly in Galileo's *Dialogue*. Vickers thinks that "the ultimate power of rhetoric in written communication was thought to reside in figures and tropes, the last stage in the elaboration of persuasive composition" (294). He demonstrates that tropic power in some of the arguments Galileo used in the *Dialogue*. But he does not distinguish rhetorical from dialectical argumentation, preferring instead to focus on the stylistic effects in praise and blame.

Pythagoreans, he replies: "I do not want to join the number of those who are too curious about the Pythagorean mysteries. But as to the point at hand, I reply that the reasons produced by Aristotle to prove that there are not and cannot be more than three dimensions seem to me conclusive; and I believe that if a more cogent demonstration had existed, Aristotle would not have omitted it" (12).

Occasionally Simplicio utters acceptable principles and argues astutely, showing that Galileo does not wish to intimate that he is lacking in intelligence, only in an independent will. Thus in response to a geometrical proof of three-fold dimensionality developed by Salviati, who wants to show that he can give better foundations for the principle than did Aristotle, Simplicio says: "I shall not say that this argument of yours cannot be conclusive. But I still say, with Aristotle, that in physical matters one need not always require a mathematical demonstration" (14).

Sagredo, acting as the "determiner," replies: "Granted, where none is to be had; but when there is one at hand, why do you not wish to use it?" To demonstrate good will and to emphasize the common ground between the new philosophy and the old, Sagredo voices the conclusions to be drawn from this argument: "But it would be good to spend no more words on this point, for I think that Salviati will have conceded both to Aristotle and to you, without further demonstration, that the world is a body, and perfect; yea, most perfect, being the chief work of God" (14).

Salviati returns immediately to his basic task of discrediting Aristotle's teachings and takes up Aristotle's classification of motion as straight, circular, and mixed, showing it to be defective and inconsistently maintained: "Moreover, it appears that Aristotle implies that only one circular motion exists in the world, and consequently only one center to which the motions of upward and downward exclusively refer. All of which seems to indicate that he was pulling cards out of his sleeve, and trying to accommodate the architecture to the building instead of modeling the building after the precepts of architecture" (16). Again, Simplicio rises ineffectually to Aristotle's defense.

Throughout the dialogue, Galileo has provided summary postils for quick assimilation of the contents, and to drive his points home. For the above he writes: "Aristotle shapes the rules of architecture to the construction of the universe, not the construction to the rules."

Having discussed the nature of the whole, Salviati moves to consider the parts: the celestial and the elemental, which in Aristotle's view, as the gloss points out, are "mutually exclusive" (14). This is a key point in the new philosophy, and one which Galileo has been able to forward through evidence provided by the telescope.

In his effort to mount more evidence that Aristotle is foolish in his opinions and fallacious in his reasoning, Galileo cites some of the Stagyrite's arguments from first principles that do not accord with contemporary ideas. For example: that straight motion characteristic of earth is imperfect because it has no terminus, whereas the circular motion of celestial bodies is perfect; it does not change and contains no contrarieties. Whatever, then, is characteristic of the earth—its propensity to change, seen in gravity and levity, generation and corruption—must be contrary to the conditions of the heavens (18). Salviati argues that if one finds such imperfections in the foundation of Aristotle's philosophy, then one must doubt the whole.

Salviati says that he now prefers to turn from the Aristotelian discussion to "discover a more direct and certain road and establish our basic principles with sounder architectural precepts." He agrees with Aristotle that the universe is perfect and contains all the dimensions noted, but he thinks the concepts of straight motion being infinite is impossible in nature, for nature would not attempt what is impossible, as the Philosopher has indicated. Unaccountably, Salviati also accepts some first principles based on logic.

We pass over the explication of Galileo's principles of motion. They have been the subject of many learned commentaries and their treatment would take us beyond the focus of this study. At the end of the discussion, which is illustrated with experiments and mathematical computations, Simplicio points out that Aristotle also thought that "sensible experiments were to be preferred above any argument built by human ingenuity, and . . . that those who would contradict the evidence of any sense deserved to be punished by the loss of that sense." But he doggedly adds that Aristotle was right in his observations about the movement of heavy and light objects.

Salviati then moves to an important implication of the dialectical topos of parts and whole. Since parts of the earth return to the earth's center and not to the center of the universe, he wonders what center should be ascribed to the universe and what center the earth would seek if it were removed from its place. He suggests that neither Aristotle nor Simplicio can ever prove that the earth is de facto the center of the universe: "if any center may be assigned to the universe, we shall rather find the sun to be placed there" (32).

Simplicio argues that it is vain to conjecture about what might happen if the sun or moon were removed from the whole because this could never happen; celestial bodies are neither heavy nor light and thus would not be pulled to a center. Salviati replies that Simplicio is guilty with Aristotle of fallacious argument. Simplicio responds indignantly: "Please, Salviati, speak more respectfully of Aristotle. He having been the first, only, and admirable expounder of the syllogistic forms, of proofs, of disproofs, of all the manner of discovering sophisms and fallacies—in short, of all logic—how can you ever convince anyone that he would subsequently equivocate so seriously as to take for granted that which is in question?" (35).

Salviati responds with a rhetorical argument designed to soften the over-throw of Aristotelian philosophy, which must take place before the new phi-losophy can be erected as the basis for the Copernican answer to the riddle of celestial motion:

Simplicio, we are engaging in friendly discussion among ourselves in order to inves-tigate certain truths. I shall never take it ill that you expose my errors; when I have not followed the thought of Aristotle, rebuke me freely, and I shall take it in good part. Only let me expound my doubts and reply somewhat to your last remarks. Logic, as it is generally understood, is the organ with which we philosophize. But just as it may be possible for a craftsman to excel in making organs and yet not know how to play them, so one might be a great logician and still be inexpert in making use of logic. Thus we have many people who theoretically understand the whole art of poetry and yet are inept at composing mere quatrains; others enjoy all the precepts of da Vinci and yet do not know how to paint a stool. Playing the organ is taught not by those who make organs, but by those know how to play them; poetry is learned by continual reading of the poets; painting is acquired by continual painting and de-signing; the art of proof, by the reading of books filled with demonstrations—and these are exclusively mathematical works, not logical ones. (35)

The teaching is a familiar one to scholastics: that logic as *utens* or use is dif-ferent from logic as *docens* or teaching.[24] Logic as the organ of demonstra-tion informs the practice of demonstration. Demonstrations about the universe that are convincing must have the sinews of perfect logic within them and properly apply mathematics to the physical; they must not be con-structed with formal principles alone, such as those criticized in earlier argu-ments resting on the topoi of perfection or of contraries.

A little later Galileo has Simplicio voice the concerns of the Peripatetics: "This way of philosophizing tends to subvert all natural philosophy, and to disorder and set in confusion heaven and earth and the whole universe. However, I believe the fundamental principles of the Peripatetics to be such that there is no danger of new sciences being erected upon their ruins" (37).

The universality of the fear expressed here is well illustrated by the senti-ment expressed in the familiar, poignant lines of John Donne's "First Anniversarie":

And new Philosophy calls all in doubt,
The Element of fire is quite put out;
The Sun is lost, and th' earth, and no mans wit
Can well direct him where to looke for it.

24. Without using these terms expressly, Galileo shows himself aware of this important distinction, known implicitly to the Greeks and much discussed by the Latins, including Thomas Aquinas, Giacomo Zabarella, and Paulus Vallius, the source from whom he appropri-ated his manuscript notes on logic. Wallace uses the distinction to structure his *Galileo's Logic of Discovery and Proof.*

And freely men confesse that this world's spent,
When in the Planets, and the Firmament
They seeke so many new; then see that this
Is crumbled out againe to his Atomies.
'Tis all in peeces, all cohaerence gone;
All just supply, and all Relation.[25]

Celestial vs. Terrestrial

Galileo continues to illustrate the inadequacies of Aristotle's account of the differences between the celestial and the earthly realms, suggesting that it should be easier to find out whether the earth moves than whether contraries actually are the source of corruption. Salviati exposes various difficulties in the latter principle, pointing to degrees of difference in generation and corruption and transmutations that stretch credulity. Sagredo returns the digression to the point, summarizing the import: if Aristotle is wrong about circular motion being inappropriate for the earth, then it might be claimed that what is characteristic of earth is also true of the heavens.

For the Peripatetic position Simplicio argues that sensible experience shows that generation and corruption occur on the earth but have not been seen in the heavens. So the two realms differ.

In an inspired analogy Salviati counters that visibility is an imperfect test, pointing out that he has not seen China or America and yet he thinks that alteration occurs there (48). Thus, from not seeing things in the heavens "you cannot deduce that there are none" (48). He comforts Simplicio with the assertion that "if Aristotle were now alive, I have no doubt he would change his opinion" on the basis of the sense observations now possible (50). Galileo must have hoped that this thought would mollify his opponents.

This exchange leads to a further discussion of Aristotelian method and of the charge that the Peripatetics (represented by Simplicio) do not understand how Aristotle actually worked. The implication is that their current methods badly invert the proper approach to nature.

Simplicio contradicts Salviati, saying that the latter is incorrect in his assumptions about Aristotle's way of philosophizing. The Stagyrite begins with a priori principles and then supports them with sense experience (50). Galileo shows the depth of his understanding of Aristotelian logic when he has Salviati respond that such is the method Aristotle uses in presenting his teachings but that it is not his method of investigating nature: "Rather, I

25. Marjorie Nicholson discusses the impact of Galileo's discoveries of 1610 in this poem composed in 1611 in her *Science and Imagination* (Ithaca: Cornell University Press, 1956, rpr. Hamden, Conn.: Shoe String, 1976), 51–57.

think it certain that he first obtained it [his doctrine] by means of the senses, experiments, and observations, to assure himself as much as possible of his conclusions. Afterward he sought means to make them demonstrable. That is what is done for the most part in the demonstrative sciences" (51). *Docens* is again contrasted with *utens,* this time recognizing Aristotle's proper use of it.

In support of the assertions that novelties have been discovered in the heavens, Salviati unabashedly mentions the appearance of comets, despite Galileo's position on the matter in his exchange with Grassi: "Excellent astronomers have observed many comets generated and dissipated in places above the lunar orbit, besides the two new stars of 1572 and 1604, which were indisputably beyond all the planets" (51). Corroborating this evidence are observations of spots on the sun. Whether Galileo changed his mind about the comets' position or whether the point simply serves the argument well is not clear. His desire to best Tycho, however, is clear.

Simplicio remarks that he thought these new opinions were very well refuted by the *Anti-Tycho* [of Chiaramonti]. To which Salviati responds: "As far as the comets are concerned I, for my part, care little whether they are generated below or above the moon, nor have I ever set much store by Tycho's verbosity" (52). Furthermore, he doubts that comets "are subject to parallax" and discounts the observations on which parallax has been based, thereby recalling to readers' minds Galileo's dismissal of Grassi's evidence. In response, Simplicio recites various opinions about the sunspots supportive of Aristotle, Scheiner's principally, concluding that even if these views are not sufficient to explain the phenomenon, "more brilliant intellects . . . will find better answers" (53).

In explaining to Simplicio how wrong are his conceptions about the method of natural philosophy, Salviati digresses and in so doing provides us with additional insights into Galileo's view of the disciplines of logic and rhetoric:

If what we are discussing were a point of law or of the humanities, in which neither true nor false exists, one might trust in subtlety of mind and readiness of tongue and in the greater experience of the writers, and expect him who excelled in those things to make his reasoning most plausible, and one might judge it to be the best. But in the natural sciences, whose conclusions are true and necessary and have nothing to do with human will, one must take care not to place oneself in the defense of error; for here a thousand Demostheneses and a thousand Aristotles would be left in the lurch by every mediocre wit who happened to hit upon the truth for himself. (53–54)

In the realm of law and the humanities—the domain of rhetoric—the aim is to persuade the will, and to do so depends upon plausible reasoning and eloquent expression. But if the aim is to discover the true nature of things,

nothing but demonstration is needed. Even the wit of Aristotle and the facility of Demosthenes can be shown up by some dolt who stumbles across
the truth of nature, and, presumably, could then demonstrate it.

The distinction Salviati makes—that proof in the realm of science is
aimed at the intellect whereas proof in rhetoric is aimed at the emotions and
the will (which is influenced by the emotions), and that to each belongs a
particular method—is a distinction known and respected by most scholars
in this period. The postil underscores the point: "In natural sciences the art
of oratory is ineffective" (54). It is edifying to read this in Galileo, but the
very nature of the *Dialogue* shows that rhetoric could, nevertheless, be very
useful to a scientist who happened not to have necessary demonstrations.

Galileo expands the argument and, ironically, does so with one of the
most brilliant rhetorical passages in the *Dialogue*. Salviati argues that since
Aristotle says that what is true of the heavens cannot be confidently concluded because of their great distance from us, and that since he also teaches
that sense experience should be preferred over reasoning, we should not be
persuaded by Aristotle's reasoning alone if sense experience provides new
data.

Sagredo, seeing Simplicio's distress at such a turn of the discussion, imagines that he might declaim as follows (the gloss entitles this "Simplicio's
declamation"):

Who would there be to settle our controversies if Aristotle were to be deposed? What
other author should we follow in the schools, the academies, the universities? What
philosopher has written the whole of natural philosophy, so well arranged without
omitting a single conclusion. Ought we to desert that structure under which so
many travelers have recuperated? . . . Should that fort be leveled where one may
abide in safety against all enemy assaults? (56)

Sagredo continues to amplify the architectural metaphor repeated in these
rhetorical questions. He compares Simplicio's predicament to that of a
wealthy aristocrat who has built and embellished a palace at great expense
and, upon finding its foundations crumbling, should attempt "to avoid the
grief of seeing the walls destroyed, adorned as they are with so many lovely
murals; or the columns fall, which sustain the superb galleries, or the gilded
beams; or the doors spoiled, or the pediments and the marble cornices,
brought in at so much cost—should attempt, I say to prevent the collapse
with chains, props, iron bars, buttresses and shores" (56–57). With un-
veiled irony Salviati reassures Simplicio that he need not fear imminent collapse, for "such a multitude of great, subtle and wise philosophers" are safe
from "one or two who bluster a bit." And in ignoring these blusterers "by
means of silence alone, they place them in universal scorn and derision." In
case the sarcasm is missed, the gloss makes it explicit: "It is the Peripatetic
philosophy that is unalterable."

Salviati concludes in frustration: "It is vanity to imagine that one can introduce a new philosophy by refuting this or that author. It is necessary first to teach the reform of the human mind and to render it capable of distinguishing truth from falsehood, which only God can do" (57).

God must reform the soul of the Peripatetic to permit intellect to dominate the will and the emotions. One can imagine Galileo's audience of supporters nodding in agreement and feeling empathy for him. He has appealed to an elite audience, an audience with minds God does not need to change. Unfortunately, in this ad hominem argument Galileo has included the audience he needs to convince.

The precision in Galileo's choice of metaphors, the climactic repetition, the beautifully wrought detail, the smug opprobrium of the grand design, these are the rhetorical weapons of the court, both ecclesiastical and secular. Galileo was a master in the fabrication and application of these arms. Unlike the heavy sarcasm and broad insults of Bruno's dialogue, Galileo knew how to destroy an opponent deftly, with grace.

Upon Sagredo's warning to stay on the subject, the interlocutors proceed to show the similitude of the moon to the earth, a task which occupies them during the rest of the first day's meeting. Simplicio can accept little of this, especially an assertion that the moon is opaque and that the earth reflects the sun and illuminates its satellite. He is unconvinced, even though he has read *The Assayer* and the *Letters on Sunspots* by "our mutual friend" and has listened to Salviati's explanation.

Differences between earth and moon are next considered. The moon appears to be composed of plains and mountains and is lacking in water. Its days are a month long, and it has extremes of heat and no rain. As to the kinds of things that may exist there, Salviati says that these are probably very different and beyond our power to conceive. He comments that this fits "the richness of nature and the omnipotence of the Creator and Ruler" (101). The pious sentiments punctuating the controversion of received philosophy found so frequently in the text are meant to reassure the theologically scrupulous.

Mind's Limitations

The palliative provokes a digression on the limitations of the human mind by Sagredo and reflective responses by the others. Its theme at first sounds quite orthodox, but in its further explication it appeared to the examiners to verge upon heresy.

Sagredo declares that he thinks it rash to make man the measure of nature: "On the contrary there is not a single effect in nature, even the least that exists, such that the most ingenious theorists can arrive at a complete understanding of it" (101).

Salviati agrees, adding that "the wisest of the Greeks . . . said openly that he recognized that he knew nothing." To Simplicio's assertion that one or the other had to be lying, Salviati replies with a dialectical argument to show that both were right:

The oracle judges Socrates wisest above all other men, whose wisdom is limited; Socrates recognizes his knowing nothing relative to absolute wisdom, which is infinite. And since much is the same part of infinite as little, or as nothing (for to arrive at an infinite number it makes no difference whether we accumulate thousands, tens, or zeros), Socrates did well to recognize his limited knowledge to be as nothing to the infinity which he lacked. But since there is nevertheless some knowledge to be found among men, and this is not equally distributed to all, Socrates could have had a larger share than others and thus have verified the response of the oracle. (101–2)

Sagredo then voices the theological commonplace that divine wisdom is more powerful and truly infinite. It far surpasses the creative abilities of man. At this juncture Simplicio confesses that he is baffled by what appears to be a contradiction: "Among your greatest encomiums, if not indeed the greatest of all, is your praise for the understanding which you attribute to natural man" (102). Yet, says Simplicio, you point out that Socrates says he has none.

These are indeed the two different themes of the preface that we recognized as discordant: The great power of some human minds, and yet the limitation of the human mind generally. Salviati proceeds to knit all up into one whole by making a philosophical distinction between two "modes" of human understanding—extensive and intensive knowledge:

Extensively, that is, with regard to the multitude of intelligibles, which are infinite, the human understanding is as nothing even if it understands a thousand propositions; for a thousand in relation to infinity is zero. But taking man's understanding *intensively*, in so far as this term denotes understanding some proposition perfectly, I say that the human intellect does understand some of them perfectly, and thus in these it has as much absolute certainty as Nature itself has. Of such are the mathematical sciences alone; that is, geometry and arithmetic, in which the Divine intellect indeed knows infinitely more propositions, since it knows all. But with regard to those few which the human intellect does understand, I believe that its knowledge equals the Divine in objective certainty, for here it succeeds in understanding necessity, beyond which there can be no greater sureness. (103)

Simplicio, surprised by the speech, declares it "bold and daring." Salviati goes on to explain that he did not intend to "detract in the least from the majesty of Divine wisdom." Yet, when man grasps the "truth" of mathematical proofs, "this is the same that Divine wisdom recognizes." God knows by intuition, a different and more excellent way, while man "proceeds with reasoning by steps from one conclusion to another" (103).

The argument from genus and differentia is qualified through its distinctions. Similarity with the divine mind is claimed only on the basis of a partial sharing of intensive knowledge. Man's mind is similar to God's when it knows with certainty. The differences Salviati notes are great; man and God differ in extensive knowledge, so much so that man can be said to know nothing; in intensive knowledge he differs in the manner of acquiring knowledge and in the fact that he knows only a small number of things intensively, while God's knowledge is infinite.

The problem for the Holy Office, whose court determined heresy, lay, apparently, in the last two lines where Galileo claims that our knowledge "equals" God's knowledge. The point noted by the examiners was that "he asserted some equality between the Divine and the human mind in geometrical matters" (477, n. 103).

The rhetorical overtones may also have influenced them. Galileo has said that brilliant powers in the human mind are possessed by only a few; he names Ptolemy and Copernicus in the dedication to the Grand Duke. Such men excel because they are philosophers. The implication is that Galileo is in the class of the greatest of these; though most of us would heartily acknowledge that today, his contemporaries were not all so agreed. His continuous disparagement of the Peripatetics' mental abilities may have made this a particularly sensitive theme. He has already stated that stubborn minds like theirs can change only if moved by God.

The end of the first day's session continues to praise the human mind, by listing its accomplishments: the statues and paintings of Michaelangelo or Raphael and Titian, musical compositions, the diversity of musical instruments, navigation, and, above all, the invention of writing. The words are made more trenchant by the passage of centuries: "But surpassing all stupendous inventions, what sublimity of mind was his who dreamed of finding means to communicate his deepest thoughts to any other person, though distant by mighty intervals of place and time! Of talking with those who are in India; of speaking to those who are not yet born and will not be born for a thousand or ten thousand years; and with what facility, by the different arrangements of twenty characters upon a page!" (105).

The Earth's Movement

The beginning of the next day's dialogue brings to the fore a discussion of the nature of the discourse in which they are engaged. Digressions such as these are interspersed throughout the dialogue.[26]

Invited by Salviati's confusion concerning the outcome of the previous day's discussion to summarize, Sagredo says that they had aired two opin-

---

26. Finocchiaro has made a tally of these and classified their subject matter in *Galileo and the Art of Reasoning*, ch. 6.

ions to determine which is "more probable and reasonable" and had concluded that the one which maintained that the earth was similar in nature to the celestial bodies was "more likely," and he suggests that they now consider whether the earth moves and if so how. Salviati stops him, determined to clarify: "I did not conclude this, just as I am not deciding upon any other controversial proposition. My intention was only to adduce those arguments and replies, as much on one side as on the other—those questions and solutions which others have thought of up to the present time (together with a few which have occurred to me after long thought)—and then to leave the decision to the judgment of others" (107).

Through adjustments like this Galileo tries to give the appearance of a dialogue in which both sides are adequately presented, the tradition of the disputation *in utramque partem*. Thus, Sagredo is forced to recognize his error in being "carried away by my own sentiments."

For his part Simplicio acknowledges that he thought the opinions "novel and forceful," but finds he is still impressed by other authorities. Sagredo, amused and frustrated by the closed mind of his friend, recounts the story of an eminent anatomist who wished to show the real source and placement of nerves in the human body, which had been greatly debated. For the benefit of a Peripatetic philosopher he traced the bundle of nerves from the brain through the body, showing its branches and the single thread that passed through the heart. On completing the demonstration he asked the philosopher if he were now convinced that the nerves had their origin in the brain and not in the heart. The philosopher after considering for a while, answered: "You have made me see this matter so plainly and palpably that if Aristotle's text were not contrary to it, stating clearly that the nerves originate in the heart, I should be forced to admit it to be true" (108).

Simplicio defends the Peripatetic approach, saying that if one knows the whole corpus of Aristotle one can combine enough passages to be able "to draw from his books demonstrations of all that can be known; for every single thing is in them" (108).

This caricature of the opposition effectively eviscerates one of the sides in the dispute by an implicit ad hominem argument.

Continuing the reductio, Sagredo provides a series of alternative texts for the kind of research Simplicio suggests—the verses of Virgil and Ovid—and, finally, recommends the most appropriate little book for the enterprise, the alphabet. He ends the digression with a lamentation: "Oh the inexpressible baseness of abject minds! To make themselves slaves willingly; to accept decrees as inviolable; to place themselves under obligation and to call themselves persuaded and convinced by arguments that are so 'powerful' and 'clearly conclusive' that they themselves cannot tell the purpose for which they were written, or what conclusion they serve to prove!" (112).

Salviati pretends to temper the mood of the discussion by suggesting that it is mainly the blind followers who are to blame, not Aristotle. His texts deserve careful study. But the remarks that follow simply amplify the invective: "And what is more revolting in a public dispute, when someone is dealing with demonstrable conclusion, than to hear him interrupted by a text (often written to some quite different purpose) thrown into his teeth by an opponent? If, indeed, you wish to continue in this method of studying, then put aside the name of philosophers and call yourselves historians or memory experts; for it is not proper that those who never philosophize should usurp the honorable title of philosopher" (113). The remarks provide new insight into the projected audience as well as the occasion for these disputes. Galileo speaks of "public dispute," indicating that the live audience for these debates was often a popular one involving others besides the academic philosophers whom he says brandish Aristotle's opinions rather than offer demonstrations. Demonstrations should be preferred by the scholarly audience. His description suggests debates in the courts of secular and ecclesiastical princes, or the salons of wealthy bankers and businessmen like Welser, or perhaps disputations open to the public at colleges and universities.

Rhetorical Construction

Returning the interlocutors to the subject at hand, Salviati again states that he has not made up his mind on the question, but because Simplicio has decided on Aristotle's view, "he shall give the reasons for his opinion step by step, and I the answers and the arguments of the other side, while Sagredo shall tell us the workings of his mind and the side toward which he feels it drawn" (113).

He begs Sagredo to serve in that capacity, for the arguments will be subtle and will need one whose wit is "acute and penetrating." Galileo here describes the rhetorical model of the audience he expects to judge the work: Sagredo, a patrician statesman educated privately under Galileo at Padua, an amateur of natural philosophy well acquainted with the major issues in astronomy and mathematics, becomes the epitome of what we might call Galileo's universal audience. The term "universal audience," invented by the philosopher and rhetorician Chaim Perelman, aptly denotes the ideal audience an author projects for philosophical matters, informed and wise people who would impartially make the same judgment regardless of time or place.[27]

27. Perelman says that even though the author assumes that the audience is universal, "each culture has its own conception." See the discussion in Perelman and Olbrechts-Tyteca, *New Rhetoric*, 31–35; I have quoted from 33.

Sagredo's response is significant: "Describe me as you like, Salviati, but please let us not get into another kind of digression—the ceremonial. For now I am a philosopher, and am at school and not at court" (114).

The intimation is that following this "digression" they can proceed to consider the matter at hand in an unbiased philosophical manner. The problem is that the interlocutors have already had their ethos constructed in such a way as to prejudice the consideration. One does not now expect much from Simplicio, and what he does put forward will be regarded only as quaint.

Salviati, evidently forgetting his announcement concerning Simplicio's role in the debate, himself lays out the problem of the earth's movement, showing, as the gloss declares: "Why the diurnal motion must more probably belong to the earth than to the rest of the universe" (115). We will look at the proofs and refutations of some of these as illustrative of the dialectical character of the arguments and for the insights periodically offered concerning Galileo's view of the argumentative process.

Arguments For and Against

Seven general arguments for the earth's motion follow; these are not presented as "inviolable laws" but simply as "plausible reasons." At the end Salviati notes that "one single experiment or conclusive proof to the contrary would overturn these and a great many other probable arguments" (122). They are simply arguing ex hypothesi and will continue to do so, "assuming both positions are equally adapted to the fulfillment of all the appearances" (124).

Salviati notes that nothing would be changed in the relations among the bodies of the heavens by either system. Only the relation of earth to those bodies is changed. Since either explanation could satisfy the appearances the simpler one would seem to be true: "who is going to believe that nature . . . has chosen to make an immense number of extremely large bodies move with inconceivable velocities, to achieve what could have been done by a moderate movement of one single body around its own center?" (117). The principle that forms the first premise is given in the postil: "Nature does not act by means of many things when she can act by means of a few" (117). The assumption lies at the base of the other six arguments as well.

Simplicio offers a scholastic refutation. He recognizes the point of the arguments given but responds that since the Mover of all has infinite power "why should not a great part of it be exercised rather than a small?" (123).

It is not a question of power but rather of simplification and ease, Salviati replies, and recites the maxim given above in the gloss, mentioning that it is Aristotle's principle.

Simplicio goes on to cite Aristotle's reasons for the earth being at rest.

First, force would be required to move the earth since it is in the center. To be moved in a circle is not natural to the earth, for such a force would have to be true of the parts of the earth. But these move with straight-line motion toward the center. Straight-line motion is forced and preternatural, and it is not eternal. But the heavens are eternal and all motions in the heavens are circular. He continues with similar objections until Salviati takes over, as the more expert in knowledge of them.

At this point Salviati rehearses the familiar arguments of bodies dropped from a tower or from a ship's mast and shows that these do not prove either system. Projectile arguments are introduced as well. Salviati is delighted that these were unknown to Aristotle because Simplicio will have no text to guide him and so might come to a new opinion. He promises: "But you will certainly see further novelties; you will hear the followers of the new system producing observations, experiments, and arguments against it [the old view] more forcible than those adduced by Aristotle and Ptolemy and the other opponents of the same conclusions. Thus you will become assured that it is not through ignorance or inexperience that they have learned to adhere to such opinions" (127–28).

The repetition of the theme of the preface, that it was not ignorance that led Italians to ban the system, develops a dangerous strain when Salviati implies that he is a "follower" of the new system and that his arguments are "more forcible." The excursus that follows amplifies the point and makes the charge that all those who have followed Copernicus first had learned the contrary opinion, "but had come over to this one moved and persuaded by the force of its arguments" (128). For such people to be converted to the other side against the opinion of the majority, he says, they "must of necessity be moved, not to say compelled, by the most effective arguments" (129).

On the other hand, Salviati reports, when he questioned Peripatetics and those who are followers of Ptolemy, he saw that none of them really studied Copernicus. "I found very few who had so much as seen it, and none who I believed understood it" (128).

To sharpen the issue, Salviati poses the question around which the whole of the *Dialogue* is constructed: "whether it should be held with Aristotle and Ptolemy that the earth alone remains fixed in the center of the universe while all the celestial bodies move, or on the other hand that the stellar sphere remains fixed with the sun in its center, the earth being located elsewhere and having the motions which appear to be those of the sun and the fixed stars?" (129). Then he asks if it is not true that one or the other of these opinions must be true or false. All agree. This brings him to a further question: "Do you believe that in dialectics, in rhetoric, in physics, metaphysics, mathematics, or finally in the generality of reasonings, there are arguments sufficiently powerful and demonstrative to persuade anyone of false no less than true conclusions?" (130). Simplicio provides the answer that validates the dialogic process: "By no means." He says that true conclusions may have many powerful demonstrations but these may not be shown to be fallacious: "I believe on the other hand that to make a false proposition appear true and convincing, nothing can be adduced but fallacies, sophisms, paralogisms, quibbles, and silly inconsistent arguments full of pitfalls and contradictions" (130).

Galileo, then, thinks that rhetoric, dialectics, and demonstration share the same truth potential. In his view both kinds of arguments would claim intelligent adherents if concerned with true arguments and true conclusions, for he thinks that man is capable of judging the validity of these arguments and would not be misled by fallacious reasoning. Sagredo, the universal audience, says: "I would have to be stupid indeed, warped in judgment, thick-witted and blind to reason, not to distinguish light from darkness, jewels from coals, truth from falsity" (131). And he adds in support of Simplicio's comment that Aristotle was the greatest master of logic, that "if Aristotle were here he would either be convinced by us or he would pick our arguments to pieces and persuade us with better ones" (131).

Salviati then stops and utters a disclaimer: "Before you go further I must tell Sagredo that I act the part of Copernicus in our arguments and wear his mask. As to the internal effects upon me of the arguments which I produce in his favor, I want you to be guided not by what I say when we are in the heat of acting out our play, but after I have put off the costume, for perhaps then you shall find me different from what you saw of me on the stage" (131).

Galileo must have hoped this ploy would silence those who might think that he himself supported Copernicus. Interestingly too, the words reflect the ambiguity implicit in the enterprise: a play depicting a disputation.

Now Salviati takes up the refutatory arguments against the earth's motion. The first is the familiar one posed by the flight of birds. How could they return to their habitats when flying high above the trees, since the earth would be turning during their flight? Would they, and clouds too, not be left far behind by the earth's movement? Other traditional arguments are introduced to an appreciative Sagredo, who praises this understanding of the Ptolemaic position.

Salviati proceeds to answer, offering a number of dialectical arguments and ingenious experiments. One of the latter, mentioned most often in the literature, is the ship's mast experiment. Salviati counters common wisdom by asserting that if a rock is dropped from the top of the mast, whether the ship is moving or not makes no difference, for the rock will fall in the same place. He inveighs against people who maintain the opposite and have never performed the experiment. But he declares, when pressed by Simplicio, that he himself did not need to perform it since he knew it could not happen otherwise. And, he says, Simplicio must know that too. "But I am so handy at picking people's brains that I shall make you confess this [view] in spite of yourself" (145). (Galileo's confidence in his own powers of persuasion is well placed.)

A lengthy consideration of projectile motion follows. Salviati is able to show that the problem with most of the arguments favoring the earth's stability is that they assume as true something that is actually in question. In the case of the stone falling from the tower, if the earth is already in motion the argument has no force. It is the same with the cannonball experiments that have been proposed. Neither the stone nor the cannonballs can be said to start from rest if they are already in motion along with the earth (174–83).

Through dialectical arguments, eloquent examples, and analogies Galileo repeatedly shows Aristotle's errors. When Simplicio suggests that Salviati's method is to argue mainly by a kind of Platonic "reminiscence," Salviati replies that he uses both words and deeds—meaning by words, dialectics and demonstration, and by deeds, experiments. Sagredo remarks at this point that when he studied logic he could never convince himself that "Aristotle's method of demonstration, so much preached, was very powerful" (191).

For Salviati the experiments often meant the application of geometry to natural phenomena. Sagredo notes a little later that "trying to deal with physical problems without geometry is attempting the impossible" (203). The implication is that this is something omitted in Aristotle.

In developing the point further, Salviati attempts to show the adjustments that must be made in applying mathematics to nature in the mixed sciences: "Just as the computer who wants his calculations to deal with sugar, silk, and wool must discount the boxes, bales, and other packings, so the mathematical scientist, when he wants to recognize in the concrete the effects which he has proved in the abstract, must deduct the material hindrances, and if he is able to do so, I assure you that things are in no less agreement than arithmetical computations. The errors, then, lie not in the abstractness or concreteness, not in geometry or physics, but in a calculator who does not know how to make a true accounting" (207–8).

This clear and eloquent statement conveys the ingenuity with which Galileo unraveled some of the great mysteries of natural science. The value of the *Dialogue* lies in the multiplicity of its revolutionary ideas. The fact that he stacks arguments and uses rhetoric to undercut his opposition is really unimportant to the science of the arguments, but it was of considerable importance to his peers in his own day.

In closing these refutations of Aristotle, Salviati returns to a theme announced in the dedicatory letter: the investigation of the universe as "one of the greatest and noblest problems in nature." This time he proposes that the investigation be turned upon another great problem: "the cause of the ebb and flow of the sea, which has been sought by the greatest men who ever lived and has perhaps been revealed by none" (210). He suggests they turn to that subject next.

Sagredo compliments Salviati for having a lofty mind occupied "with the highest meditations," but states that others are not so high-minded. He would be happy to hear further reflections. Salviati replies:

I have always taken great joy in the things I have found out, and next to this greatest pleasure I rank that of discussing them with a few friends who understand them and show a liking for them. Now, since you are one of these, I shall loosen the reins a little on my ambition (which much enjoys itself when I am showing myself to be more penetrating than some other person noted for his acuity) and I shall for good measure add to the last discussion one more fallacy on the part of the followers of Ptolemy and Aristotle. (210–11)

Contemporary Authors Refuted

In the last part of the second day, Galileo takes up two arguments made by his contemporaries against the motion of the earth. Simplicio introduces them, saying that what he has thus far heard from Salviati has somewhat disinclined him to accept the traditional views he had entertained. Recently, however, he has read refutations of Copernicus that are difficult to dismiss. Galileo here draws attention to two authors he wants to discount but does not explicitly identify. Drake notes that one is a pupil of Scheiner, Locher, and the other is Chiaramonti, who until the *Dialogue* considered himself a friend of Galileo.[28]

Locher

Simplicio introduces the objections made by Locher, which we will simply summarize, noting the general attack and its refutation. These Simplicio tries to summarize from memory, and not always very reliably. The first argument against the motion of the earth is supported by computations about the very long time it would take a falling body dropped from the moon's orbit to fall to the earth. Salviati is able to confute the conclusions based on the evidence of experiments on falling bodies by "our friend the Academician" (221). This new authority counterposed against Simplicio's is extolled as having disclosed in an unpublished manuscript a "whole new science,"

---

28. Drake, (*Dialogue,* note for p. 91 on 476, note for p. 52 on 473. Drake (473) points out that Urban VIII is said to have appointed Chiaramonti to examine the *Dialogue.*

one until now completely unknown. Salviati will not be deflected from his task, in spite of Sagredo's entreaties to tell him more of the new science, but instead proceeds to offer his own mathematical calculations concerning falling bodies.

Salviati continues to reveal fallacies in the objections made by Simplicio's authorities, using the Academician's opinions regarding motion and the acceleration of falling bodies to do so. His reasoning is supported with examples taken from common experience, the motion of pendulums and cannon balls, and mathematical proofs. Sagredo is so convinced that he cannot understand why Salviati terms this sort of argument "probable": "I wish to Heaven that in the whole body of ordinary philosophy there could be found even one proof this conclusive" (230).

Being held to greater accuracy in his account of the opponent's arguments, Simplicio, now supplied with the book, quotes from the theses of Scheiner's disciple. Often the arguments are simply stated and summarily refuted. Occasionally these are supported by dialectical arguments and thought experiments, which Salviati refutes with counter-reasoning and examples. The refutations are elaborately polite and at times satirical. At the end Salviati says his ears are weary and that if he thought he would hear "nothing cleverer from that other author, I don't know but what I should decide to go and take the air in a gondola" (247). Thus, again Galileo has fanned the flames of Jesuit indignation with his contemptuous dismissal of Scheiner's pupil.

Chiaramonti

Simplicio then begins to treat Chiaramonti's arguments against the diurnal motion of the earth.[29] The initial proposition claims that if Copernicus is believed, all the criteria by which science is constructed will be abrogated, for our senses and experiences cannot be our guide (248).

Salviati concedes the charge that we must disavow the testimony of our senses, but in support of Copernicus he proposes a number of experiments to show that our senses may deceive us regarding motion if we are participating in that motion, as in a ship, for example: "It is therefore better to put aside the appearance, on which we all agree, and to use the power of reason either to confirm its reality or to reveal its fallacy" (256). He illustrates the point with a brilliant analogy comparing our inability to perceive the earth's motion to a common illusion: "This event is the appearance to those who travel along a street by night of being followed by the moon, with

29. Wallace discusses the significance of these arguments in *Galileo and his Sources*, 303–6; Finocchiaro (38–39) analyzes and summarizes the content.

steps equal to theirs, when they see it go gliding along the eaves of the roofs. There it looks to them just as would a cat really running along the tiles and putting them behind it; an appearance which, if reason did not intervene, would only too obviously deceive the senses" (256).

Simplicio turns next to Chiaramonti's consideration of arguments against the Copernican view based on genus, on similarities, and on differences. Continuing the dialectical refutation, Chiaramonti notes that Copernicus proposes to place the earth, "this sink of all corruptible material," among those perfect, pure celestial bodies, such as Venus and Mars (268). The "sink" image Galileo had used in *Sidereus nuncius* to contrast the nobility given to the earth by Copernicus. Simplicio goes on to proclaim the greater propriety of the opposite view: "How much superior a distribution, and how much more suitable it is to nature, indeed to God the Architect Himself—to separate the pure from the impure, the mortal from the immortal, as all other schools teach, showing us that impure and infirm materials are confined within the narrow arc of the moon's orbit, above which the celestial objects rise in an unbroken series!" (268).

Salviati observes that indeed there are great "disturbances" to the old system created by the new, but that the three friends are speaking of what is really the case in the universe. Calling Chiaramonti's conclusion a "rhetorical deduction," he goes on to make one of his own:

What is more vapid than to say that the earth and the elements are banished and sequestered from the celestial sphere and confined within the lunar orbit? Is not the lunar orbit one of the celestial spheres, and according to their consensus is it not right in the center of them all? This is indeed a new method of separating the impure and sick from the sound—giving to the infected a place in the heart of the city! I should have thought that the leper house would be removed from there as far as possible. (268)

Then Galileo recalls the elegant rhetorical argument of Copernicus:

Copernicus admires the arrangement of the parts of the universe because of God's having placed the great luminary which must give off its mighty splendor to the whole temple right in the center of it, and not off to one side. As to the terrestrial globe being between Venus and Mars, let me say one word about that. You yourself, on behalf of this author, may attempt to remove it, but please let us not entangle these little flowers of rhetoric in the rigors of demonstration. Let us leave them rather to the orators, or better to the poets, who best know how to exalt by their graciousness the most vile and sometimes even pernicious things. (268–69)

For Galileo it appears that rhetoric is acceptable in this discourse, so long as it is carefully kept out of demonstrations. Poetics, on the other hand, is too far removed from reality to be enlisted in the undertaking.

Galileo introduces at this stage an argument of Kepler in favor of Copernicus, one of the few times Galileo displays Kepler's views to advantage.

At the close of the day's discussion Simplicio pronounces himself impressed by the arguments, but says he does not feel "entirely persuaded to believe them." They simply show that the proofs for the earth's stability are not necessary demonstrations. He points out the obvious problem: "But no demonstration on the opposing side is thereby produced which necessarily convinces one and proves the earth's mobility" (274).

Salviati utters another disclaimer, saying that he had not intended to persuade Simplicio but simply to make clear that those who think otherwise "were not blindly persuaded of the possibility and necessity of this." They came to that opinion after examining the contrary view. In keeping with this approach it is only fitting that the group examine tomorrow the views of Aristarchus and Copernicus that the earth circles the sun in a year's time.

The Earth's Annual Revolution

At the beginning of the third day's discussion Sagredo reflects on the arguments, saying that he realizes that those for the Copernican system are much stronger, but because of their difficulty and novelty the view has few followers, while that of the Peripatetics, held for so long, has many. Hoping to hang on to this view, some of its partisans advance absurd opinions.

Salviati agrees, immediately noting that when arguments are brought against the Peripatetics they are scornful and angry: "Beside themselves with passion, some of them would not be backward even about scheming to suppress and silence their adversaries. I have had some experience of this myself" (277). Sagredo replies significantly: "No good can come of dealing with such people especially to the extent that their company may be not only unpleasant but dangerous." In these sentiments Galileo may intend a reference to the growing animosity of the Jesuits reported to him. As if to reinforce the impression by the contrast, he remarks that Simplicio is not such a one, being "a man of great ingenuity and entirely without malice."

Salviati then offers further reactions to Chiaramonti's arguments against Tycho. He says that he was greatly astonished at what he read, but thinks the author wrote the work in order to stay in the good graces of others. Thus he should expect it to bring only censure from the learned.

Sagredo agrees, adding that those who might be able to see his errors would be far outnumbered by the multitudes who favor Peripateticism: "And even the few who do understand scorn to make a reply to such worthless and inconclusive scribbles" (279). The remark seems inappropriate for Sagredo, who has supposedly not read Chiaramonti's book.

Salviati declares that silence would be the best response, but echoing the reason given for the *Dialogue* he thinks this is not desirable now:

One reason is that we Italians are making ourselves look like ignoramuses and are a laughing-stock for foreigners, especially for those who have broken with our religion; I could show you some very famous ones who joke about our Academician and the many mathematicians in Italy for letting the follies of a certain Lorenzini appear in print and be maintained as his views without contradiction. But this also might be overlooked in comparison with another and greater occasion for laughter that might be mentioned, which is the hypocrisy of the learned toward the trifling of opponents of this stripe which they do not understand. (279)

Now Galileo castigates the rest of his informed peers who do not stand up for what they know is true. His stance of the preface is amplified: he will show that Italy can boast intelligent, informed scholars.

Parallax and Observations

Salviati first has to treat the problem of the absence of stellar parallax, which Chiaramonti, following Aristotle and Ptolemy, had maintained would not be lacking if the earth revolved around the sun. Galileo argues astutely that the nature of our instruments prevents us from discerning parallax since the stars are at such great distances from us. In connection with this argument he examines the data of a number of astronomers to support his contention. In his attacks on the author's views he frequently accuses him of deliberate attempts to fudge his results, which even if true could not win Galileo many friends among the Peripatetics. Throughout, his tone is contemptuous. He terms one of Chiaramonti's explanations a "refuge" both "miserable" and "ridiculous," and accuses him of grasping at "spiderwebs," of proffering "unhappy and beggarly excuses" (311–17).

Salviati then moves to establish the revolution of the earth around the sun as the center of the universe, contending that this opinion is "deduced from most obvious and therefore most powerfully convincing observations" (321). He reviews his conclusions about the orbits of Venus and of Mercury, and with Simplicio assigns orbits to the rest of the planets. Simplicio without noticing it has come to see that the most sensible arrangement necessitates the earth's being placed among the planets and revolving with them around the sun.

Admitting that the sense arguments offered against the heliocentric view are very persuasive, Salviati intimates that the reason he has been able to rise above them is "the existence of a superior and better sense than natural and common sense to join forces with reason." In the clause that follows he says a "clearer light than usual has illuminated me," intimating that he has been illuminated by divine light to see the better answer (328).

The Sunspot Proof

The next important argument introduced in favor of the earth's revolution is taken from the Academician's observation of the movement of the sunspots. The appearances can best be explained, maintains Salviati, by the movement of the earth around the sun. He then defends the worth of telescopic observations in the face of Simplicio's doubts. After Salviati has explained the significance of the observations of the phases of Venus, Sagredo is moved to exclaim: "O Nicholas Copernicus, what a pleasure it would have been for you to see this part of your system confirmed by so clear an experiment" (339).

To which Salviati replies: "Yes, but how much less would his sublime intellect be celebrated among the learned." He then notes also how the satellites of Jupiter offer an analogous model for the moon's revolutions around earth and both around the sun.

Referring to the letters of Apelles, Scheiner's pseudonym, Salviati says that the Academician showed just how "vain and foolish" were the explanations given there and that he had correctly predicted that Apelles would come around to his opinion (346). Furthermore, since our Academician thought that in his *Letters* to Welser he had "looked into and demonstrated" everything pertinent, "if not everything that human curiosity might seek and desire," he ceased these observations and turned to other things.

Nevertheless Simplicio voices objections:

Step gently, my friend; perhaps you have not got so far as you think you have. For although I have not entirely mastered the content of Salviati's discourse, still, when I consider the form of the argument, I cannot see that my logic teaches me that this mode of reasoning necessarily forces me to any conclusion in favor the Copernican hypothesis. . . . Unless you first demonstrate to me that such an appearance cannot be accounted for when the sun is made movable and the earth fixed, I shall not change my opinion, nor believe that the sun moves and the earth remains at rest. (352)

Salviati makes an attempt to answer his charge, offering plausible reasons, but admits at the end that the arguments are neither conclusive nor inconclusive. Simplicio pronounces himself incapable of making a decision about the matter, but states that he prefers to remain neutral until such time as his "mind will be freed by an illumination from higher contemplations than these of our human reasoning and all the mists which keep it darkened will be swept away" (355).

The neutral determiner Sagredo, having heard all the arguments put forward so far, observes:

I have not, among all the many profundities that I have ever heard, met with anything which is more wonderful to my intellect or has more decisively captured my mind (outside of pure geometrical and arithmetical proofs) than these two conjectures, one of which is taken from the stoppings and retrograde motions of the five planets, and the other from the peculiarities of movement of the sunspots. And it appears to me that they yield easily and clearly the true cause of such strange phenomena, showing the reason for such phenomena to be a simple motion which is mixed with many others that are also simple but that differ among themselves. (356)

Sagredo is convinced by the proofs and the promise of necessary proof. His allusion to geometric and arithmetical proofs strengthens what has been offered, since mathematical proofs are the most certain possible. In the realm of nature, however, true causes are the foundation for necessary demonstration. But his conviction that the arguments have revealed "the true cause" rests on dialectical arguments amplified by rhetoric, for the Academician's predictions have not been thoroughly verified, nor have other possible explanations been ruled out. Through the rhetoric of Sagredo's response Galileo hopes to lead the audience to look upon the Copernican solutions as nearly certain.

In order that the discussion be evenhanded, Simplicio is asked to give some opinions from the other side. He does so referring to the booklet authored by Scheiner's pupil, wherein the opinion of Copernicus is derisively presented as proposing "sublime inanities." Among these are the necessity that Christ would have to rise to hell and descend into heaven when he neared the sun (357). None of the company take the arguments seriously. Salviati observes that the author mentions at the outset that he is not well informed about the position of Copernicus: "This is a poor beginning for gaining the confidence of the reader" (357).

The evidence of other astronomers cited by the author of the booklet is shown to be untrustworthy, leading Salviati to mention the "delusions of a number of others"; among these he names "al-Fergani, al-Battani, Thabit ben Korah, and more recently Tycho, Clavius, and all the predecessors of our Academician" (360). In strong terms he notes: "Nor can these men be excused for their carelessness" (360).

The Argument from the Tides

We turn from the other ingenious refutations to the most persuasive proofs Salviati offers. In a dramatic introduction he points out that, for all of the arguments concerning the earth in relation to the celestial bodies, either explanation—the earth's mobility or its stability—could equally serve. In addition, all the terrestrial events discussed could be shown to accord with either side except for one: the movement of the waters of tidal bodies. This we have to recognize as caused by the earth's motion. "No ebb and flow if the terrestrial globe were immovable," the postil informs us (417).

The Primary Cause

The question of what Galileo really thought about the proof is unresolvable, but certainly in the dialogue itself Salviati, his surrogate, implies that the argument is conclusive. The language used is the language of demonstration and would signal to knowledgeable readers that he plans to offer one. Nevertheless, Galileo cannot have Salviati term it a demonstration absolutely, since he has been told by Riccardi and the pope that this will not be acceptable under the terms of the decree against Copernicanism. Consequently, he has Salviati say that he has not had much chance to think about this possibility, but is simply proposing "a key to open portals to a road never before trodden by anyone" in hopes that others with more penetrating minds will carry it further (418). This proof is what Galileo called an ingenious fantasy (*una fantasia ingegnosa*) in the preface.

The quandary he is in demands equivocation. In the next statement Salviati notes that although other seas in remote areas have not been observed, if "the reason and the cause" can be verified by the events seen in the Mediterranean, these will hold true for other similar cases: "For ultimately one single true and primary cause must hold good for effects which are similar in kind" (418).

He lays out the plan of attack for the day: "I shall, then, tell you the story of the effects which I know to exist, and assign to them the cause that is believed by me to be true; and you, gentlemen, shall produce others noticed by you in addition to these of mine, and then we shall see whether the cause I am about to adduce can account for them also" (418).

To accomplish his purpose Galileo first seeks to dispose of counter-theories for the tides. Simplicio introduces them. The first based on Aristotle assigns the cause to variations in the depths of the seas. Another suggests that the moon attracts the waters, although how it does this when it has set is not explained. Some say that the moon heats the water and thus lifts it.

These are dismissed in turn by Salviati, who comments: "Let us just say that there are two sorts of poetical minds—one kind apt at inventing fables, and the other disposed to believe them" (420). Simplicio replies that no one would actually believe a fable if he knew it to be one, but that, frankly speaking, the cause Salviati has attributed to the tides seems "no less fictitious than all the rest." Foreshadowing what will be the last argument he raises in the *Dialogue*, Simplicio remarks: "If no reasons more agreeable to natural phenomena were presented to me, I should pass on unhesitatingly to the belief that the tide is a supernatural effect, and accordingly miraculous and inscrutable to the human mind—as are so many others which depend directly upon the omnipotent hand of God" (421). Salviati is not offended, but declares that Simplicio reasons "prudently" just as Aristotle does in the *Mechanics* when he assigns to miracles causes that remain hidden. But this cause need not remain hidden, he argues adroitly, for it can be replicated by experiments, which he proceeds to describe.

Arguing from the analogy of the movement of water in a barge as it floats in the sea, and what would happen if it were started and stopped or suddenly accelerated, he explains that in a similar manner the tide is produced by diurnal rotation when this is added to and subtracted from the annual motion of the earth. The combination of these movements causes the unevenness of motion in various parts of the earth. This is the "fundamental and effective cause of the tides, without which they would not take place" (428).

Secondary Causes

Galileo also introduces secondary causes that explain some accidental effects: the way in which water moves and stabilizes, the configuration of the basin in which the water lies, its depth. He uses thought experiments to illustrate his points and gives examples of motion in various seas.

At the end of these elaborate proofs, Simplicio evaluates his effort:

I do not think it can be denied that your argument goes along very plausibly, the reasoning being *ex suppositione*, as we say; that is, assuming that the earth does move in the two motions assigned to it by Copernicus. But if we exclude these movements, all the rest is vain and invalid; and the exclusion of this hypothesis is very clearly pointed out to us by your own reasoning. Under the assumption of the two terrestrial movements, you give reasons for the ebbing and flowing; and vice versa, arguing circularly, you draw from the ebbing and flowing the sign and confirmation of those same two movements. Passing to a more specific argument, you say that on account of the water being a fluid body and not firmly attached to the earth, it is not rigorously constrained to obey all the earth's movements. From this you deduce its ebbing and flowing. (436)

The interesting point regarding Simplicio's comment is that Galileo has here tried to disarm those who would probably raise similar objections to his argument. He knows how to reason like the Peripatetics, having studied the *Posterior Analytics* with a Jesuit commentary, as we noted in the first chapter. Nevertheless, the criticism is not sufficiently answered: that Galileo has assumed what he would prove, and that the same reasoning can be used to produce effects contrary to fact.[30] But to have Simplicio voice the objection robs it of rhetorical force; he makes it appear pedantic.

Kepler's view must be discounted if Salviati's explanation is to be triumphant. That Kepler could have attributed the tides to the moon Salviati says left him "more astonished" than that others might. Rather than addressing the argument, Galileo prefers to address the man: "Despite his open and acute mind, and though he has at his fingertips the motions attributed to the earth, he has nevertheless lent his ear and his assent to the moon's dominion over the waters, to occult properties, and to such puerilities" (462).

In concluding his exposition Salviati once more echoes the self-effacing disclaimer of the preface: "I do not claim and have not claimed from others that assent which I myself do not give to this invention, which may very easily turn out to be a most foolish hallucination and a majestic paradox" (463). The words do not ring true, however, given the manner in which the explanations have been extolled and opposing opinions damned. Galileo declares at the beginning of the day's discussion that he believes he has found the "true cause" for the effects, and he then invokes "the fundamental and effective cause" and posits their secondary causes. These terms indicate a demonstration, despite the final protestations to the contrary. It is no wonder that the Inquisitors viewed the disclaimer as being tacked on.

Simplicio is used to voice the doubts Galileo is sure the Peripatetic audience must continue to cherish concerning the four days' results: "I admit that your thoughts seem to me more ingenious than many others I have heard. I do not therefore consider them true and conclusive . . . " (464). This part of Simplicio's remarks undergird the view that Galileo has interjected at various points and reaffirms at the end, that the opinions offered are purely conjectural. These statements would seem to satisfy Riccardi's request concerning the preface and the end of the book and would simply reiterate a stance he had taken in the letter to Ingoli.

Perhaps all would have been well had Galileo let Simplicio stop there. But, instead, he has given to him the opinion expressed by Urban VIII in his audience with Galileo. It is the opinion that Simplicio had introduced to deaf ears earlier that day. Had Sagredo now embraced this opinion, even though it would have needed considerable preparation for him to do so and stay "in character," then perhaps Urban would not have reacted as he did. Instead Simplicio voices the opinion in the rest of the sentence quoted incompletely above:

. . . indeed, keeping always before my mind's eye a most solid doctrine that I once heard from a most eminent and learned person, and before which one must fall silent, I know that if asked whether God in His infinite power and wisdom could have

30. See the discussion of the point in Wallace, *Galileo and his Sources*, 310–11.

conferred upon the watery element its observed reciprocating motion using some other means than moving its containing vessels, both of you would reply that He could have, and that He would have known how to do this in many ways which are unthinkable to our minds. From this I forthwith conclude that, this being so, it would be excessive boldness for anyone to limit and restrict the Divine power and wisdom to some particular fancy of his own. (464)

Simplicio has uttered too many absurdities and has too often been ridiculed for his blind acceptance of authority to ask the reader to accept this, his last opinion, as a sage corrective to Salviati's ingenious proof. But Galileo has acquiesced to the pope's wishes and ameliorated the tidal argument with a summary objection. What he privately thought of it is obvious by his assigning the speech to Simplicio.

Salviati's response to that sentiment, on the other hand, does seem out of character unless we take it as sarcasm: "An admirable and angelic doctrine, and well in accord with another one, also Divine, which while it grants to us the right to argue about the constitution of the universe (perhaps in order that the working of the human mind shall not be curtailed or made lazy) adds that we cannot discover the work of His hands" (464). The last part of the sentence, however, has the effect of resolving the two paradoxical themes of the preface. The "marvelous powers of the human mind" have been nullified so that a safer motif may linger in reader's minds.

The *Dialogue* ends with the promise of another discourse, a treatise by the Academician on local motion.

Reflections on the Copernican Debate

The most significant immediate effect of the *Dialogue* was to provoke the wrath of the pope and to prompt him to appoint a commission to examine the work. From our review of the dialectical and rhetorical strategies of the *Dialogue*, we can see that most of the charges made by the papal commission are far from groundless. Drake summarizes the eight offending points:[31]

1. That the imprimatur of Rome was put on the title page without proper authority.
2. That the preface was printed in different type and thus vitiated, that the closing argument was put in the mouth of a simpleton, and that it was not fully discussed.
3. That Galileo often treated the motion of the earth as real and not hypothetical.
4. That he treated this subject as undecided.

31. Drake, *Dialogue*, 477; the report of the commission of qualifiers is given in *Opere* 19:327.102–3.

5. That he contemned opponents of the Copernican opinion.
6. That he asserted some equality between the Divine and the human mind in geometrical matters.
7. That he represented it to be an argument for the truth that Ptolemaics become Copernicans, but not vice versa.
8. That he ascribed the tides to motion of the earth which was nonexistent.

The Charges against Galileo

The first objection is not relevant to our discussion but has reference to the transfer of authority for the imprimatur from Rome to Florence. We have mentioned the Commission's dismay at seeing the preface in type different from the text. It is true that without the preface the work would seem to challenge the Church's wisdom in suppressing the Copernican opinion. The fact that the examiners link the preface with the reference to the closing argument "in the mouth of a simpleton" underscores the pope's interest in both the preface and conclusion. Galileo's error in this regard was not simply that he made a poor rhetorical choice; he wanted to air the pope's opinion but he really could not countenance the objection himself and thought it best voiced by the conservative Simplicio. It was a tactical error but an honest judgment. To better the rhetorical effect he would have had to do an injustice to the personas he had invented and to his own opinion. The claim that he "often treats the motion of the earth as real and not hypothetical" is undeniable. In further development of the point the examiners said that Galileo had "retreated from his hypothesis in asserting absolutely the mobility of the earth and the stability of the sun and either sustains the argument as based on demonstration and necessity or treats the negative side as impossible."[32] The reference to his treating the earth's motion as "undecided" recalls for us the fact that the Church held it to be decided.

Galileo's claim that the human mind had the ability to achieve knowledge equal in intensity to the divine mind bordered on blasphemy in the examiners's view. The repeated airing of "the remarkable powers of the human mind" theme no doubt helped to magnify the importance of the statement.

The commission saw Galileo's discussion of the conversion of Ptolemaics to Copernicanism and the absence of the opposite case as a distortion of the truth. Sagredo implied that Ptolemaics remained so because they gave inadequate consideration to the other side.

The final point has specific reference to the decree of 1616 in which the motion of the earth was declared to be "false and altogether opposed to the Holy Scripture," and, of course, it was the prohibition against defending and holding the doctrine that prompted the other charges concerning it.

32. *Opere* 19:326.95–98.

We need not rehearse the trial that ensued in 1633, since a historical account is not the aim of our study. The trial has been extensively and expertly aired in the mid-twentieth century by Giorgio de Santillana and later by James Langford. More recently, numerous interpretations and analyses of the documents have added to our knowledge of the events and deepened our understanding of the positions taken by Galileo and the Church.[33]

Galileo's Intentions

Obviously Galileo meant to convince people of the superiority of the Copernican system. He said so in the preface, and the language of the rhetorical and dialectical arguments bear this out. As he promised, he used every artifice in the effort. One of these was the proof based on the tides. Did he think he had offered a demonstration, as he had so long wanted to have? The answer has, of course, broad implications, both for the history of science and for our understanding of the man himself.

The careful development, the terminology employed, and the reasoning all imply a demonstration, and the temptation is to believe that he thought it was one. But the objection raised by Simplicio and Salviati's earlier acknowledgement that complete verification for the premises needs further work cloud its status. Besides, Salviati renders the presentation ambiguous by noting near the end of the fourth day that he does not regard the proofs he has introduced as "true and conclusive." Galileo seems then to have thought the argument was dialectically convincing but not completely demonstrated.

But why did he advance the case so vigorously? It may simply have been a matter of pride. Even though the answer given in the preface is often discounted, it may offer the best explanation: he wanted to show that Italian scientists knew all the possible proofs in order to quell the jibes of those who thought Italians ignorant. Galileo gave that same excuse to Ingoli.

In support of this view we might recall also the fact that Galileo claimed at the trial that he had not held the Copernican view since the decree of 1616 was announced, which meant that he did not maintain it as a viable explanation. Perhaps he told the truth, although most people prefer to believe that he had his fingers crossed.

The *Dialogue* may have been a case of déjà vu. He said in the preface that he wanted to rehearse the arguments heard in Rome before the ban. In the

33. Santillana, *Crime of Galileo;* Langford, *Galileo, Science and the Church;* Wallace, "Galileo's Science and the Trial of 1633"; R. S. Westfall, *Essays on the Trial of Galileo* (Vatican City: The Vatican Observatory, 1989); the documents and analysis by scholars of the Roman Catholic Church are in *I Documenti del Processo*, cited above; Finocchiaro's English translation of pertinent documents is *The Galileo Affair*. An excellent nineteenth-century account is Karl von Gebler's, cited in n. 21 above.

composition of the book, he may have so thoroughly imagined himself back in time that he was oblivious to the effect this might have.

Antonio Querengo, a diplomat, poet, and priest who was a friend of Galileo, observed him in debate at Rome. In early 1616 he wrote a description of him that makes this explanation of his later statement to the inquisitors credible:

> We have here Sig. Galileo, who, often, in gatherings of men of curious mind, bemuses many concerning the opinion of Copernicus that he holds for true. . . . He discourses often amid fifteen or twenty guests who make hot assaults upon him, now in one house, now in another. But he is so well buttressed that he laughs them off; and although the novelty of his opinion leaves people unpersuaded, yet he convicts of vanity the greater part of the arguments with which his opponents try to overthrow him. Monday in particular, in the house of Federico Ghislieri, he achieved wonderful feats; and what I liked most was that, before answering the opposing reasons, he amplified them and fortified them himself with new grounds which appeared invincible, so that, in demolishing them subsequently, he made his opponents look all the more ridiculous.[34]

After the decree, Querengo writes: "The disputes of Signor Galileo have dissolved into alchemical smoke, since the Holy Office has declared that to maintain this opinion is to dissent manifestly from the infallible dogmas of the Church. So here we are at last, safely back on a solid Earth, and we do not have to fly with it as so many ants crawling around a balloon."[35]

It is no wonder that Galileo would wish to remember those exciting times, when he could allow his mind to range over all the possible arguments and show the system to best advantage. His rhetoric, too, is more understandable, given these circumstances. In memory he was giving as good, and better, than he got. The taunts of his opponents always haunted him, and in his writing he could annihilate them as he had in those encounters in Rome. His ego could not let it be supposed that his opponents raised better arguments.

Only piety, as the preface notes, was responsible for his relinquishing the joy of teaching that opinion. Galileo was a pious Christian, as Olaf Pedersen has argued in a convincing essay.[36] As complex a man as the father of modern science must have been, it is not unthinkable that he could have bracketed that opinion until the evidence was complete and he had a demonstration. The words written on a flyleaf of the *Dialogue* in his own hand

34. Santillana, *Crime of Galileo*, 112–113.
35. Ibid., 124.
36. Pedersen, "Galileo's Religion."

attest to his frustration at not having one. We cited these words in chapter 7, but they bear repeating here: "Beware, theologians, that in wishing to make matters of faith of the propositions attendant on the motion and stillness of the sun and the earth, in time you probably risk the danger of condemning for heresy those who assert the earth stands firm and the sun moves; in time, I say, when sensately and necessarily it will be demonstrated that the earth moves and the sun stands still."[37] If he had had a demonstration, then he could have, and probably would have, penned most boldly that such was the case.

Perhaps he had hopes when he resumed work on the book in 1627 that the new pope would relax the decree or even reverse it. This might explain his having been carried away by the remembrance of the earlier debates. When the pope's intentions were made clearer during the negotiations before publication, he knew he had to continue to "bracket" the opinion and to tone down the insinuations in the claims he had initially drafted.

Concerning the effects of the work on the audience, and especially the elements of dialectic and rhetoric of it, a number of other inferences can be made with more confidence. The effect on the audience of the day was mixed. Those who had no prior intellectual allegiances found it quite persuasive. The rhetorical arguments, appealing to intellectual elitism and denigrating the conservatism of the opponents to Copernicanism, struck an answering chord in the breasts of many readers. Intellectuals already inclined to the view would have had their opinions reinforced by the compelling dialectical proofs, while the artistry of the examples delighted their imaginations.

The effects of Galileo's rhetoric on those who already had allegiances to the Tychonian system could not have been salutary. The jaundiced review of their work and the contempt shown the authors surely increased the indignation of the Jesuits. The Dominicans who had raised doubts about Galileo's orthodoxy before the decree of 1616 could not have been comforted by the weight given the Copernican side in the *Dialogue*. The Peripatetics in the universities knew that their opinions were those most ridiculed by the work. The progressives among them would find their views ignored or inadequately conveyed. Antonio Rocco, an Aristotelian philosopher, must have echoed the opinion of these when he said: "But come on, if there is a necessary truth and conclusion such that it is also evident as you

37. Westfall quotes from a letter of Galileo to Elia Diodati written on 15 January 1633 that echoes this sentiment. Galileo says of the efforts of Froidmont to support the ban on Copernicanism: "When Froidmont and others have established that to say the earth moves is heresy, while demonstrations, observations, and necessary conclusions show that it does move, in what a swamp will he have lost himself and the Holy Church" (*Essays*, 24, citing *Opere* 15:25).

say, show the evidence, bring in the reasons and the causes, leave persuasion to rhetoric, and no one will contradict you."[38]

The rhetoric of the work was probably most effective in its impact on the opposition; it strengthened and unified them. The damage done by it to the cause of science in Rome and in the Italian universities at the time was far greater than was its aid. If Galileo had stuck to dialectical arguments alone or if he had used rhetoric simply to open minds to his novelties as Copernicus had, he might have come closer to convincing those who were opposed to him. Most importantly, had he employed the regressus argument from the tides more carefully and considered alternative causes seriously, he would not have presented the earth's movement as the "true cause" so indubitably. Then his opponents would have been forced to recognize the strength of his arguments or he could have justly called them to account. As it was, those who were against him were inflamed by the book, and even the pope, who seems to have been at least inclined to entertain the opinions, was enraged by the rhetorical slant of the arguments. The many instances of rhetoric woven into the dialectical arguments of the piece perhaps make more understandable the reactions of the conservatives regnant in the Church and the universities.

38. Quoted in Carugo and Crombie, "The Jesuits and Galileo's Ideas of Science and of Nature," 24, citing *Opere* 7:629.

CHAPTER TEN

Galileo Interpreted for Englishmen

Galileo's *Dialogue* became available in bookstores on 21 February 1632. In recording the reactions of its readers Robert Westman has translated a letter of Tommaso Campanella, which discloses how the rhetoric of Galileo's work was perceived by that sympathetic reader:

Everyone plays his part marvelously: Simplicio as the laughing stock of this philosophical comedy, who, at the same time, shows the foolishness of his sect—the manner of speaking, the insecurity, the stubbornness and what not. Clearly we need not envy Plato. Salviati is a great Socrates who causes things to be born that are not yet born and Sagredo is a free intellect who, not corrupted by the schools, judges all with great wisdom. . . . You have done what I wished when I wrote to you from Naples [many years ago], to wit, that you ought to put your teachings into [the form of] a dialogue in order to assure reaching all, etc.[1]

In defining the work as a "philosophical comedy" Campanella gives no sign of realizing just how seriously it might be taken by conservatives in the Church. Viewing the book as a comedy, he is not concerned that its biting sarcasm might be seen as something more than playful teasing. His attitude probably mirrors that of many scholars resentful of the Simplicios who dominated the intellectual forum. Campanella's allusion to an earlier letter gives rise to the conjecture that he may have first given Galileo the idea of turning his treatise on the tides into a dialogue.

Others wrote similar accolades. But the response was not all positive. Castelli informed Galileo about the reaction of Father Scheiner reported to him by a bookseller. Scheiner supposedly encountered another priest at a bookstore and when the latter recommended the *Dialogue* as "the greatest book ever published," Scheiner's color changed, his hands shook, and he said that he would pay "ten gold ducats" for a copy in order to reply to it "extremely quickly."[2]

A year and a half later, on 16 June 1633, the book was prohibited, its sales curtailed, and all professors of philosophy and mathematics were to be read the Church's admonition against it. The effect, of course, was to publicize

1. Westman, "The Reception of Galileo's Dialogue," *Novità Celesti e Crisi del Sapere*, 334.
2. Ibid., 335.

the work even more. Westman reports that the book circulated mainly among a nonacademic audience, which included many of those whom Galileo had hoped to reach, knowledgeable men and those of political importance. It soon spread to the rest of Europe and England. But copies were scarce in England.

Galileo in English Translation

Although Galileo's writings were known in the original Italian, English translations were slow in coming. The *Dialogue* was finally available in English in 1661, when Thomas Salusbury's translation was published in his two-volume work, *Mathematical Collections and Translations*. In the first volume he included his translation of the *Letter to Madame Christina of Lorraine* and the exegetical writings of Kepler, Zuñiga, and Foscarini.[3] Salusbury discloses in the preface to that massive volume that he worked on the translations while living on the Continent during the Commonwealth period, 1649–60. His sympathies were obviously with the Royalists, and remarks in the preface and the text indicate that he probably was a crypto-Catholic.

In the preface to the first tome of the first volume, for which the *Dialogue* is the focus, the Englishman remarks that mathematical literature, especially that "nobler and sublimer part," astronomy, is quite sparse in England. The exegetical writings are included, he says, to reassure persons who might have pious objections to that controversial book of Galileo. Of the Roman Church's actions he says:

I shall not presume to Censure the Censure which the Church of Rome past upon his Doctrine and its Assertors. But, on the contrary, my Author having bin indefinite in his discourse, I shall forbear to exasperate, and attempt to reconcile such persons to this Hypothesis as devout esteem for Holy Scripture, and dutifull Respect to Canonical Injunctions hath made to stand off from this Opinion. . . . And least what I have spoken of the prohibiting of these Pieces by the Inquisition may deterre any scrupulous person from reading of them, I have purposely inserted the Imprimatur by which that Office licenced them.

The preface and marginalia disclose more than the usual deference to the Catholic position on Copernicus. Having himself been forced into exile by religious oppression during the Protectorate, Salusbury was particularly

3. The pieces described make up the first tome, part one. The second tome, which is extremely rare, contains three pieces on the measurement of flowing waters by Benedetto Castelli and an additional work by D. Corsinus on the floods in Bologna and Ferrara. It also contains four of Galileo's books on mechanics and writings by Descartes, Torricelli, and others. For details of Salusbury's life see Drake's preface to Galileo, *Discourse on Bodies in Water* (Urbana: University of Illinois Press, 1960), xxii–xxiv.

sensitive to the theological difficulties engendered by the Copernican system. The Salusbury translation of the *Dialogue,* however, seems not to have had an extensive influence, possibly because the volume was not printed in great numbers. It did not have a second printing.[4]

Wilkins and Puritan Science

A much earlier and more widely known effort to make Galileo's writings known to Englishmen was that of John Wilkins. It occurred to him to publicize the discoveries of *Sidereus nuncius* as early as 1638, while he was a young tutor at Oxford. The success of the first led to a second book two years later in which he defended the Copernicanism of the *Dialogue* with theological and philosophical arguments.

These two works of Wilkins permit an examination of how scholars in England, reportedly dominated by the outlook of Francis Bacon and the Royal Society, understood the place of rhetoric, dialectics, and demonstration in scientific discourse, and, more narrowly, they allow speculation on the effect of Galileo's views on a scientist of a different religious and national ambience. Wilkins's writings are important because they were very popular and his influence was profound. He has been termed "the most dynamic force in seventeenth-century England."[5]

Wilkins's stated objective was to make the discoveries of Galileo and their implications concerning the Copernican thesis more acceptable to his countrymen. Later Bishop of Chester, and a founding member of the Royal Society, Wilkins was one who understood how to appeal to the public he chose to address: Englishmen of varying degrees of education and wide interests.[6] Many of these were Puritans and what would today be called Protestant fundamentalists. But Wilkins was not, strictly speaking, a Puritan or a fundamentalist.

4. In his edition of the *Dialogue* Drake notes (xxv) that few copies of volume one survive and suggests that some may have been destroyed in the great fire of London. Santillana also published an edition of the Salusbury work under the title *Dialogue on The Great World Systems* (Chicago: University of Chicago Press, 1953). He says (viii) that Salusbury's translation was "badly printed, exceedingly unreliable, and it was often obscure." Yet he found the style so particularly well-suited to Galileo's baroque prose that he decided to undertake this edition of it.

5. Barbara Shapiro, *John Wilkins 1614–1672: An Intellectual Biography* (Berkeley: University of California Press, 1969), 2. Shapiro says that Grant McColley shared this estimate.

6. The major study of Wilkins is Shapiro's; see also Grant McColley, "The Ross-Wilkins Controversy," *Annals of Science* 3 (1968): 153–89. The biography by P. A. Wright Henderson, *The Life and Times of John Wilkins* (Edinburgh and London: William Blackwood and Sons, 1910) is unreliable in its discussion of Wilkins's writings. The author admits (79) he has not read his subject's books "save in the most hurried and even careless way, except two of them, the 'Real Character' and 'Natural Religion'."

Let us first look at Wilkins's work in relation to English science. Over the years much has been made of the influence of Protestantism on England's culture and science. As Barbara Shapiro points out, Robert Merton and those who have found a formative Puritan influence in the burgeoning of English science in the seventeenth century, especially Christopher Hill, apply the term Puritan too broadly. In effect they assume that English science is Puritan science, that the particular religious and social ethic of Puritanism explains the achievements of English science. Shapiro examines but does not accept Dorothy Stimson's similar but more nuanced view of the influence of Puritanism. Stimson, who distinguishes among kinds of Puritans, thought the moderate Puritans to be the most productive scientists. This progress Stimson attributed to their espousal of the "right to private judgment and their independent interpretation of Scripture."[7] To Shapiro "moderate" and "Puritan" are contradictory terms, for the Puritan was anything but moderate in his views, as the Civil War evidences. Scientists such as John Wilkins were not "Puritans," but rather stood somewhere in between the High Anglican Churchmen and the Puritans; they were moderate Protestants in matters of religion, not moderate Puritans. Wilkins, Shapiro thinks, always remained a moderate in the face of the great controversies he endured, espousing "religious purity and religious unity, parliament and monarch, religion and science."[8]

7. Shapiro, *John Wilkins,* 7. The Merton thesis was based on Robert K. Merton, "Science, Technology and Society in Seventeenth Century England," *Osiris* 4 (1938): 360–631. (The thesis was the subject of a unique conference in which Merton participated, entitled Fifty years of the Merton Thesis, Tel Aviv and Jerusalem, 16–19 May 1989. Some of the discussion in the present chapter was prepared for that conference. Selected papers and Merton's engaging response to the participants were published in *Science in Context* 3, no. 1 [1989].) Dorothy Stimson's early study laid the basis of some of Merton's ideas, "Puritanism and the New Philosophy," *Bulletin of the Institute of the History of Medicine* 3:(1935): 321–34, and also for the continuing discussion in *Past and Present* from 1964 through 1968. Rupert Hall's treatment of the thesis on the twenty-fifth anniversary of Merton's work is particularly insightful, "Merton Revisited, or Science and Society," *History of Science* 2 (1963): 1–16. Richard Westfall discusses the role of Protestantism and Calvinism in particular in the remarkable growth of science in England. He makes a more qualified assessment of its importance than Merton and Stimson, but also finds that Protestantism did have a beneficial effect in that it was more accepting of science than was Catholicism; see his *Science and Religion in Seventeenth-Century England* (New Haven: Yale University Press, 1953), 7–8. Perhaps Westfall means that in England there was a wider acceptance of heliocentrism.

8. Shapiro, 29, emphasis hers. The observations and evidence offered by Shapiro (4–9) and others regarding the imprecision of the use of the term Puritan are certainly convincing. Thomas Kuhn suggests that the real point of Merton's work is that there *was* something distinctive that made English science so productive and different from the Continental variety following the Scientific Revolution, not during it. He suggests that a revised Merton thesis might well hold the key to understanding the great achievements of eighteenth-century English mechanical science and experimentation; see his *The Essential Tension: Selected Studies in*

One can see more similarities than differences in Wilkins's and Galileo's conception of science and the roles accorded to sense experience and reasoning. Wilkins, like Galileo, does disparage some scholastic and Aristotelian teachings in natural philosophy, but he praises Aristotelian logic. The logic of Aristotle had continued to be an important element of university education, especially at Oxford after its revival there in the late sixteenth century, as Charles Schmitt has convincingly demonstrated.[9] Wilkins's epistemology and methodology is Aristotelian, and he, as did Galileo, maintained that sense knowledge unmediated by reason was not to be trusted in matters of astronomy.

The foundation of knowledge leading to the proofs of dialectic and of demonstrative or certain knowledge was a subject of special interest to Wilkins also. He treats it explicitly in his *Of the Principles and Duties of Natural Religion* (1675), published posthumously. The principles he expresses there guided his early writings as well. Since the problem of knowledge and certainty, and discourse about them, has been such a prominent concern in the discussion of the debates on the Copernican question, a brief look at his conception of methodology is instructive, especially since Wilkins as one of the founding members of the Royal Society was very much concerned with its scientific methodology.[10]

In his work on natural religion Wilkins states that men gain knowledge or belief outside of revelation by evidence of the senses, understanding, and a combination of the two. Sense knowledge he divides into two types, one derived from the outward senses, provided by the external sense organs, the other from the inner senses, provided from within us and furnishing a con-

---

Scientific Tradition and Change (Chicago: University of Chicago Press, 1977), 58–59, 136–137. Leo F. Solt has made a good case for a view opposite to Merton's: Roman Catholicism and Arminian Anglicanism promoted individual responsibility by their emphasis on good works, whereas Lutheranism and Calvinistic Puritanism "annihilated" the individual by their sole reliance on grace and God's arbitrary election. He states: "This extreme emphasis upon the sovereignty of God disposed Puritans to accept authority in politics, science or business that had its sanctions from above, not from below" ("Puritanism, Capitalism, Democracy and Science," *American Historical Review* 73 [Oct. 1967]: 20).

9. Charles B. Schmitt, *John Case and Aristotelianism in Renaissance England*, 29–43. Unlike our popular notion of the strength of Ramist revisionist logical teaching at the English universities, Schmitt shows that Aristotelianism, shorn of its elaborate late scholastic intricacies, was in the ascendency during the last quarter of the sixteenth and continued to expand through the early seventeenth century. It reached its height in the Laudian statues of 1636 that made Aristotle central to the curriculum. He points out (43–44) that this was a step forward rather than backward since the new curriculum now contained, besides its solid instruction in Aristotelian philosophy, added emphasis on mathematics, astronomy, and botanical science.

10. Wilkins wrote a text on language in which he attempted to create a more objective and efficacious method of notation, reducing words to a universal minimum of univocal references: *Essay towards a Real Character and a Philosophical Language* (London 1668).

sciousness of data derived from our outward senses and from thinking. Understanding is that by which we "apprehend the objects of knowledge"; these may be general or particular objects and may be either present or absent.

Evidence, he says, comes "from the nature of things in themselves" and from a "Congruity or Incongruity betwixt the Terms of a Proposition, or the Deductions of one Proposition from another, as doth either satisfie the mind, or else leave it in doubt." Evidence can also be based upon the testimony of others "when we depend upon the credit and relation of others for the truth or falsehood of anything." This last is especially important when there is no other way of knowing, such as in "matters of fact," and "accounts of persons and places at a distance." Mixed evidence relates "both to the Senses and Understanding, depending upon our own observation and repeated trials of the issues and events of Actions or Things, called Experience."[11]

After his discussion of epistemology, Wilkins distinguishes between two kinds of assent we may give to evidence: knowledge or certainty and opinion or probability. The first, certainty, he further subdivides into physical, mathematical, and moral. Physical certainty is dependent upon sense evidence and is the "first and highest kind of Evidence, of which humane nature is capable." Mathematical certainty includes not only mathematics but "simple abstracted beings" that must be recognized, such as that the whole is greater than its parts. He continues: "Connexion betwixt the Terms of some Propositions and some Deductions are so necessary as must unavoidably enforce our assent. There being an evident necessity that some things must be so, or not so, according as they are affirmed or denied to be, and what supposing our faculties to be true, they cannot possibly be otherwise, without implying a Contradiction."[12]

Moral certainty is less simple, he says, and depends more upon mixed circumstances; that is, it contains evidence from several sources: observation, experience, and understanding. He goes on to discuss propositions that are self-evident, those incapable of further proof, and first principles, those that cannot be proved a priori. When he speaks of moral certainty, Wilkins refers to reasoning in the realm of moral philosophy, not to moral certainty about natural philosophy.

Continuing in this scholastic vein, Wilkins says that necessary deductions made from these principles possess the same kind of certainty. He characterizes physical and mathematical certainty as infallible and moral certainty as indubitable. In a pointed reference to Catholic tradition, which ascribes in-

11. Wilkins, *Of the Principles and Duties of Natural Religion*, 3–5.
12. Ibid.

fallibility to the pope and Church Councils, he denies that any man can "pretend to such a perfect unerring judgment on which the divine power it self could not impose" in physical and mathematical matters. Such a claim would be "no less than a blasphemous arrogance." Absolute infallibility is an "Incommunicable Attribute." He explains: "But I mean a Conditional infallibility, that which supposes our faculties to be true, and that we do not neglect the exerting of them. And upon such a supposition there is a necessity that some things must be so as we apprehend them, and that they cannot possibly be otherwise."[13]

As his discussion makes clear, necessary demonstrations founded upon observation and sense evidence provide certainty in Oxford as well as in Rome.

The Plurality of Worlds

Wilkins's conception of scientific proof is announced initially in the preface to his first book touching on the Copernican hypothesis, *The Discovery of a World in the Moone or, A Discourse Tending to Prove that 'tis Probable there may be another Habitable World in that Planet* (1638).[14] He explains that this is not an "exact, accurate Treatise" since it was "but the fruit of some lighter studies and hurriedly composed," written in a matter of "some few weeks." Instead, it contains "onely probable arguments for the proofe of this opinion" (that the moon may be habitable) so that each conclusion does not have "an undeniable dependency" nor can "the truth of each argument . . . be measured by its necessity." He grants that the appearances might be solved otherwise than he has done, "But the thing I aime at is this, that probably they may so be solved, as I have here set them downe." This is the language of Aristotelian dialectics. It shows that Wilkins shares common ground with

13. Ibid., 7–9. My reading of Wilkins's conception of certainty differs in some points from Shapiro's, who sees his mathematical certainty as less certain than his physical. She does not note that Wilkins's use of "moral certainty" is confined to moral philosophy but seems to think that the last variety of certainty applies to science as well. Her discussion is in *Probability and Certainty in Seventeenth-Century England* (Princeton: Princeton University Press, 1983), 29–32. Concerning truth claims in general and the scientific revolution see Benjamin Nelson, "The Early Modern Revolution in Science and Philosophy," *Boston Studies in the Philosophy of Science* 3 (1964–66): 1–40; W. A. Wallace reviews some of the changing attitudes towards certitude in this period in "The Certitude of Science in Late Medieval and Renaissance Thought," *History of Philosophy Quarterly* 3 (July 1986): 281–91.

14. Wilkins's *The Discovery* was published in three editions. We will cite the first one (1638) here; a facsimile of this edition contains an introduction by Barbara Shapiro (New York: Scholars' Facsimiles and Reprints, 1973). The third edition includes an additional proposition in which Wilkins discusses the possibility of man's being able to journey to the moon, which is treated at the end of this chapter.

Galileo in his understanding of what constitutes acceptable proof in the realm of natural philosophy.

Dialectical Argument

Wilkins's plea to his readers in *The Discovery* makes use of familiar distinctions: "Let me then advise thee to come unto it with an equall minde, not swayed by prejudice, but indifferently resolved to assent unto that truth which upon deliberation shall seeme most probable unto thy reason, and then I doubt not, but either thou will agree with mee in this assertion, or at least not thinke it to be as farre from truth, as it is from common opinion."

He will attempt to provide dialectical arguments and hopes these will be met with rationality. Just as he asks the audience to approach the issues with minds free of prejudice, he implies that he will not try to sway his audience by rhetorical pleas. His desire to separate the two kinds of persuasion is very different from Galileo's intent in the preface to the *Dialogue,* where he vows to use every artifice to persuade his readers.

Wilkins's preface is intended to prepare the ground for the acceptance of rational argument. He keeps his promise; the arguments he presents in the text are generally syllogistic, even though these are presented in an informal conversational style. Emotional and ethical appeals are not developed to gain assent to his proofs, but he does use these and poetic touches occasionally to provide a lively context for a scientific problem. His is an attempt to popularize some difficult new philosophical conceptions for an audience who may not be deeply conversant with the sciences involved, but still he prefers to clarify by means of the deep structure of scientific argument employed by an intellectual elite. The language is folksy but the method is precise. From the preface one can see that Wilkins expects his audience to respond to principles of rational argument. He could presume for most of his readers a common schooling in grammar, rhetoric, and logic, whether this was acquired in a formal setting or in textbooks readily available on these subjects. Regardless of whether students were taught within the traditional scholastic logic and Ciceronian rhetoric, or the logic and rhetoric reforms of Ramus, the more highly educated readers would recognize arguments designed to stir the emotions and would think these inappropriate to dialectic and the investigation of nature.[15]

The Discovery is a kind of colloquial disputation. Scholastic in its arrange-

15. Even in Ramus's reform, wherein dialectical and rhetorical techniques of finding arguments are combined and one method of logic proclaimed for all discourse, Ramus distinguishes between types of discourse. That for scientific and learned audiences emphasizes the subject and logical proofs, and that for a popular audience must indulge in flattery. See the discussion of Ramus's reforms in Howell, *Logic and Rhetoric in England,* 146–281, esp. 164.

ment, the first edition of the book contains thirteen propositions taking up objections and airing supportive arguments, but these are always expressed in an unaffected manner in ordinary language. In the construction of proofs Wilkins often uses the revelations of the *Sidereus nuncius* and arguments from the *Dialogue*. He seems to have had in mind a more general audience than that envisioned for his later work, *A Discourse Concerning a New Planet: Tending to Prove That ('tis probable) our Earth is one of the Planets* (1640), where more advanced mathematical arguments and glosses point to more learned readers.

Wilkins argues in *The Discovery* that the moon's having land, water, atmosphere, and seasons indicates the strong probability that the moon could be inhabited or habitable. Wilkins shows himself to be a convinced Copernican who is very well read in the relevant literature. Besides Galileo, he cites Kepler, Tycho, Scheiner, Maestlin, Blancanus, among others, and, of course, Aristotle.

His initial task is to overcome both the religious fears and the philosophical objections of his audience that impede their acceptance of Galileo's discoveries about the moon and the new philosophical tenets concerning the heavens that these discoveries require. In an attempt to accomplish this he turns to rhetoric at times. Under the first proposition, "That the strangeness of this opinion is no sufficient reason why it should be rejected," Wilkins takes up complaints about the novelty of the discovery that the moon is a planet. He urges his readers not to reject the new theory of the heavens just because it is novel and received opinion is against it. Using the commonplace analogy of the similarities between Galileo's discoveries and those of Columbus, he argues that the celestial ones should be acknowledged as real also. The analogy between Columbus and Galileo, essentially a rhetorical argument, plays off the familiar discoveries of Columbus against the strange vistas of Galileo, in hopes of dispelling the fears of his audience by inducing a comparison between the wonderful resources of the new world beyond the sea to the marvelous possibilities of the novelties in the heavens.

Sounding much like Salviati, he maintains that it is generally the case that new truths are derided by the ignorant or by those "whose perverseness ties them to contrary opinion" and by "men whose curious pride will not allow any new thing for truth which they themselves were not the first inventors of" (1–4). Carrying the point further with obvious enjoyment, he then points out that a number of truths rejected by very wise men as ridiculous were later shown by "sense or demonstration" to be so. He cites the Antipodes, which scholars said could not exist (for how could men walk around with their "heels higher than their heads?"), and the superstitions of some ancient Romans who, thinking that an eclipse of the moon was caused by the moon's being in labor, made loud noises and held up torches to help ease her labor and to make sure she did not fall asleep (8–13).

Quite absurd opinions have commonly been held to be true by those who "know not the causes of things." As he explains, "nothing is in its selfe strange, since every naturall effect has an equall dependance upon its cause, and with the like necessity doth follow from it" (19). In each of these cases he shows that sense evidence or reason, or a combination of both, controverted the received opinion.

Wilkins observes that it is not the commonness of an opinion that can "priviledge it for a truth" (18). He admits that probable reasoning rests on opinion, but this is the opinion of an informed people, adopted after investigation of causes or held in light of propositions that are self-evident. Wilkins seems to wish to move the argument from the rhetorical level, where emotions would help to make the decision, to the dialectical, where decisions would be made on logical grounds alone.

Religious Objections

In treating the second proposition, "That a plurality of worlds doth not contradict any principle of reason or faith," Wilkins first must dispose of the philosophical basis for religious objections to the earth's motions, that founded on Aristotle's views. He begins by questioning the Stagyrite's relevance for matters of faith, noting that he was reputed to have read the books of Moses and praised them for their "majesticke stile as might become a God, but withall hee censured that manner of writing to be very unfitting for a Philosopher because there was nothing proved in them, but matters were delivered as if they would rather command than perswade beliefe" (24–25).

This is not an ad hominem argument of vilification, but an argument questioning the validity of a pagan philosopher's opinion on religious matters. His mention of Aristotle's reading of Moses, as well as some of the erroneous opinions he lists, seem to be based on similar passages in Campanella's *Apologia*. Unlike Campanella, though, he does not suggest that the pagan philosophy of Aristotle be replaced by Christian thought.

Wilkins wants to show respect for Aristotle but at the same time intends to judge him at the bar of reason. He goes on to point out the weaknesses of the philosopher's arguments, using other points of his doctrine to disprove them. Concerning the singularity of the world, he says, if there were "such necessary proofe as might confirme it," Aristotle would have supplied them in *De caelo* (26). He cites Campanella's *Apologia pro Galileo* on the futility of accepting Aristotle on all points (30). The Oxford tutor's position on Aristotle's cosmology is the same as his views of the contentions of others: they are to be believed when the arguments are demonstrative, based upon true causes, and when the evidence they put forward is not contradicted by present knowledge. He thus judges them on the basis of their conformity to Aristotelian principles of reasoning.

In an amusing comment that underscores his benign intent, Wilkins says he disagrees with both St. Vincentius and Serafinus de Firmo, who were quoted as saying that Aristotle was the vial of "God's wrath" poured out on the waters by the third angel in the book of Revelation. Although he believes "the world is much beholden to Aristotle for all its sciences," he thinks it a shame to hold that all we can know has been given to us by our forefathers and that "wee are set upon their shoulders, not to see further then they themselves did." It is, he says, "a superstitious, a lazie opinion to thinke Aristotles workes the bounds and limits of humane invention" (32–33).

Taking up one of the problems debated in the literature concerning the subject matter of *De caelo,* Wilkins clarifies the meaning of 'world'. There are two senses: The first denotes the universe and the second denotes another inferior elemental body. He thinks that there are no other worlds in the first sense but that there are in the second.

In considering arguments from Scripture in this second proposition of *The Discovery,* Wilkins approach is similar to Galileo's in the *Letter to Christina.* Wilkins had probably not read it at this point, for he does not cite the letter here, but he does refer to it in *A Discourse.* The hermeneutic principles he employs in this earlier work, however, are the traditional approaches to exegesis.

To say Wilkins reads Scripture with an independent mind as Stimson would expect of a Puritan is not quite accurate. The divine origin of Scripture is unquestioned and the traditional authorities are accorded great respect, although not necessarily consent. Accordingly, he reviews the opinion on the plurality of worlds in Scripture and in the writings of Church and secular authorities, mentioning Moses, St. John, Aquinas, Cardinal Baronius (on Virgilius who was excommunicated for this belief), and Julius Caesar la Galla, all of whom were against the view.

Wilkins offers the two main principles of scriptural interpretation we have seen in preceding exegetical works: the Scriptures are relevant to faith and morals, not to natural philosophy, and the language and thought of the word is accommodated to the common man. Augustine's well-known *De Genesi ad litteram* furnishes the main principles of interpretation for Wilkins, just as it had for Galileo. The Englishman maintains that opinions drawn from scriptural statements are not relevant to astronomical issue since these are not matters of religious doctrine. Wilkins cites Aquinas, who said the reason Moses did not write of the air was because people could not see it and would not know whether it existed or not. And similarly, St. Augustine thought that this was why Moses did not describe the creation of angels. Furthermore, the fact that Scripture does not tell us that there are other worlds cannot be brought against the proposition, for the Scripture does not speak of planets. The New and Old Testaments are not concerned with philosophy but with history, exegesis, and prophecy.

One would expect, following the drift of the Merton thesis, that Protestant writers would not cite authorities as often as do Catholic authors, and that the Fathers of the Church, important sources for Catholics, would find little place in English writings. As we have noted, Dorothy Stimson thought that what made the Puritan scientists so enterprising was a different habit of mind evinced in their belief in the "right of private judgment." Wilkins does demonstrate an independent reading of authorities in both of the books we are examining here. Yet he is more inclined than Galileo to cite strings of authorities in approbation of opinions. But he always assesses these on logical and scientific grounds.

In his conclusions, Wilkins is more adventuresome than most Catholic writers would be at this time. Although Cardinal Cusanus supported the plurality of worlds, so did Bruno, who had been burned at the stake. Reflecting as usual the Jesuit view, Carbone in discussing the question in a manuscript commentary on *De caelo* states that philosophically speaking the unity of the world can be maintained only with probability, but from the standpoint of theology the view is erroneous if not heretical.[16] Thus for Roman Catholics the teaching would be termed heretical if the Scripture clearly stated the opposite, but it could be judged erroneous on the basis of the account of creation alone, for this does not mention other worlds.

The New Cosmology

The third proposition maintains that the heavens are not composed of pure uncorruptible matter. As we have noted, this principle of Galileo's new philosophy confutes one of the basic tenets of Aristotle's cosmology. Wilkins provides new support for Galileo's position, citing Peter Lombard, St. Ambrose, and Bede, all of whom spoke of three and four elements as compositive of heavenly bodies. The rhetorical use of authority here shows that the opinion has been held by reputable theologians.

Aristotle himself had not observed any changes, but, Wilkins points out, Scheiner reported some in his *Rosa Ursina* (51). Tycho's view receives attention here. Wilkins reveals even at this early point in his career that he has an independent mind in things scientific, but it was not Galileo's reasoning on the sunspots but Scheiner's that convinced him. Wilkins had excellent

16. Carbone's view is described in my "Ludovico Carbone's Commentary on Aristotle's *De caelo*," in Daniel O. Dahlstrom, ed., *Nature and Scientific Method* (Washington, D.C.: Catholic University of America Press, 1990), 179.

sources for his up-to-date scientific knowledge: the Bodleian library's science collection and the Savillian lectures delivered during his stay at Oxford from the late 1620s to 1637.[17]

In the propositions that follow, Wilkins treats the nature of the moon, considers whether it is solid and opaque, whether it shines with reflected light, and whether there is "a world in the moon" (54–93). In treating these questions he draws not only on Galileo but on Tycho, Kepler, and Julius Caesar la Galla. He takes up the myths of the ancients and uses the moderns to refute them. His arguments for the propositions are dialectical and often supported by sense observations, either those of the telescope or by sight alone, and sometimes by mathematical computations. He occasionally incorporates Galileo's diagrams, changing only the identifying letters. Wilkins was also a mathematician and later wrote a work on its practical applications in machines, *Mathematical Magick or the Wonders That May be Performed by Mechanical Geometry* (1648).

Under Proposition 8 Wilkins considers whether land and water exist on the moon, concluding that both do. In this regard he cites Kepler as much as Galileo. Continuing to expose the latest opinions regarding the surface of the moon, in Proposition 9 he argues that there are high mountains on the moon, as may be inferred from their shadows. Using an illustration and diagram from *Sidereus nuncius,* he combines telescopic evidence with mathematical proofs, declaring "the observation of Galilaeus, whose glasse can shew this truth to the senses a proofe beyond exception and certaine that man must needs be of a most timerous faith who dares not believe his own eye" (133–34).

The manner in which Wilkins builds his dialectical arguments on sense evidence is well illustrated in the proofs for Proposition 10, which states "that there is an Atmo-sphaera, or an orbe of grosse vaporous aire, immediately encompassing the body of the Moone." After noting that Maestlin, Kepler, Galileo, Baptista Cisatus, and Scheiner all think that such an atmosphere is present, he develops their argument as follows: "'Tis observed, that so much of the Moone as is enlightened, is alwaies part of a bigger circle then that which is darker. Their frequent experience hath proved this, and an easie observation may quickely confirme it. But now this cannot proceede from any other cause so probable, as from this orbe of aire, especially when we consider how that planet shining with a borrowed light, doth not send forth any such rayes as may make her appearance bigger than her body" (139–40).[18]

17. Shapiro, *John Wilkins,* 30.

18. This point seems to have been based on remarks by Galileo in *Sidereus Nuncius;* see Drake, *Discoveries,* 39.

In like manner he continues to make other points, beginning with sense observation and reasoning from it to the probable cause. No matter that the conjectured cause was later found not to be the moon's atmosphere but our own that produced the appearances, the reasoning is still that of the re- gressus, which figured so frequently in Galileo. But since Wilkins could not show that only this cause could produce the effect, the reasoning has only probable force.

In urging the habitability of the moon in Proposition 13, Wilkins founds his argument on all of the evidence earlier presented dialectically. He states that "tis probable there may be inhabitants in this other World, but of what kinde they are is uncertaine" (186).

He wonders whether these inhabitants could be Adam's seed, the ques- tion that was of most concern to the Church. Adam's sin, or original sin, he thinks, would not in fact apply, but other sin might and if so Christ's sacrifice could save them. Scriptural passages are cited to buttress the point. In a final reflection meant to encourage the incredulous reader he declares: "Provi- dence does not enlighten us all at once, but prefers to lead us from one thing to another."

Wilkins's prophecy at the end of the book is moving in its prescience: "So, perhaps, there may be some other meanes invented for a conveyance to the moone, and though it seem a terrible and impossible thing ever to passe through the vaste spaces of aire, yet no question there would bee some men who durst venture this as well as the other" (208).

He concludes his work succinctly saying, "that tis possible there may be, and tis probable there is another habitable world in that Planet. And this was that I undertooke to prove" (209–10).

In the two years between the popular book's first two editions (both pub- lished in 1638) and the third of 1640, the idea of travel to the moon must have generated much discussion and further reflection, for Wilkins adds more than thirty pages to consider these possibilities and additionally a fourteenth proposition: "That tis possible for some of our posteritie, to finde out a conveyance to this other world; and if there be inhabitants there, to have commerce with them."

Although he conceives of a "flying Chariot," he did not envision space suits to supply the warmth and oxygen needed for the trip. He speculates, rather, that the atmosphere might not be as cold when we travel further and if food is insufficient our bodies might adapt and be able to take nourish- ment from "smells" or live on air, like certain plants described by Aristotle and others (224–25).

The only discussion dwelling upon "possibility" rather than probability in either work, the fourteenth proposition is pure speculation based mainly on analogy. Wilkins must have been influenced to some degree by Kepler's speculations about moon creatures in his response to Galileo's discoveries. The flight Wilkins proposes is reminiscent of the German astronomer's description of a journey to the moon in the *Somnium,* published in 1634.

The Scientific Style of the *Discovery*

If we compare the presentation of scientific discoveries and the arguments about them in Wilkins's *Discovery* to that which inspired it, Galileo's *Sidereus nuncius,* we find a number of differences engendered by the nature and purposes of the works. Galileo employs rhetoric, dialectic, and demonstration in a manner more complex and subtle than does Wilkins. Apart from the great differences in their talents, their professions, and their audiences, both obviously had different aims: Galileo was intent on imparting his discoveries, not in defending them philosophically or theologically, while Wilkins was concerned to dispel fears resulting from those revelations and the corroboration of the Copernican system they seemed to promise.

Galileo's account of his observations with the telescope in *Sidereus Nuncius* contains a much freer, more graceful mingling of observation with dialectical deductions than does Wilkins's recounting of them. Not only does Galileo provide vivid narrative, but he occasionally embellishes it with poetic language and rhetorical flourishes, maintaining throughout a kind of reverent wonder at all he has seen. The contrast in their approaches is well illustrated by the way Wilkins treats Galileo's argument concerning the moon's secondary light. Galileo posits a rhetorical question with poetic elegance: "Yet what is so remarkable about this? The earth, in fair and grateful exchange, pays back to the moon an illumination similar to that which it receives from her throughout nearly all the darkest gloom of night."[19]

While Wilkins repeats the personification, he amplifies it in a different way in the *Discovery:* "And as loving friends equally participate of the same joy and griefe, so doe these mutually partake of the same light from the Sunne, and the same darkenesse from the eclipse" (153–54). Wilkins's choice of figure cannot match the charm and power of Galileo's style. The cadence of a preacher already emerges in Wilkins's prose.

Nor does Wilkins employ the sharp thrusts of ridicule for which the Italian was renowned. Wilkins instead turns a whimsical sense of humor on absurd beliefs and superstitious practices, which must have disarmed his readers and gently induced them to call their own beliefs into question. His rebuttals to more serious objections are dialectical and often grounded in sense experience or in general principles. Occasionally he argues from final cause. Rhetorical artifice he seldom uses other than to allay fear. By this time

19. Ibid., 44.

in England the notion that one should make a conscious effort to avoid *copia* and "the flowers of rhetoric" has been voiced by Francis Bacon and would soon become an article of faith for the Royal Society. For Verulam, the role of rhetoric was "to apply reason to the imagination for the better moving of the will."[20]

As is appropriate to an Oxford tutor and a future bishop, Wilkins's aim is to teach, and judging from the content and style of his writing he must have hoped to reach not only the educated laity in his wide audience but also clerics of little learning who were often influential in areas beyond their competence. At the same time he wished to persuade those who were knowledgeable and to make sure that his arguments could stand up against their objections.

Galileo on the other hand spoke to an audience composed of educated laymen, learned ecclesiastical officials, scholars in religious orders, and academicians who would appreciate and expect from him, the scion of Florentine humanist society, a literary eloquence that characterized the Renaissance man, even though he spoke of scientific matters.

The differences we see in content and style in the two works may be attributed to the authors' varied aims, temperament, and talents. What is significant, and surprising, is their agreement on the logical methodology that constitutes proof in science and their divergence in the role granted to rhetoric in such discourse. This divergence becomes more apparent in the next work of Wilkins to be considered.

Wilkins's Defense of Copernicus

In *A Discourse* Wilkins defends the Copernican thesis in general, and, as in *The Discovery*, he approaches the subject from both a theological and scientific standpoint. Again, dividing the work into propositions in the scholastic manner, he devotes the first half of his ten propositions to theological issues and the rest to those of natural philosophy or astronomy. Besides Galileo's *Letter to Christina*, which he must have read in the Latin-Italian edition,[21] he draws heavily upon Galileo's *Dialogue*. But Galileo is only one of a number of authorities to whom he turns in the development of his arguments.

In his consideration of the basis of knowledge in *Of Principles and Duties*

20. The translation is from Kennedy, *Classical Rhetoric*, 217. Much has been written on the effect of Bacon and the Royal Society on scientific language. A good summary of the discussion is in Brian Vickers's introduction to *English Science: Bacon to Newton* (Cambridge: Cambridge University Press, 1987). The definition of rhetoric appears in Bacon's *Advancement of Learning* 2.18.2.

21. Wilkins cites *Nov-antiqua*, the first words of its long title: *Nov-antiqua Sanctissimorum Patrum*. . . .

of Natural Religion, discussed above, Wilkins spoke of the testimony of others as yielding understanding, but he carefully circumscribes what he will accept: *matters of fact* and accounts of *persons* and *places* at a distance, when there is no other way of learning these things ourselves, and when the witnesses are credible. We have seen that in the first part of *The Discovery* he urges his readers not to reject the new theory of the heavens just because it is novel and that received opinion is against it.

Authority and Prejudice

In the beginning of *A Discourse* Wilkins explicitly takes up the problem of authority, since that constitutes the major obstacle standing in the way of the new opinion. In theology one must seek divine authority, but in philosophy, he argues, it would be "a preposterous course to begin at the testimony and opinion of others, and then afterwards to descend unto the reasons that may bee drawne from the Nature and essence of the things themselves: because these inartificiall Arguments (as the Logicians cal them) doe not carry with them any cleere and convincing evidence; and therefore should come after those that are of more necessary dependence, as serving rather to confirme, than resolve the Judgment" (2).[22]

The purpose of *A Discourse,* Wilkins states in the preface, is to remove "common prejudices" that prevent men from accepting new opinions. He tells his readers that he will not trouble them "with an *Invective* against those multitudes of Pamphlets which are every day prest into the World; or an *Apology,* why this was published amongst the rest." His hope is that through this book men might see "a greater comelinesse and order in this great Fabricke of the World, and more easily understand the appearances of Astronomy."

As if to underscore the fact that the work is dialectical and not rhetorical, Wilkins states that the manner he chooses to present his ideas will not proceed "with such heate and religion, as if every one that reads it, were presently bound to yeeld up his assent: But as it is in other Warres where *victory* cannot bee had, men must be content with peace." He notes that if some of "our hot adversaries" had not been so vehement in opposing "the Persons" but had invested their vigor in "confuting the cause" they would have achieved better effects. He adds an admonition that reveals his own intent: "'Tis an excellent rule to bee observed in all disputes, that Men should give soft Words and hard Arguments, that they would not so much strive to vex, as to convince an Enemy."

22. The logicians are the Ramist logicians who took over Aristotle's distinction in the *Rhetoric* between artificial proofs—the arguments invented by the author, and inartificial proofs—testimony of witnesses, contracts, oaths, etc. See Aristotle *Rhetoric* 1.2; Loeb edition, 15–16.

Wilkins's aim thus requires him to use both rhetorical and scientific arguments. In attempting to remove prejudice he will use "soft words" and treat those stumbling blocks that most prevent acceptance of the Copernican system. But when he enters into matter scientific or theological, he plans to leave the domain of rhetoric and enter into discussions of necessary and probable truths.

Wilkins observes these scholastic distinctions and uses pathetic appeals only when emotional barriers require it. Appeal from his own authority or that of others is employed not to persuade, but to confirm. These were not Galileo's methods, as we have witnessed, and one wonders if Wilkins meant to offer a mild corrective to the Italian whom he admired so much.

In line with his rhetorical objective, Wilkins first refutes common opinions that find the Copernican view both too novel and too singular, for only a few have supported it. The space that Wilkins expends on allaying fears of novelty in both books may seem odd to the modern reader. But in England as well as in Italy novelty was regarded with suspicion. Galileo himself was quick to point out that the *novità celesti* were revealed by sense experience. Before him Copernicus had tried to overcome imputations of rashness for the novelty of his position by carefully noting the ancients who shared similar views of heliocentrism.

In response to the common question about how men for five thousand years could not have seen that the earth moves, Wilkins says that this "novelty" has been urged only by a few: "some fabulous Pithagorians, and of late Copernicus." Not even the ancients are infallible, he notes, and yet we build on earlier knowledge. He then quotes Bacon's memorable answer: "Truth is the Daughter of Time" (6).[23] Concerning the more serious question of silence on these matters by such figures as Moses, Job, David, and Solomon, Wilkins reiterates the familiar argument, saying that the Holy Spirit enlightened the Old Testament figures about spiritual matters, and in some cases gave them special human knowledge, but left them ignorant of philosophy.

As to its singularity, he mentions that others besides Pythagoras and the Pythagoreans have held this view, notably Aristarchus Samius, and a number of others mentioned by Plutarch. More recently, besides Copernicus, Cardinal Cusanus and most astronomers—including Joachinus Rheticus, Christopherus Rothman, Maestlin, and Erasmus Reinholdus—think the new theory to be true. Even more importantly, Gilbert, Kepler, and Galileo have "confirmed this Hypothesis, with their new inventions." After citing the testimonies of these experts, he concludes that now "it is a greater argument of singularitie to oppose it" (18).

23. The quotation is in *Novum Organum*, Aphorism 85.

Wilkins observes that even Ptolemy cautioned men not to think that his was the true picture of the heavens but that his aim was to present an hypothesis to measure and explain the appearances and to permit us to calculate their motions. Showing that he discounts Osiander's preface, Wilkins says Copernicus's intent was to present "the true naturall Causes of these severall Motions, and Appearances." Pythagoras meant merely "to settle the Imagination," while Copernicus wished "to satisfy their judgment." Certainly then Ptolemy would have no reason not to assent to this hypothesis if he knew the grounds for it. He, too, reports the reaction of the illustrious mathematician Clavius upon hearing of Galileo's discoveries with the telescope to the effect that now we must consider a hypothesis other than Ptolemy's to explain the appearances (20–21).

Pointing out that those on the Copernican side in the controversy are just as knowledgeable concerning Aristotle and Ptolemy as those opposed, he remarks that on the other hand those opposed have not bothered to read Copernicus. Wilkins, then, accepts the argument Galileo had made very forcibly in the *Dialogue*, which the Papal Commission thought an unjustified claim.

Wilkins posits three reasons for men's reluctance to accept the hypothesis. First, people are naturally partial to their own inventions. Thus, Tycho, whom Wilkins otherwise often praises, was desirous of spreading his own opinion. Second, "a servile and superstitious feare of derogating from the authoritie of the antients, or opposing that meaning of Scripture phrases" intrudes. Finally, people judge by sense evidence rather than by "discourse and reason," and they adhere to the letter of Scripture and ignore "all those grounds and probabilities in Astronomie, upon which this opinion is bottomed." This last, he says, is the major reason scholars, who are otherwise reputable, and also the common people rage against it (25–27). Except for the reservations due to belief in the literal meaning of Scriptures, Wilkins has summarized the reasons given in the *Dialogue*. Galileo, we know, did not there take up opposition based on scriptural grounds, as advised.

Scriptural Difficulties

In the following several propositions Wilkins rehearses familiar principles of biblical interpretation, but he adds some new explanations of his own. For example he phrases the next proposition: "The Holy Ghost could have enlightened us, but he left us something to trouble us and distract us from our sins." Moving from general objections to specific scriptural passages at odds with heliocentrism, Wilkins first notes that the difficult passages are of two kinds: those that "imply a motion in the Heavens," and "those that seem to express a rest and immobilitie in the Earth" (30). He explains these with the accommodation principle introduced in *Discovery*, this time clearly basing the argument upon Galileo's *Letter to Christina*. But the second refutation is his extrapolation: if the Holy Ghost did speak in the Scripture of astronomical phenomenon as they really are, then people might begin to doubt the spiritual truths it is designed to teach. And if the motions of the heavens were discussed men, might ignore more important things (31–32).

Wilkins then analyzes the texts that in their literal sense appear to support the sun's rather than the earth's movement, among them the problematic verse from Joshua that Galileo accepts literally and tries in the *Letter to Christina* to establish as a passage supportive of Copernicanism: "Sun stand thou still upon Gibeon." Wilkins disagrees with Galileo and asserts that the Holy Ghost only intended to speak of the appearances, how ordinary men would perceive it. He explains that the sun was probably low in the sky and seemed to stand still on the peak of Gibeon (35–40). But Wilkins thinks that Galileo and Zuñiga were correct in finding possible corroboration for the earth's motion in the passage from Job, "Who moveth the earth from its place?" (103).

In developing further the principle of the Scripture's accommodation to the language and understanding of man, Wilkins draws his confirmatory authorities from both Catholic and Protestant camps. He cites Mersenne as one who is opposed to Copernicus but does invoke scriptural grounds because he too accepts the principle of accommodation. Wilkins goes on to cite Calvin and Clavius, both of whom give examples of hyperbole in Scripture.

He notes in the fourth proposition the absurdities that ensue from looking to the Scriptures for philosophical knowledge and argues in the fifth that if the proper interpretations were applied we would see that Scripture does not teach the immobility of the earth; it speaks, rather, of the stability of the universe.

In the sixth and final proposition touching on religious objections to the Copernican view, Wilkins contends boldly that no arguments from Scripture, nature, and astronomical observations support the earth's being at the center of the universe (105). He examines the problem of the location of heaven and hell within the Copernican system but suggests no solution. He says that the objections raised in this regard are all founded upon "uncertainties," and about these no answers are forthcoming (107).

It is surprising in this theological section of the book that, although Wilkins ranges widely through the Scripture and cites Galileo's *Letter to Christina* in many instances, he does not develop one of Galileo's major arguments, that two truths cannot contradict each other. Galileo had referred to the truths given to us by God in the Book of Nature and in the Book of Life, the Scriptures. Since Wilkins was both a theologian and a scientist, this would seem to be a particularly attractive argument.

Philosophical Objections

In the next section of the book Wilkins turns to natural philosophy to refute objections to the earth's being removed from its position at the center of the universe. In Proposition 6 he again uses proofs familiar to us from Galileo's treatment in the *Dialogue*. Regarding the first objections, that the earth must be removed from the heavens because of the vileness of its matter, he says that the nobility and incorruptibility of the heavens are assumptions that have not been proved. He finds the second and third arguments, those treating of the necessity of earth being at the center of the cosmos because of its heaviness and because it is the center of gravity, to be faulty. He prefers to follow Galileo in his opinion of Aristotle's proof: "Though Aristotle were a Master in the art of Syllogismes, and he from whom we received the rules of disputation; yet in this particular, 'tis very plain that hee was deceived with a fallacie, whilst his Argument do's suppose that which it do's pretend to proove" (110). Wilkins wonders how it actually can be proved that heavy bodies descend to the center or light ones rise to the circumference of the heavens. We have only experience with the earth and the air above it, which is only "an insensible point," bearing as much relation to the whole as "a grain of sand to the earth." He concludes these considerations: "Wherefore it were a sencelesse thing, from our experience of so little a part, to pronounce any thing infallibly concerning the scituation of the whole" (111).

The arguments from appearances in favor of the earth's stability he takes up in scholastic fashion, stating objections and responding to each. Adapting some of the proofs in the *Dialogue,* he proposes mathematical demonstrations that position the earth in relation to the signs of the Zodiac, to the position of the axis and the equator, and, finally, to the unchanging size of the stars.

In the last four propositions of the book Wilkins turns to the major arguments for Copernicanism from natural philosophy and astronomy. The seventh asserts that "'Tis probable that the Sun is in the center of the World" (133). Wilkins argues that the Copernican explanation frees Nature from the deformity Tycho's system would give it and the inconveniences of the Ptolemaic system. These are arguments enough to confirm the Copernican view, he says, but other probable arguments likewise support it. All of the proofs introduced here are dialectical, but they are not always those advanced by Galileo. Wilkins has summarized the ones he finds most compelling, most of which are grounded in philosophical principles. One, based on final cause, holds that the sun is better able to distribute its light and heat from the center; others argue from the topos of more and less—what would be better or what would be more fitting. One of the arguments from what is more fitting is taken from the harmonic proportionality that a central position for the sun would obtain. Wilkins here praises Kepler's argument in the *Mysterium cosmographicum,* a proof Galileo did not find relevant. If we grant the system, "an excellent Harmonie will exist in the number and the distance of the planets, . . . for then the five Mathematicall bodies so much spoken of by Euclid, wil beare in them a proportion answerable to the severall distances of the Planets from one another." He then describes the geometrical relationships of cube, tetrahedron, dodecahedron, icosahedron, and octohedron conjectured by Kepler (139).

Wilkins argues in Proposition 8 that the negative cannot be proved: that there is "not any sufficient reason to prove the Earth incapable of those motions which Copernicus ascribes unto it" (142). In developing the point Wilkins accords respectful attention to the arguments of Aristotle and Ptolemy, "which beare in them a great shew of probabilitie" because these men of "excellent parts and deep judgements, did ground upon them, as being of infallible and necessarie consequence" (142–43).

Wilkins then considers one of the foremost difficulties engendered by the system, why we cannot discern that the earth moves. He concludes his consideration by placing the blame on our *sensus communis* that makes the eye to seem immobile and not notice the effects of motion on the body (144–45). In this part of the work Wilkins repeatedly counters the arguments of Alexander Ross, presented in *Commentum de terrae motu circulari* (1634). Without consciously intending to, Ross, in his espousal of Peripatetic positions, actually mirrors the qualities of the fictitious Simplicio of the *Dialogue.* Generally Wilkins is restrained in his retorts, but he occasionally permits his true feelings to show.

In replying to claims that the earth's motion would topple tall buildings, Wilkins says that the earth's movement would not be sudden but equal, and he compares the effect to that on a glass of beer which stands upright on a ship moving upon a smooth stream. He then cites Ross's objection: suppose that motion were natural to the earth; it would still not be natural to towns and buildings, "for these are artificiall." Wilkins's response is short: "To which I answer: Ha, ha, he" (148–49). Such is the extreme of Wilkins's use of humor to vanquish an opponent.

Additional Arguments

Wilkins takes arguments from Gilbert's treatise on the magnet to argue analogously that clouds and birds, arrows and cannonballs, may be bound to the earth in the same way as magnetized particles to the loadstone. In still another analogy, he points out that we do not see the motion of each drop of water, yet the sea ebbs and flows. So a body considered by itself may seem to have only up and down motion, yet in reference to the "whole Frame of which it is a part" may also move in another way "as naturall unto it" (162–63).

In a homespun but apt illustration, Wilkins counters one last objection to the diurnal motion of the earth, namely, that two distinct motions of the earth are difficult to conceive. If one considers that both motions tend in the same direction from West to East they can more easily be understood, he says: "Thus a Bowle [a bowling ball] being turned out of the hand, ha's two motions in the Aire; one, whereby it is carried round; the other, whereby it is cast forward" (188).

In the most engaging and convincing passage Wilkins argues in the ninth proposition, again dialectically, that it is more probable that the earth moves than the sun or the heavens. He begins by explaining that if the earth is held to be stationary the heavens would have to move at least 4,529,538 German miles an hour to circle it in twenty-four hours. He makes the import more concrete by relating that Cardan has said that a star would then have to travel 1,132 miles for each pulse beat, while Tycho maintains it would move 732 miles. Wilkins explains that these are the smallest estimates, for some, like Clavius, say things would move even faster, that at the equator every star moves 42,398,437 ½ miles in an hour. So if a man traveled 40 miles a day, he would not be able to travel in 2904 years as far as a star does in one hour (189–90). Bringing the point closer to home in a humorous aside, he observes that if a bird could fly this fast it would go around the world seven times in the time one could say *"Ave Maria, gratia plena, Dominus tecum"* (191). Wilkins adds that the speed considered here is that conjectured for the eighth sphere only, so that for the primum mobile the speed would be much faster. He quotes from Gilbert's *De magnete* in this regard, translating from the Latin: "A man may more easily conceive the possibilitie of any Fable or Fiction how Beasts and Trees might talke together, than how any materiall Body should bee moved with such a swiftnesse" (192).

The counter-explanation that God has the absolute power to accomplish such movement if he chooses, the kind of argument favored by Urban VIII, does not confound Wilkins. He says that of course God could turn the spheres even faster if he wished, but the question is "not what can bee done, but what is most likely to be don according to the usuall course of Nature. 'Tis the part of a Philosopher, in the resolution of naturall events, not to fly unto the absolute Power of God, and tell us what he can doe; but what according to the usuall way of Providence, is most likely to be done, to find out such causes of things, as may seem most easy & probable to our reason" (193). Galileo may have thought this himself, but it would have been too dangerous to have said so.

Brightening the philosophical discussion with some humorous arguments from analogy, he recounts Kepler's comparison of the motion of the stars around the earth to the cook who roasts his meat by turning the fire around it, and Galileo's illustration of the man who after climbing a high tower would save himself the trouble of turning his head to survey his surroundings by having them revolve around him (203–4). Again the humor is not pointed at adversaries, rather at the absurdity of the arguments.

The cause of the motions of the planets Wilkins considers in detail, citing Aristotle, Aquinas, Dur, Soncinas, and other Schoolmen. He does not think that such motion can be attributed to angels or intelligences, for he cannot imagine how the will can move an object, or how orbs can perceive the intelligence or the will. Kepler's view he finds not "very improbable," namely, that the sun is responsible. To the question of how this power can act at a distance, he proposes the action of light and heat and "those other secret influences, which work upon Minerals in the Bowels of the Earth" (215). If the moon "according to common Philosophy may move the Sea, why then may not the Sun move this Globe of Earth?" (215). Could the sun not provide the source of the earth's movement?

The a fortiori argument reveals that Wilkins, following Kepler, assumes that the moon, and not the earth's movement, is responsible for the motion of the tides. Thus, in spite of the force Galileo attributed to his argument from the tides, it did not impress Wilkins, who was intent upon interpreting Galileo for his readers. He was not convinced by the dialectical proofs or the rhetorical appeals Galileo marshalled in its favor.

At the end of these speculations Wilkins turns to a familiar answer to the puzzling problems that continue to confound philosophers, the verse from Ecclesiastes 3.11: "That no man can find out the Works of God, from the beginning to the end" (215).

The last proposition declares that "this Hypothesis [the Copernican] is esactly agreeable to common appearances." In support of this contention he takes up the phenomena of the seasons, the years, and day and night. Concerning the last, the problem of the sun's rising and setting, he cites Aristotle's *De caelo* to the effect that the appearances will remain the same regardless of whether the eye or the object is moved. Since this sage observation appears in Aristotle, Wilkins cannot understand why the Peripatetic Alexander Ross concluded his critique of Copernicanism with the argument that the earth cannot be moving, for if it did the shadow on the sundial would be altered (218).

In explaining the appearances of the months and the solstices, Wilkins draws upon diagrams and arguments from two authors not mentioned by Galileo, Lansbergius and Fromundus. He quotes the latter, translating from the Latin: "There is not any more probable Argument to prove the annuall motion of the Earth, than it's agreeablenesse to the *station, direction,* and *regression* of the Planets" (229). Copernicus, says Wilkins, furnishes also the best explanation of the variation in size of the planets and of eclipses.

The Value of Astronomy

At the end of his treatise Wilkins includes a discussion of the value of astronomy in "earthly contentments" (234). This is an original and entirely different reflection from anything we have seen in Galileo's *Dialogue*. The high moral tone is reminiscent of Ptolemy's preface to the *Almagest*, but it is at the same time a product of the pervasive concern of English scientists who, inspired by Bacon, demanded that knowledge be both practical and edifying.

He thinks that the *main* utility of the science is that it conduces man to religion, to an understanding of the immensity of the universe and the point of earth in relation to it, and, most importantly, to a recognition of the fact that his soul is more important than all else. At the end of these considerations Wilkins briefly mentions the utility of astronomy for commerce and travel. It has created "one commonwealth," he says. But for Wilkins commercial or political utility of science is not its primary end; rather it is a secondary benefit. Thus, his assessment of the reasons for undertaking the study of astronomy differ markedly from that usually attributed to the English Protestant scientists of the period.

In a final encomium to astronomy, Wilkins describes the pleasure that attends this study: "there cannot be any fairer prospect then to view the whole Frame of nature, the fabrick of this great Universe, so to discern that order and comlinesse which there is in the magnitude, situation, motion of the severall parts that belong unto it; to see the true cause of that constant variety and alteration which there is in the different seasons of the yeare. All which must needs enter into a mans thoughts, with a great deale of sweetnes and complacency" (244).

He ends by citing the eminent men who spent much of their time in such work: "Ptolomey, Julius Caesar, Alphonsus King of Spain, the noble Tycho, &c" (surprisingly he does not name Galileo or Kepler), remarking that they have found more lasting monuments than the pyramids—"the monuments of learning are more durable than the Monuments of Wealth or Power" (246).

The Significance of Wilkins's Writings

Despite their derivative nature and intent, Wilkins's contributions to the Copernican debate are impressive works through which the author assays to remove doubts from the minds of a variety of objectors. As arguments for the new philosophy and for the Copernican system, both of his books are remarkable for the calm yet determined manner in which Wilkins goes about his task. Apparently Wilkins attempted consciously to return the debate to the realm of dialectic, to a disputational mode in which consideration of the issues could be conducted dispassionately. Both sides were painstakingly heard, propositions and objections fully explored dialectically. The motivation for the corrective he applied came not only from a concern for methodological rectitude but also from the recognition that eloquence could become a two-edged sword, as had been patently illustrated by the case of Galileo.

Wilkins was much too realistic to attempt to remove rhetoric entirely from the debate, however, for he knew it could help to change the attitudes of those on the opposing side. He understood the powers of positive rhetoric much better than anyone writing in the controversy, while few saw the dangers of negative rhetoric so clearly. He incorporated the canons of dialectic in his answer to the Copernican question, but remembered that a large part of his audience were untutored in astronomy and so pitched his discourse to their level. Neither Wilkins's natural talent nor his experience prepared him to invest his writings with the creative and stylistic qualities of Galileo's. The Englishman was gifted with soft words, hard reason, and good humor, but not with eloquence.

In the end, the elements that impressed Wilkins in Galileo's writings were the sense evidence recorded, the mathematical calculations, and the dialectical proofs developed from these. In his own writings he uses the Tuscan astronomer's observations and his dialectical arguments repeatedly, even his diagrams. But he does not mimic the vituperative passages nor does he accept Galileo's valuations of Tycho and Kepler. He must have thought that these passages would not persuade, might even harden the defenses of opponents.

The quasi demonstration that Galileo claimed for his argument from the tides does not seem to have won many to the cause. It is not even mentioned by Wilkins, for he recognized the superior arguments of Kepler and others in this regard.

On that score we might note the reaction to that argument by another contemporary Englishman, Thomas White, a Roman Catholic priest. In 1642 White published a dialogue modeled on Galileo's entitled *De Mundo,* but he did not use the tidal theory as a persuasive proof.[24] He does, however, offer further confirmation that the methodology of the *Posterior Analytics* continued to furnish the grounds for scientific claims. In a later

---

24. Russell describes the work in "The Copernican System in Great Britain," 222–23. The author does not express his own views except through the interlocutor, Asphalius (Steadfast). Simplicio is called Andabata, the term for the gladiators in Rome who engaged in combat while blindfolded. Ereunis, the Searcher, takes the role of Sagredo.

work, published in English under the title *Peripateticall Institutions* (1656), White emphasizes the compelling force of probable arguments for Copernicanism and says that in this way "astronomers prove these motions of the Earth: because otherwise greater motions of greater bodies must be supposed." He acknowledges that he does not have a perfect demonstration but that in its absence dialectical proofs command assent.[25] Unaccountably, White seems never to have been pursued by Rome for his espousal of heliocentrism.

Wilkins's interest in the more scholastic format of disputation rather than the dialogue form reflects an irenic disposition. He was not a combative person. In fact, he was often faulted for what some saw as a too compliant nature. He could be tolerant of both Roman Catholics and radical Puritans in an age when strong opinions were demanded by contemporaries. One biographer calls him a "Trimmer," but, writing at the beginning of the twentieth century when those controversies were long past, he finds Wilkins's ability to compromise a virtue.[26] And so it was in the Copernican debate. His mildness permitted him to apply himself to answering the other side with an appreciation of what it meant to give up the old ideas. That approach bore fruit. His writings were instrumental in bringing about a change of opinion in England. John Russell, himself a Jesuit, believes that Copernicanism had "effectively won by 1650."[27]

Should we conclude that dialectic was a more effective weapon in the cause of science and religion than was rhetoric? Among most scholars in the disciplines of natural philosophy or theology, only dialectical argument could be officially recognized as persuasive in the absence of demonstration. But, on the other hand, it is impossible to say just how much rhetoric may have contributed to induce minds to receive dialectical proofs and may have thereby fostered assent to the thesis.

The ethos of men like Galileo and Kepler surely prepared the way for a favorable hearing for Copernicus. And even though Wilkins may not have been aware of it, their ethos worked on him too, as did his own on his audience when he asked them to read his probable arguments with an open mind. If asked whether they were certain that the system was true, the learned among them would have had to reply in the negative; however, they might then have added "but 'tis probable." Those untrained in, or unconcerned with, the niceties of scientific methodology would be swayed by the rhetoric of a Galileo or a Ross when it appealed to their interests. If they wanted to be numbered among the most elite intelligentsia of the day, they

25. Ibid., 223.
26. P. A. Wright Henderson so describes him in *The Life and Times of John Wilkins*, 127.
27. Russell, 223.

would embrace the thesis. If they thought the arts and sciences would indeed be thrown into confusion, or if they imagined the fires of hell burned brighter around Copernicans, then they would cling to the old worldview, ignoring the force of the probable arguments presented.

The Rhetorical Revolution Assessed

The rhetorical revolution led by Galileo may have blurred the scope and validity of the scientific proofs supportive of the Copernican System, but no doubt it won many supporters among those who found his arguments ingenious, the probabilities persuasive, and opposing views too ridiculous to countenance. But in shaking the foundations of the Peripatetic edifice in this way he brought down upon his head the weight of the establishment's dependence on it. Too much was founded upon the old philosophy to give it up so quickly for the reasons advanced. When the novelties in the heavens paled with frequent sightings, the prospect seemed less frightening and the new explanation could be cautiously entertained. By the time Wilkins wrote about Galileo's discoveries a quarter century later, their reality was little questioned, although the significance was still debated.

But the very sighting of these novelties was what inspired Galileo to use all the means of persuasion at his command to proclaim that here was proof indeed of what Copernicus had declared. The depth of his knowledge of natural philosophy allowed him to grasp the vast implications of his discoveries. He knew that he was standing on the threshold of a new era, and he had the courage to lead the way.

For Galileo, nurtured by eloquence, living and working in a society where something was only worth saying if it were said well, rhetoric and communication were synonymous. Since the advent of humanism, eloquence had come to dominate Italian intellectual life and the patrician social ambience as well. Although the major battle between rhetoric and philosophy had taken place almost two centuries before, the victory won by rhetoric permitted it to extend its hegemony over all forms of public communication. We have seen it in the humanized disputation of Grassi and glimpsed it even at the edges of Wilkins's more scholastic version in England.

But the victory was not total, for though rhetoric dominated social and political discourse, it had not vanquished the private investigations and the deliberations of philosophy. When questions usually reserved to the private arena of philosophy and theology were argued in the open forum, scholars felt the need to suit the discourse to the audience. Truth might still be the aim of the speaker, but truth would now be judged not only by scholars but by a public, and in the light of its interests. It was this occasion that permitted rhetoric to enter so naturally and unobtrusively into debates on scientific questions.

Galileo, then, was moved by two divergent impulses: above all, the desire to show himself a singularly accomplished heir to the Florentine humanistic tradition and, at the same time, to manifest his membership in an elite community of scholars extending through time and over continents. To the latter group a precise methodology determined the value of the discourse. To the former, elegant persuasion suited to the moment was the key. Galileo succeeded brilliantly in fulfilling the first desire, for he was surely the most eloquent man of that time and place. As to the second, the community of scholars, he showed himself to have honored their traditions and to have applied them in ways no one had dreamed they could be. If he failed in the eyes of some members of that community, it was because they were so incensed by the rhetorical overtones that they could not see the value of his contribution. But the conflict of interests that led him to trespass in the expression of the science did not intrude in the making of that science. The wonder is that his extraordinary gifts of expression did not carry him farther away from the mark. Yet the point of this book is not to judge the father of the rhetorical revolution. It is simply to understand why and how that revolution could have taken place in the context of the Copernican controversy.

POSTSCRIPT

Dialectic and Rhetoric in Modern Science

Today, dialectic is no longer recognized as a tool of scientific investigation; it seems to have been replaced by rhetoric. If we are to accept the recent interpretations of a number of scholars, scientists are ever conscious of audience and approach problems with an aim to persuade, not as their predecessors did to find the most probable answer to the problem. Now scientists are said to have become rhetoricians. Bruno Latour and Steve Woolgar argue that researchers are continuously engaged in rhetoric, seeking to gain assent to their findings by massing their evidence in persuasive ways.[1] The very impressive instruments used and the "inscriptions" these yield have rhetorical force. The old canon of delivery, transformed in the print medium to "presentation," was never employed to more advantage. Other scholars note the rhetorical techniques of invention, arrangement, and style also at work in scientific discourse: the use of idealized narration of the research to delete intransigent data, the careful selection of data, of authorities, or of research reports to show similarities with the findings of the author and omission of those that show the opposite, the placement of the report within a desired historical context.[2]

1. Bruno Latour and Steve Woolgar, *Laboratory Life: The Social Construction of Scientific Facts* (London/Beverly Hills, Calif: Sage, 1979); Bruno Latour, *Science in Action* (Cambridge: Harvard University Press, 1987); and Steve Woolgar, "Discovery: Logic and Sequence in a Scientific Text" in Karin D. Knorr et al., eds., *The Social Process of Scientific Investigation* (Dordrecht: Reidel, 1981), 239–68.

2. See the earlier work of Stephen Shapin and Simon Schaffer on sociological influences on the history of science, *Leviathan and the Air-Pump: Hobbes, Boyle, and the Experimental Life* (Princeton: Princeton University Press, 1985), and the recent study by Alan G. Gross analyzing the rhetorical dimensions, *The Rhetoric of Science* (Cambridge: Harvard University Press, 1990). See also the rhetorical analysis of Charles Bazerman, *Shaping Written Knowledge: The Genre and Activity of the Experimental Article in Science* (Madison: University of Wisconsin Press, 1988); W. Weimar, "Science as a Rhetorical Transaction," *Philosophy and Rhetoric* 10 (1977): 1–29; Joseph Gusfield, "The Literary Rhetoric of Science, *American Sociological Review* 41 (1976): 16–34; R. Allen Harris, "Assent, Dissent, and Rhetoric in Science," *Rhetoric Society Quarterly* 20 (Winter 1990): 13–37.

On the other hand, another of the characteristics of modern scientific discourse noted by commentators would seem to indicate that dialectic has not been retired completely, except explicitly. One reason Latour and Woolgar find the scientific enterprise rhetorical is that it is "agonistic": scientists battle each other, anticipate objections that might be made and try to overcome these in their presentations. But that agonistic quality is the fundament of scientific inquiry, the dialectical process that we have traced throughout the debate on the Copernican question. It is what Aristotle claimed lay at the heart of scientific investigation: the pro and con questioning that goes on in the mind of the investigator as he tries to see a problem from all angles, a questioning that once found expression in the disputation as the inquirer strove to come to an ultimate resolution of what is the case. That activity is not what makes the science rhetorical. What does is the scientist's preoccupation with audience to the extent that it dominates his research and the presentation of it. It seems hardly credible that all scientific research proceeds with one eye on a client or an audience. If this is so and researchers aim mainly to gain funding or acclamation, then the charge hurled against the Sophists centuries ago, that they made the weaker cause the stronger, might understandably occur to evaluators of research today. It seems more likely, however, that rhetoric typically enters at a later stage in the scientific process. Scientific research probably proceeds most often dialectically in its early phases, as a scientist looks at a problem, considering one explanation and then its opposite. When at the final stage the researcher prepares a report for colleagues, he or she aims not to put a "spin" on the findings but to find the best answer to the problem through the evaluation of other scientists. In that view of the enterprise, rhetoric would take a more prominent role in the discourse when the presentation of the data had to be made to a general public or to an agency for funding. Surely in most instances the integrity of the science can be assumed. Otherwise how can one explain the outrage vented at scientists who fudge results? Is outrage only expressed by disgruntled persons who are envious of others' rewards and who employ negative rhetoric to gain them?

It can be argued, of course, as many do, that all knowledge is the result of a rhetorical effort in that it creates a communal field in which conversation occurs, that all is subjective and there is no way out of that solipsistic mirror that must operate within a field of mirrors. That was the skeptical position of Gorgias. But in my brief excursus on modern scientific discourse I have assumed that a realist epistemology still moves most working scientists to investigate a world they think is "out there."

The concomitant debate about whether knowledge is socially constructed and whether the scientific "conversational" context determines the nature of the investigation might also be partially illuminated by reviewing the nature of dialectical investigation as it was understood in the period of our study. Dialectic rests on informed opinion, and few would question the fact today that informed opinion does determine the nature of scientific inquiry, the questions being asked. Whether one should go further and say that what is real is only what is agreed upon as "fact" and approved by the relevant community is a conception of reality that would never have occurred to scientists in the late Renaissance. Probably it does not occur to most in our own day, except in the areas most removed from observation—the very small and the very large.[3]

In the case of the subject of this book, I have approached the scientific discourse of our authors from a contextual standpoint, describing their writings within the constructs they employed. To see their dialectical and demonstrative efforts as rhetorical in spite of what they maintained, and my own view as a rhetorical reflection of theirs, would require me to move through vertigo and beyond, which is more than I think I could do.

The texts have been read "simplistically," without an attempt to deconstruct them linguistically or rhetorically beyond what the authors themselves seem to have intended. They "depersonalize" and "normalize" their discoveries by means of dialectics and demonstration, but personalize them through rhetorical elements. This interweaving of the personal with the impersonal I have taken on face value. The aim has been to permit a brighter reminiscence of the living practice of science and rhetoric in the period.

I have claimed that Galileo, as the father of the rhetorical revolution in science, led the way in demonstrating that rhetoric could indeed invade the scientific domain and spearhead the cause of science and of a reformed theology. In this he can be said to have fostered the approach of those scientists, historians, and philosophers who do see science as a rhetorical enterprise. He did so, as I have observed earlier, because his audience was greatly enlarged beyond an elite, and the outcome of the debate was not confined to the classroom but would touch the worldview and the faith of a greater public.

3. Well-argued support for an approach similar to that I have taken in this study is that of J. E. McGuire and Trevor Melia in "Some Cautionary Strictures on the Writing of the Rhetoric of Science," *Rhetorica* 7 (Winter 1989): 87–99. A discussion of current approaches in the history and philosophy of science is developed by Jan Golinski in "The Theory of Practice and the Practice of Theory: Sociological Approaches in the History of Science," *Isis* 81 (September 1990): 492–505.

BIBLIOGRAPY

Aristotle. *The "Art" of Rhetoric*. Loeb Classical Library. 1975.

———. *Topica*. Loeb Classical Library. 1976.

Armstrong, C. J. R. *French Renaissance Studies*. Edinburgh: Edinburgh University Press, 1976.

Ashworth, Elizabeth J. "Traditional Logic." In *The Cambridge History of Renaissance Philosophy*, ed. Charles B. Schmitt et al. Cambridge: Cambridge University Press, 1988.

Baldwin, Charles Sears. *Ancient Rhetoric and Poetic*. New York: Macmillan, 1924. Reprint. Gloucester, Mass.: Peter Smith, 1959.

———. *Renaissance Literary Theory and Practice: Classicism in the Rhetoric and Poetic of Italy, France, and England, 1400–1600*. Ed. Donald Leman Clark. New York: Columbia University Press, 1939.

Baron, Hans. *The Crisis of the Early Italian Renaissance: Civic Humanism and Republican Liberty in an Age of Classicism and Tyranny*. 2 vols. Princeton: Princeton University Press, 1955.

Baumgardt, Carola. *Johannes Kepler: Life and Letters*. New York: Philosophical Library, 1951.

Bazerman, Charles. *Shaping Written Knowledge: The Genre and Activity of the Experimental Article in Science*. Madison: University of Wisconsin Press, 1988.

Berti, Enrico. "Ancient Greek Dialectic as Expression of Freedom of Thought and Speech." *Journal of the History of Ideas* 39 (1978): 347–70.

———. *Logica Aristotelica e Dialectica*. Bologna: L. Cappelli, 1983.

Bird, Otto. "The Tradition of the Logical Topics: Aristotle to Ockham." *Journal of the History of Ideas* 23 (1962): 307–23.

Black, Robert. "Italian Renaissance Education: Changing Perspectives and Continuing Controversies." *Journal of the History of Ideas* 52 (1991): 315–34.

Bonansea, Bernardino M. "Campanella's Defense of Galileo." In *Reinterpreting Galileo*, ed. William A. Wallace, 205–39.

———. *Tommaso Campanella: Renaissance Pioneer of Modern Thought*. Washington: The Catholic University of America Press, 1969.

Bruno, Giordano. *The Ash Wednesday Supper*. Trans. Stanley Jaki. The Hague: Mouton, 1975.

Burke, Kenneth. *A Rhetoric of Motives*. Berkeley and Los Angeles: University of California Press, 1969.

Butterfield, Herbert. *The Origins of Modern Science*. Rev. ed. New York: Macmillan, 1967.

Campanella, Tomasso. *Apologia pro Galilaeo, mathematico florentino*. Ed. Salvatore Femiano. Milan: Marzorati, 1971.

Camporeale, Salvatore. "Giovanmaria dei Tolosani O.P.: 1530–1546, Umanesimo, Riforma e Teologia controversista." *Memorie Domenicane* 17 (1986): 145–252.

Carugo, Adriano, and Alistair Crombie. "The Jesuits and Galileo's Idea of Science and Nature." *Annali dell'Istituto e Museo di Storia della Scienza di Firenze* 8 (1983): 1–69.

Cicero. *De inventione*. Loeb Classical Library. 1949.

Cicero (Pseudo). *Ad Herennium*. Loeb Classical Library. 1981.

Codina Mir, Gabriel. *Aux sources de la pédagogie des Jésuites. Le "Modus Parisiensis."* Rome: Institutum Historicum S.I., 1968.

Copernicus, Nicolai. *On the Revolutions*. Trans. Edward Rosen. Baltimore: Johns Hopkins University Press, 1978.

Dahlstrom, Daniel O., ed. *Nature and Scientific Method*. Essays in Honor of William A. Wallace. Washington: The Catholic University of America Press, 1990.

Dear, Peter. *Mersenne and the Learning of the Schools*. Ithaca: Cornell University Press, 1988.

Dobryzcki, Jerzy, ed. *The Reception of Copernicus' Heliocentric Theory*. Papers from a Symposium hel ' in Torun, Poland, 1972. Dordrecht: Reidel, 1972.

Drake, Stillman. *Discoveries and Opinions of Galileo*. Garden City: Doubleday, 1957.

———. *Galileo against the Philosophers*. Los Angeles: Zeitlin & Ver Brugge, 1976.

———. *Galileo at Work: His Scientific Biography*. Chicago: University of Chicago Press, 1978.

———. "Reexamining Galileo's *Dialogue*." In *Reinterpreting Galileo*, ed. William A. Wallace, 155–75.

Drake, Stillman, and I. E. Drabkin. *Mechanics in Sixteenth-Century Italy*. Madison: University of Wisconsin Press, 1969.

Drake, Stillman, and C. D. O'Malley. *The Controversy on the Comets of 1618*. Philadelphia: University of Pennsylvania Press, 1960.

Duhem, Pierre. *To Save the Phenomena*. Trans. E. Dolan and C. Maschler. Chicago: University of Chicago Press, 1969.

Erasmus, *De duplici copia rerum et verborum*. Ed. and trans. Craig R. Thompson. *Collected Works of Erasmus*, vol. 2, 284–659. Toronto: University of Toronto Press, 1978.

Feyerabend, Paul K. *Against Method*. London: Redwood Burn, 1978.

Feldhay, Rivka. "The Jesuits' Educational Ideology: From *'officium docendi'* to *'ministerium'*." Typescript, 1985.

———. "Knowledge and Salvation in Jesuit Culture." *Science in Context* 1 (1987): 195–213.

Finocchiaro, Maurice A. "Commentary: Dialectical Aspects of the Copernican Revolution." In *The Copernican Achievement*, ed. R. S. Westman, 204–12. Berkeley and Los Angeles: University of California Press, 1975.

———. *The Galileo Affair: A Documentary History*. Berkeley and Los Angeles: University of California Press, 1989.

———. *Galileo and the Art of Reasoning*. Boston Studies in the Philosophy of Science, vol. 61. Dordrecht and Boston: Reidel, 1980.

———. "The Methodological Background to Galileo's Trial." In *Reinterpreting Galileo*, ed. William A. Wallace, 241–72.

———. "Varieties of Rhetoric in Science." *History of the Human Sciences* 3 (1990): 177–93.

Firpo, Luigi. *Apologia di Galileo*. Turin: Unione Tipografica Editrice Torinese, 1968.
———. "Cinquant'anni di studi sul Campanella (1901–1950)." *Rinascimento* 6, no. 483 (1955): 300.
Flynn, Lawrence J. The "De Arte Rhetorica." Ph.D. diss. University of Florida, 1955.
———. "The *De Arte Rhetorica* of Cyprian Soarez, S.J." *Quarterly Journal of Speech* 42 (Dec. 1956): 367–74.
———. "Sources and Influence of Soarez' De Arte Rhetorica." *Quarterly Journal of Speech* 43 (Oct. 1957): 257–65.
Foscarini, Paolo Antonio. *Lettera del R.P.M. Paolo Antonio Foscarini Carmelitano Sopra l'Opinio de' Pittagorici e del Copernico, della Mobilità della Terra e Stabilità del Sole, e del Nuovo Pittagorico Sistema del Mondo*. Naples, 1615.
Fumaroli, Marc. *L'Age de L'Eloquence*. Geneva: Librairie Droz, 1980.
Funkenstein, Amos. "The Dialectical Preparation for Scientific Revolutions." In *The Copernican Achievement*, ed. R. S. Westman, 165–203.
Galilei, Galileo. *Dialogue Concerning the Two Chief World Systems*. Trans. Stillman Drake. Berkeley and Los Angeles: University of California Press, 1953. Rev. ed., 1962.
———. *Dialogue on the Great World Systems*. Trans. Thomas Salusbury. Ed. Giorgio de Santillana. Chicago: University of Chicago Press, 1953.
———. *Discourse on Bodies in Water*. Trans. Thomas Salusbury. Ed. Stillman Drake. Urbana: University of Illinois Press, 1960.
———. *Le Opere di Galileo Galilei*. Ed. Antonio Favaro. 20 vols. in 21. Florence: G. Barbèra, 1890–1909; rpt. 1968.
———. *Sidereus nuncius or The Sidereal Messenger*. Trans. Albert van Helden. Chicago: University of Chicago Press, 1989.
———. *Tractatio de praecognitionibus et praecognitis* and *Tractatio de demonstratione*. Transcribed from the Latin autograph by William F. Edwards, with Notes and Commentary by William A. Wallace. Padua: Editrice Antenore, 1988.
Gallego, F. Jordan. "La Metafisica de Diego de Zuñiga (1536–1597) y la Reforma Tridentina de los Estudios Ecclesiasticos." *Estudio Agustiniano* 9 (1974): 3–60.
Galluzzi, Paolo, ed. *Novità Celesti e Crisi del Sapere*. Atti del Convegno Internazionale di Studi Galileiani. Florence: Istituto e Museo di Storia della Scienza, 1983.
Garin, Eugenio. "Alle origini della polemica anticopernicana." Colloquia Copernicana, vol. 2. *Studia Copernicana* 6:31–42. Cracow: Ossolineum, 1975.
———. *Italian Humanism: Philosophy and Civic Life in the Renaissance*. Trans. Peter Munz. Oxford: Blackwell, 1965.
———. *Rinascite e rivoluzioni: Movimenti culturali dal XIV al XVIII secolo*. Bari: Laterza, 1976.
———. *Science and Civic Life in the Italian Renaissance*. Trans. Peter Munz. Garden City: Doubleday, 1969.
Geymonat, Ludovico. *Galileo Galilei: A Biography and Inquiry into his Philosophy of Science*. Trans. Stillman Drake. New York: McGraw-Hill, 1965.
Gilbert, Neal. "The Italian Humanists and Disputation." In *Renaissance Essays in Honor of Hans Baron*, ed. A. Molho and J. A. Tedeschi, 205–26.

————. *Renaissance Concepts of Method*. New York: Columbia University Press, 1960.

Gillispie, Charles C., ed. *Dictionary of Scientific Biography*. 16 vols. New York: Scribners, 1970–80.

————. *The Edge of Objectivity*. Princeton: Princeton University Press, 1960. Reprint. 1990.

Gingerich, Owen. "The Censorship of Copernicus' *De Revolutionibus*." *Annali dell'Istituto e Museo di Storia della Scienza di Firenze* 6 (1981): 45–61.

————. "From Copernicus to Kepler: Heliocentrism as Model and Reality." *Proceedings of the American Philosophical Society* 117 (1973): 513–22.

————. "Kepler, Johannes." *Dictionary of Scientific Biography*, ed. C. C. Gillispie, 7 (1983): 289–312.

————. "Ptolemy, Copernicus, and Kepler." *The Great Ideas Today 1983*. Chicago: Encyclopaedia Britannica, 1983, 137–80.

Goldstein, Bernard R. *The Arabic Version of Ptolemy's Planetary Hypotheses*. Transactions of the American Philosophical Society, vol. 57, pt. 4, 1967.

Golinski, Jan. "The Theory of Practice and the Practice of Theory: Sociological Approaches in the History of Science." *Isis* 81 (Sept. 1990): 492–505.

Grafton, Anthony and Lisa Jardine. *From Humanism to the Humanities: Education and the Liberal Arts in Fifteenth- and Sixteenth-Century Europe*. Cambridge, Mass.: Harvard University Press, 1986.

Gray, Hanna. "Renaissance Humanism: The Pursuit of Eloquence." *Journal of the History of Ideas* 24 (1963): 497–514.

Green-Pedersen, Niels J. *The Tradition of the Topics in the Middle Ages*. Munich: Philosophia Verlag, 1984.

Grendler, Paul. *Schooling in Renaissance Italy: Literacy and Learning, 1300–1600*. Baltimore: Johns Hopkins University Press, 1989.

Grimaldi, William M. A. *Aristotle, Rhetoric I, A Commentary*. New York: Fordham University Press, 1980.

————. *Studies in the Philosophy of Aristotle's Rhetoric*. Wiesbaden: Franz Steiner, 1972.

Gross, Alan G. *The Rhetoric of Science*. Cambridge, Mass.: Harvard University Press, 1990.

Gusfield, Joseph. "The Literary Rhetoric of Science." *American Sociological Review* 41 (1976): 16–34.

Hall, A. Rupert. "Merton Revisited, or Science and Society." *History of Science* 2 (1963): 1–16.

Harré, Rom. *The Philosophies of Science*. London and Oxford: Oxford University Press, 1972.

Harris, R. Allen. "Assent, Dissent, and Rhetoric in Science." *Rhetoric Society Quarterly* 20 (Winter 1990): 13–37.

Harris, Stephen J. "Transposing the Merton Thesis: Apostolic Spirituality and the Establishment of the Jesuit Scientific Tradition." *Science in Context* 3 (1989): 29–65.

Helbing, Mario Otto. *La Filosofia di Francesco Buonamici, professore di Galileo a Pisa*. Pisa: Nistri-Lischi Editore, 1989.

Horace. *Satires, Epistles and Ars Poetica*. Loeb Classical Library. 1978.

Howell, Wilbur Samuel. *Logic and Rhetoric in England, 1500–1700*. Princeton: Princeton University Press, 1956.

Jardine, Lisa. "Humanistic Logic." *The Cambridge History of Renaissance Philosophy*, ed. Charles B. Schmitt et al., 173–98.

Jardine, Nicholas. *The Birth of History and Philosophy of Science: Kepler's "A Defence of Tycho against Ursus" with Essays on its Provenance and Significance*. Cambridge: Cambridge University Press, 1984.

Kennedy, George A. *Classical Rhetoric and its Christian and Secular Tradition from Ancient to Modern Times*. Chapel Hill: University of North Carolina Press, 1980.

Kepler, Johannes. *Astronomia nova*. Trans. William Donahoe, Owen Gingerich, and Ann Wegner. *The Great Ideas Today 1983*, 309–341. Chicago: Encyclopaedia Britannica, 1983.

———. *Mysterium cosmographicum*. Trans. A. M. Duncan, commentary by Eric Aiton. New York: Abaris, 1981.

———. *Kepler's Somnium: A Dream or Posthumous Work on Lunar Astronomy*. Trans. Edward Rosen. Madison and Milwaukee: University of Wisconsin Press, 1967.

Kinneavy, James L. "*Kairos:* A Neglected Concept of Classical Rhetoric." In *Rhetoric and Praxis*, ed. Jean Dietz Moss, 79–105.

Knoll, Paul W. "The Arts Faculty at the University of Cracow at the End of the Fifteenth Century." In *The Copernican Achievement*, ed. Robert S. Westman, 137–56.

Koestler, Arthur. *The Sleepwalkers: A History of Man's Changing Vision of the Universe*. London: 1959. New York: Macmillan, 1968.

Koyré, Alexandre. *The Astronomical Revolution*. Trans. R. E. W. Maddison. Ithaca: Cornell University Press, 1973.

———. *Metaphysics and Measurement: Essays in the Scientific Revolution*. Cambridge, Mass.: Harvard University Press, 1968.

Kristeller, Paul Oskar. *Renaissance Thought and its Sources*. Ed. Michael Mooney. New York: Columbia University Press, 1979.

Kuhn, Thomas S. *The Copernican Revolution*. Cambridge, Mass.: Harvard University Press, 1957.

———. *The Essential Tension: Selected Studies in Scientific Tradition and Change*. Chicago: University of Chicago Press, 1977.

Langford, James J. *Galileo, Science and the Church*. Ann Arbor: University of Michigan Press, 1971.

Latour, Bruno. *Science in Action*. Cambridge, Mass.: Harvard University Press, 1987.

Latour, Bruno, and Steve Woolgar. *Laboratory Life: The Social Construction of Scientific Facts*. London and Beverly Hills: Sage, 1979.

Lechner, Sister Joan Marie. *Renaissance Concepts of the Commonplaces*. New York: Pageant, 1962.

Leff, Michael C. "The Topics of Argumentative Invention in Latin Rhetorical Theory from Cicero to Boethius." *Rhetorica* 1 (Spring 1983): 23–44.

Lindberg, David C., and Ronald L. Numbers, eds. *God and Nature: Historical Essays on the Encounter between Christianity and Science.* Berkeley and Los Angeles: University of California Press, 1986.

Lohr, Charles H. "Jesuit Aristotelianism and Sixteenth-Century Metaphysics." *Paradosis* 32 (1976): 203–20.

———. *Latin Aristotle Commentaries.* Vol. 2, *Renaissance Authors.* Florence: Olschki, 1988.

Mahoney, Michael. *Vico in the Tradition of Rhetoric.* Princeton: Princeton University Press, 1985.

Marsh, David. *The Quattrocento Dialogue.* Cambridge, Mass.: Harvard University Press, 1980.

McColley, Grant. "The Defense of Galileo of Thomas Campanella." *Smith College Studies in History* 22 (April–July 1937): i–xliv, 1–93.

———. "The Ross-Wilkins Controversy." *Annals of Science* 3 (1968): 153–69.

McGuire, J. E., and Trevor Melia. "Some Cautionary Strictures on the Writing of the Rhetoric of Science." *Rhetorica* 7 (Winter 1989): 87–99.

McNulty, Robert. "Bruno at Oxford." *Renaissance News* 13 (1968): 300–305.

Merton, Robert K. "Science, Technology and Society in Seventeenth-Century England." *Osiris* 4 (1938): 360–631.

Miller, Carolyn. "Aristotle's 'Special Topics' in Rhetorical Practice and Pedagogy." *Rhetoric Society Quarterly* 17 (Winter 1987): 61–70.

Molho, A., and J. A. Tedeschi, eds. *Renaissance Essays in Honor of Hans Baron.* Florence: Sansoni, 1971.

Moss, Jean Dietz. "Dialectics and Rhetoric: Questions and Answers in the Copernican Revolution." *Argumentation* 5 (1990): 17–37.

———. "Galileo's *Letter to Christina*: Some Rhetorical Considerations." *Renaissance Quarterly* 36 (Winter 1983): 547–76.

———. "Ludovico Carbone's Commentary on Aristotle's *De caelo*." In *Nature and Scientific Method*, ed. Daniel O. Dahlstrom, 169–92.

———. "Newton and the Jesuits in the *Philosophical Transactions*." In *Newton and the New Direction in Science,* ed. George V. Coyne et al., 117–34. Vatican City: Vatican Observatory, 1988.

———, ed. *Rhetoric and Praxis: The Contribution of Classical Rhetoric to Practical Reasoning.* Washington: The Catholic University of America Press, 1986.

———. "The Rhetoric Course at the Collegio Romano in the Latter Half of the Sixteenth Century." *Rhetorica* (Spring 1986): 137–51.

———. "The Rhetoric of Proof in Galileo's Writings on the Copernican System." In *Reinterpreting Galileo,* ed. William A. Wallace, 179–204.

Murphy, James J. *Rhetoric in the Middle Ages.* Berkeley and Los Angeles: University of California Press, 1974.

———, ed. *Renaissance Eloquence.* Berkeley and Los Angeles: University of California Press, 1983.

Nelson, Benjamin. "The Early Modern Revolution in Science and Philosophy." *Boston Studies in the Philosophy of Science* 3 (1964–66): 1–40.

Nelson, Norman E. "Peter Ramus and the Confusion of Logic, Rhetoric, and Poetic." *Contributions in Modern Philology* 2 (April 1947): 1–22.

Nicholson, Marjorie. *Science and Imagination*. Ithaca: Cornell University Press, 1956. Reprint. Hamden, Conn: Shoe String, 1976.

Olivieri, Luigi, ed. *Aristotelismo Veneto e Scienza Moderna*. 2 vols. Padua: Editrice Antenore, 1983.

O'Malley, John W. *Praise and Blame in Renaissance Rome: Rhetoric, Doctrine and Reform in the Sacred Orators of the Papal Court, c. 1450–1521*. Durham: University of North Carolina Press, 1979.

Ong, Walter J. *Ramus, Method, and the Decay of Dialogue*. Cambridge, Mass.: Harvard University Press, 1958.

Ornstein, Martha. *The Role of Scientific Societies in the Seventeenth Century*. Chicago: University of Chicago Press, 1928.

Pagano, Sergio M., ed. *I Documenti del Processo di Galileo Galilei*. Vatican City: Pontifical Academy of Sciences, 1984.

Pedersen, Olaf. *A Survey of the "Almagest."* Acta Historica Scientiarum Naturalium et Medicinalium, 30. Odense: Odense University Press, 1974.

———. "Galileo's Religion." *The Galileo Affair: A Meeting of Faith and Science*, ed. George V. Coyne et al., 75–102. Vatican City: Vatican Observatory, 1985.

Pera, Marcello, and William R. Shea, eds. *Persuading Science*. Canton, Mass.: Science History Publications, Watson Publishing, 1991.

Perelman, Chaim, and L. Olbrechts-Tyteca. *The New Rhetoric*. Notre Dame: University of Notre Dame Press, 1971.

Purnell, Frederick. "Jacopo Mazzoni and Galileo." *Physis* 3 (1972): 273–94.

Quintilian. *Institutio Oratoria*. Loeb Classical Library. 1980.

Rauh, Sister Miriam Joseph. *Shakespeare's Use of the Arts of Language*. New York: Columbia University Press, 1947, 1949. Abridged and reprinted as *Rhetoric in Shakespeare's Time*. New York: Harcourt, 1962.

Redondi, Pietro. *Galileo Heretic*. Trans. Raymond Rosenthal. Princeton: Princeton University Press, 1987.

Riccobono, Antonio. *De natura rhetoricae*. Venice, 1579.

Righini Bonelli, Maria Luisa. "Le Posizioni Relative di Galileo e dello Scheiner nelle Scoperte delle Macchie Solari nelle Pubblicazioni Edite entro il 1612." *Physis* 12 (1970): 405–12.

Rosen, Edward. "Copernicus, Nicholas." *Dictionary of Scientific Biography*, ed. C. C. Gillispie, 3 (1971): 401–11.

———. *Kepler's Conversation with Galileo's Sidereal Messenger*. New York and London: Johnson Reprint, 1965.

———. *Kepler's Somnium: The Dream or Posthumous Work on Lunar Astronomy*. Madison and Milwaukee: University of Wisconsin Press, 1967.

———. *Three Copernican Treatises*. New York: Dover, 1959.

Russell, John L. "The Copernican System in Great Britain." In *The Reception of Copernicus' Heliocentric Theory*, ed. Jerzy Dobryzcki, 189–239. Dordrecht: Reidel, 1972.

Russo, François. "Lettre à Christine de Lorraine Grande-Duchesse de Toscane (1615)." *Revue d'histoire des sciences* 17 (1964): 330–66.

Salusbury, Thomas. *Mathematical Collections and Translations*. 2 vols. London, 1661.

Santillana, Giorgio de. *The Crime of Galileo*. Chicago: University of Chicago Press, 1955.

Scaglione, Aldo. *The Liberal Arts and the Jesuit College System*. Amsterdam and Philadelphia: John Benjamins, 1986.

Schmitt, Charles B. *Cesare Cremonini un Aristotelico al tempo di Galileo*. Quaderni 16. Venice: Centro Tedesco di Studi Veneziane, 1980.

———. *John Case and Aristotelianism in Renaissance England*. Kingston: McGill-Queens University Press, 1983.

———. "The Rise of the Philosophical Textbook." In *The Cambridge History of Renaissance Philosophy*, 792–804.

Schmitt, Charles B., Quentin Skinner, et al., eds. *The Cambridge History of Renaissance Philosophy*. Cambridge: Cambridge University Press, 1988.

Seigel, Jerrold. *Rhetoric and Philosophy in Renaissance Humanism*. Princeton: Princeton University Press, 1968.

Shapin, Stephen, and Simon Schaffer. *Leviathan and the Air-Pump: Hobbes, Boyle, and the Experimental Life*. Princeton: Princeton University Press, 1985.

Shapiro, Barbara. *John Wilkins 1614–1672: An Intellectual Biography*. Berkeley and Los Angeles: University of California Press, 1969.

———. *Probability and Certainty in Seventeenth-Century England*. Princeton: Princeton University Press, 1983.

Shea, William R. *Galileo's Intellectual Revolution*. New York: Science History Publications, 1972.

Singer, Dorothea Waley. *Giordano Bruno: His Life and Thought, with Annotated Translation of his Work "On the Infinite Universe and Worlds."* New York: Henry Schuman, 1950.

Solt, Leo F. "Puritanism, Capitalism, Democracy and Science." *American Historical Review* 73 (Oct. 1967): 18–27.

Stimson, Dorothy. "Puritanism and the New Philosophy in Seventeenth-Century England." *Bulletin of the Institute of the History of Medicine* 3 (1935): 321–34.

———. *The Gradual Acceptance of the Copernican Theory of the Universe*. New York: Baker and Taylor, 1917.

Struever, Nancy S. *The Language of History in the Renaissance: Rhetoric and Historical Consciousness in Florentine Humanism*. Princeton: Princeton University Press, 1970.

———. "Lorenzo Valla: Humanist Rhetoric and the Critique of the Classical Languages of Morality." In *Renaissance Eloquence*, ed. James J. Murphy, 191–206.

Stump, Eleonore. *Boethius's "De Topicis Differentiis."* Ithaca: Cornell University Press, 1978.

———. *Boethius's "In Ciceronis Topica."* Ithaca: Cornell University Press, 1988.

Swerdlow, Noel. "The Derivation and First Draft of Copernicus's Planetary Theory: A Translation of the *Commentariolus* with Commentary." *Proceedings of the American Philosophical Society* 117 (1973): 445–50.

Thorndike, Lynn. *The Sphere of Sacrobosco and its Commentators*. Chicago: University of Chicago Press, 1949.

Van Helden, Albert. *Measuring the Universe: Cosmic Dimensions from Aristarchus to Halley*. Chicago and London: University of Chicago Press, 1985.

Vasoli, Cesare. *La dialettica e la retorica dell'Umanismo*. Milan: Feltrinelli, 1968.

Vickers, Brian. *English Science: Bacon to Newton*. Cambridge: Cambridge University Press, 1987.

———. "Epideictic Rhetoric in Galileo's Dialogo." *Annali dell'Istituto e Museo di Storia della Scienza* 8 (1983): 69–102.

———. *In Defence of Rhetoric*. Oxford: Clarendon, 1988.

Villoslada, Riccardo G. *Storia del Collegio Romano dal suo inizio (1551) alla soppressione della Compagnia di Gesù (1773)*. Rome: Gregorian University, 1954.

Von Gebler, Karl. *Galileo Galilei and the Roman Curia*. Trans. Mrs. George Sturge. London: C. K. Paul & Co., 1879.

Wallace, William A. "Aristotelian Influences on Galileo's Thought." In *Aristotelismo Veneto e Scienza Moderna*, ed. Luigi Olivieri, 1:349–78. 2 vols. Padua: Editrice Antenore, 1983.

———. *Causality and Scientific Explanation*. 2 vols. Ann Arbor: University of Michigan Press, 1972, 1974.

———. "The Certitude of Science in Late Medieval and Renaissance Thought." *History of Philosophy Quarterly* 3 (July 1986): 281–91.

———. *Galileo and his Sources: The Heritage of the Collegio Romano in Galileo's Science*. Princeton: Princeton University Press, 1984.

———. "Galileo's Early Arguments for Geocentrism and his Later Rejection of Them." In *Novità Celesti e Crisi del Sapere*, ed. Paolo Galluzzi, 31–40.

———. *Galileo's Logic of Discovery and Proof: The Background, Content, and Use of his Appropriated Treatises on Aristotle's "Posterior Analytics."* Boston Studies in the Philosophy of Science, vol. 137. Dordrecht: Kluwer Academic Publishers, 1992.

———. *Galileo's Logical Treatises: A Translation, with Notes and Commentary, of his Appropriated Latin Questions on Aristotle's "Posterior Analytics."* Boston Studies in the Philosophy of Science, vol. 138. Dordrecht: Kluwer Academic Publishers, 1992.

———. "Galileo's Science and the Trial of 1633." *The Wilson Quarterly* 7 (Summer 1983): 154–64.

———. "The Problem of Apodictic Proof in Early Seventeenth-Century Mechanics: Galileo, Guevara, and the Jesuits." *Science in Context* 3 (1989): 67–87.

———. "Randall *Redivivus*: Galileo and the Paduan Aristotelians." *Journal of the History of Ideas* 49 (1988): 133–49.

———, ed. *Reinterpreting Galileo*. Studies in Philosophy and the History of Philosophy, vol. 15. Washington: The Catholic University of America Press, 1986.

Weinberg, Bernard. *A History of Literary Criticism in the Italian Renaissance*. 2 vols. Chicago: University of Chicago Press, 1961.

Weimar, W. "Science as a Rhetorical Transaction." *Philosophy and Rhetoric* 10 (1977): 1–29.

Westfall, Richard S. *Essays on the Trial of Galileo*. Vatican City: The Vatican Observatory, 1989.

———. *Science and Religion in Seventeenth-Century England*. New Haven: Yale University Press, 1953.

Westman, Robert S. "The Copernicans and the Churches." In *God and Nature*, ed. D. C. Lindberg and R. L. Numbers, 76–113.

————, ed. *The Copernican Achievement*. Berkeley and Los Angeles: University of California Press, 1975.

————. "Magical Reform and Astronomical Reform: The Yates Thesis Reconsidered." In *Hermeticism and the Scientific Revolution*, by Robert S. Westman and J. E. McGuire, 1–91. Los Angeles: William Andrews Clark Memorial Library, University of California, 1977.

————. "La Préface de Copernic au Pape: Esthétique humaniste et réforme de l' Eglise." *History and Technology* 4 (1987): 365–84.

————. "Politics, Poetics, and Patronage." In *Reappraisals of the Scientific Revolution*, ed. David C. Lindberg and Robert S. Westman, 167–205. Cambridge: Cambridge University Press, 1990.

————. "The Reception of Galileo's *Dialogue*." In *Novità Celesti e Crisi del Sapere*, ed. Paolo Galluzzi, 329–371.

White, Hayden V. "The Tropics of History: The Deep Structure of the New Science." In *Giambattista Vico's Science of Humanity*, ed. Giorgio Tagliacozzo and Donald P. Verene, 65–85. Baltimore: Johns Hopkins University Press, 1973.

Wilkins, John. *A Discourse Concerning a New Planet: Tending to Prove that (tis probable) our Earth is one of the Planets*. London, 1640.

————. *Essay towards a Real Character and a Philosophical Language*. London, 1668.

————. *The Discovery of a World in the Moone or, A Discourse Tending to Prove that 'tis Probable there may be another Habitable World in that Planet*. London, 1638. Reprint. New York: Scholars' Facsimiles and Reprints, 1973.

————. *Of the Principles and Duties of Natural Religion*. London, 1675.

Witt, Ronald. "Medieval 'Ars Dictaminis' and the Beginnings of Humanism: A New Construction of the Problem." *Renaissance Quarterly* 35 (Spring 1982): 1–35.

Woolgar, Steve. "Discovery: Logic and Sequence in a Scientific Text." In *The Social Process of Scientific Investigation*, ed. Karin D. Knorr et al., 239–68. Dordrecht: Reidel, 1981.

Wright Henderson, P. A. *The Life and Times of John Wilkins*. Edinburgh and London: William Blackwood and Sons, 1910.

Wrightsman, Bruce. "Andreas Osiander's Contribution to the Copernican Achievement." In *The Copernican Achievement*, ed. R. S. Westman, 213–43.

Wyss-Morigi, Giovanna. Contributo allo Studio del Dialogo all'Epoca dell' Umanismo e del Rinascimento. Ph.D. diss. University of Bern, 1947.

Yates, Frances. *Giordano Bruno and the Hermetic Tradition*. Chicago: University of Chicago Press, 1964.

————. "Giordano Bruno's Conflict with Oxford." *Journal of the Warburg Institute* 2 (1938–39): 227–42.

Young, Richard E., Alton L. Becker, and Kenneth L. Pike. *Rhetoric: Discovery and Change*. New York: Harcourt, Brace, and World, 1970.

INDEX

Abbot, George, 180n
Accademia dei Lincei (Lyncean Academy), 97, 115, 137, 222; publisher of Galileo's *Assayer,* 241–43
accommodation, principle of, 131, 132, 138, 159, 175, 189, 311, 319, 320
Acquaviva, Claudio, 125n, 197, 211, 258
Adami, Tobias, 257
Agricola, Rudolf, 16n, 19, 20, 67
Agrippa, Cornelius, 172n; *De occulta philosophia,* 172n
Albert the Great, Saint, 157, 172
Alexander of Aphrodisias, 162
Alfarabi, 159
Alfonsine Tables, 46
Alphonsus, King of Spain, 325
Ambrose, Saint, 154, 162, 166, 312
analogy and metaphor, 110, 140, 143, 177, 190, 225, 234, 245, 275, 309, 324
anaphora, 90
ancients and moderns, 136, 242, 311, 313
angels, 129
Antipodes, 309
Apelles. *See* Scheiner, Christopher
Apollonius, 33
appropriateness *(prepon),* 60, 73, 95
argument(s), argumentation, vii, ix, xiv, 96; *a fortiori,* 74, 109, 324; *ad hominem,* 276, 279, 310; dialectical, xi, 47, 49, 51, 56, 58, 74, 82, 109, 112, 117, 129, 137, 142, 147, 187, 210, 226, 238, 250, 254–55, 281, 300, 308–10; *ex hypothesi,* 281; from authority, 115, 131, 147, 159, 164, 188, 199, 312, 313, 317; induction and deduction, 4, 11, 199; mathematical, 115; mathematical-physical, x; probable, 112, 203, 281, 307, 321, 327; Rogerian, 53n. *See also* reasoning; demonstration(s); dialectic
Ariosto, *Orlando Furioso,* 77
Aristarchus, 62, 318
Aristotelians, 113; conservative, 98; Paduan, 252–53; progressive, 200, 267
Aristotle, xii, 3, 8, 47, 49, and passim;

biology of, 32; commentaries on the *Rhetoric,* 3n, 10, 11n, 20; heaviness and lightness in, 55; logic of, viii; science of, ix; turned against the Aristotelians, 113
—works: *Analytics,* 2, 3; *Categories,* 3, 9, 267; *Ethics,* 76; *On the Heavens (De caelo),* 32, 41, 140, 158, 267, 310, 324; *On Interpretation,* 3; On the soul *(De anima),* 267; *Metaphysics,* 158; *Meteorology,* 41, 229; *Organon,* 3–4, 20, 32; *Physics,* 3–4, 41, 267; *Poetics,* 20, 76; *Posterior Analytics,* 6, 41, 113, 265, 293; *Rhetoric,* 10–11, 16, 37n, 76, 265; *Sophistical Refutations,* 23; *Topics,* 4, 265
arrangement, rhetorical, 17–18
ars dictaminis (art of letter writing), 37, 101n
Ashworth, E. J., 4n
Assayer, The (Il Saggiatore), 241–54; endorsements and injustices, 242–45; Grassi's fictional authorship, 245–47; optical arguments and experiments, 248–52; Galileo's atomism, 252–54, 257
astronomy, 30, 34, 42; as highest endeavor of man, 30, 325; insufficiently developed, 157; moral edification of, 30, 50, 118, 325; new kind of, 88; value of, 325; written for astronomers, 46. *See also* mixed science
Attavanti, Gianozzo, 216
audience, viii, xii, xiii, 1, 9–11, 53, 85, 111, 227–29, 308, 309, 326, 328; Galileo's, 75, 81, 85, 101, 106, 119, 251, 265, 302, 316; modern views of, 22–23, 53–54, 101n, 280, 283; primary and secondary, shadow, 191–93, 243
Augustine, Saint, 135, 157, 162, 165, 167, 196; *De Genesi ad litteram,* 160, 195, 206, 311
Averroes, 159, 177
Avicenna, 157, 159

Bacon, Francis, 147, 303, 316, 318
Bandini, Cardinal Ottavio, 183